AF388252

Contributions to Operator Theory and its Applications

The Tsuyoshi Ando Anniversary Volume

Edited by

T. Furuta
I. Gohberg
T. Nakazi

Springer Basel AG

Volume Editorial Office

Raymond and Beverly Sackler Faculty of Exact Sciences
School of Mathematical Sciences
Tel Aviv University
IL-69978 Tel Aviv
Israel

A CIP catalogue record for this book is available from the Library of Congress, Washington D.C., USA

Deutsche Bibliothek Cataloging-in-Publication Data
Contributions to operator theory and its applications: the
Tsuyoshi Ando anniversary volume / ed. by T. Furuta ... – Basel
; Boston ; Berlin : Birkhäuser, 1993
   (Operator theory ; Vol. 62)
   ISBN 978-3-0348-9690-0   ISBN 978-3-0348-8581-2 (eBook)
   DOI 10.1007/978-3-0348-8581-2
NE: Furuta, Takayuki [Hrsg.]; Ando, Tsuyoshi: Festschrift; GT

Camera-ready copy prepared by the authors
Printed on acid-free paper produced from chlorine-free pulp
Cover design: Heinz Hiltbrunner, Basel

ISBN 978-3-0348-9690-0

9 8 7 6 5 4 3 2 1

Contents

TSUYOSHI ANDO

Operator Theory:
Advances and Applications, Vol. 62
© 1993 Birkhäuser Verlag Basel

BIOGRAPHY OF T. ANDO

Tsuyoshi Ando was born on February 1, 1932 and grew up in Sapporo, the provincial capital of the northern island Hokkaido in Japan. His father was a middle-ranking city official. Tsuyoshi has two elder brothers.

The second world war broke out when he was in elementary school. In the spring of 1945, after finishing the first year of a junior high school, he entered a military cadet school in Sendai with the hope of becoming a military officer in the Imperial Japanese Army. After half a year, however, Japan surrendered to the allied nations, and he returned to the junior high school. At that time Japan was in great confusion.

He entered Hokkaido University, which was transformed from an old imperial university into an American style university. His primary interest was in physical science. But just when he was about to decide his major, Dr. Hideki Yukawa won the Nobel prize for physics as the first Japanese Nobel Laureate, and that created a boom in physics among university students. Many wanted to choose physics as their major. Ando did not wish to be swept away by a popular trend so he chose mathematics. Among the few who chose mathematics was Tetsuya Shimogaki, who was destined to make important contributions to the interpolation theory and passed away at the young age of 39, in the year 1970.

The Department of Mathematics in 1950 had four chairs; algebra, real analysis, complex analysis and geometry. But when Ando chose mathematics as his major, the department was in chaos because professors were at odds with each other. Many faculty members left the department and Dr. Akitsugu Kawaguchi was the only remaining full professor. Kawaguchi was a famous geometer, but was sick and could deliver only 3 or 4 lectures a year. Other mathematics courses were taught by visiting professors.

When Ando was a junior, Dr. Hidegoro Nakano was invited from the University of Tokyo to be a full professor for the chair of real and functional analysis. Ando was attracted to the lectures of Nakano, and after receiving a B.S. degree, he together with Shimogaki entered the newly opened graduate school and began to study functional analysis under Nakano.

Professor Nakano had at that time already made significant contributions in the abstract treatment of measure theory in regard to vector lattices. During the most difficult years of wartime he had successfully completed three volumes of measure theory (in Japanese) and was devoting himself to an axiomatic treatment of spaces of Orlicz type, which he named "modulared semi-ordered linear spaces."

Ando's lifelong interest in order structure is a reflection of the influence of Nakano. But, he said, yet he could have little personal contact with the professor because of Nakano's aristocratism. Ando seemed to be significantly influenced, however, by Dr. Amemiya, at that time a young lecturer. He recalls that Dr. Amemiya appeared to have a vacant look to his eyes, but his ability to grasp the essence of mathematical problems and to discover astounding solutions was truly amazing.

In his first year of graduate school Ando together with Shimogaki read the famous book of S. Banach "Théorie des opérations linéaires" and later the newly published volume of Nakano "Semi-ordered linear spaces." He was very impressed by the original ideas of Nakano. He says, however, that it is quite regrettable that Professor Nakano used his original notations in all of books, which prevented the writings as widely accepted in the mathematics world as they should have been.

As his own field of research Ando took up the problem of investigating aspects of Perron-Frobenius and Kreĭn-Rutman theory in regard to order complete Banach lattices. In 1958 he was awarded Ph.D. in Science for his thesis entitled "Positive linear operators in semi-ordered linear spaces." Because he was the first doctorate in the new system of graduate study, many news papers acknowledged his accomplishment.

Subsequently Ando was awarded a junior academic position at the Division of Applied Mathematics in the Research Institute of Applied Electricity of Hokkaido University. At that times this research institute consisted of several divisions; electronics, physics, chemistry, physiology and mathematics, and had been developing researches of interdisciplinary character.

In this setting Ando sought to do research into a direction deviated from the ever far established theory of Nakano. His first important contribution concerned with finitely additive measures. This was an extension of the theorem of Nikodým on equicontinuity of countably additive measures. He showed that the Yosida-Hewitt decomposition is continuous with respect to sequential convergence. Then he turned his attention to the study of predual characterization in the theory of ordered Banach spaces. One of his most significant contributions, known sometimes as a Kreĭn-Ando theorem, gives predual characterization for lattice property of the dual space in terms of "Riesz interpolation property."

During this same period he started investigation of Hilbert space operators. His interest was in the theory of unitary dilation, discovered by the famous Hungarian mathematician, B. Sz.-Nagy. In this theory there remained open a problem as to whether a pair of commuting contractions admits a pair of commuting unitary dilations. Ando solved this problem affirmatively. This result was refined by several mathematicians in a form of the "commutant lifting theorem", which is now a key tool in the mathematical theory of linear system theory.

Ando made another important discovery with I. Amemiya. They settled that when T_i ($i = 1, 2, \ldots, N$) are positive contractions, any products $T_{i_1} T_{i_2} \cdots T_{i_n}$ converges weakly as $n \to \infty$, where i_n are chosen randomly from $\{1, 2, \ldots, N\}$. This result is now recognized as one of the mathematical basis of CT (computer tomography).

To distinguish metric difference between Hilbert space L^2 and other classical Banach lattices L^p ($p \neq 2$), Ando proved that any contractive linear projection in L^p ($p \neq 2$) is essentially a conditional expectation operator. This work was connected with his later metric characterization of L^p spaces.

He acknowledges that his mathematical ability is not so much in developing a large theory but rather in solving special problems by original ideas. But in the later years of his career he together with F. Kubo, a research associate, developed a theory of operator means, which is based on a deep analysis of the Loewner theory of operator monotone functions. The theory of operator means has a strong connection with realization of resistive linear electrical networks. An initial concept of this theory, he said, originated from his joint research on operator ranges with K. Nishio, a research associate. In connection with this theory he established many operator inequalities, including the concavity theorem of Dyson-Yanase. Having published such inequalities in a journal of matrix theory, he has been recognized as a matrix expert among matrix theorists.

On several occasions Ando resided abroad and has established a large circle of colleagues in the field of mathematical research. In early 1960's he spent a year with Professors W.A.J. Luxemburg and A.C. Zaanen at the California Institute of Technology. In the late 1960's he worked with Professor H.H. Schaefer at the University of Tübingen in Germany. In the middle 1970's he worked with Professor B. Sz.-Nagy at the Szeged University in Hungary.

In 1969 Ando was promoted to full professor as a division chief of the interdisciplinary institute, he has been interested in systems and networks from the standpoint of operator means and inequalities. From the analytic standpoint he has been involved in Toeplitz and Hankel operators. His work in analytic direction is now being developed by his former students, T. Nakazi and K. Takahashi.

For more than twenty years Ando has been running a functional analysis seminar and inspiring students and young colleagues. He himself has been a frequent speaker, presenting new ideas and introducing new results. Among his former students and research associates are M. Takaguchi, K. Nishio, T. Nakazi, M. Uchiyama, K. Okubo, F. Kubo, F. Hiai and others, who are now active mathematicians in their respective fields.

Ando has been a member of the faculty of the graduate school of mathematics of the university, and is indeed an inspiring educator. He has presented a one semester special course every other year. Surprisingly he focused on a different topic every course, ranging from general operator theory, operator algebra, Hankel operators, Kreĭn spaces, de Branges spaces, etc. Apparently he prepared his lectures so well that despite the complexity of the subject his lectures were always clear and instructive. He says "I could have been a lecturer like I. Schur." Most of his lectures have been published in the form of lecture notes from the institute.

Since 1988 Ando has been the director of the research institute. Japanese universities are now under severe pressure to reform. Research institutes have particularly been requested to reform. As the institute director, he has devoted a great amount time to designing the future of the institute. From the academic year of 1992 his institute has changed its name to the Research Institute for Electronic Science, and its Division of Applied Mathematics is now known as the Laboratory of Information Mathematics. He is still serving as the director of the new research institute.

Although Ando has precious little time for research these years, he is still very active in mathematics. He is on the editorial board of five international journals, and in 1991 he organized the International Workshop on Operator Theory and Complex Analysis in Sapporo.

We all express our sincere congratulations to Professor Ando on this occasion of his sixtieth anniversary and we hope that he will be even more active and productive than he has been before sixty as was Mark G. Kreĭn.

Sapporo, May 1992   Takahiko Nakazi

[36] (with Nishio, Katsuyoshi) Convexity properties of operator radii associated with unitary ρ-dilations. Michigan Math. J. **20** (1973), 303–307. MR 48#12091

[37] Closed range theorems for convex sets and linear liftings. Pacific J. Math. **44** (1973), 393–410.   MR 48#6888

[38] A theorem on nonempty intersection of convex sets and its application. Collection of articles dedicated to G. G. Lorentz on the occasion of his sixty-fifth birthday. J. Approximation Theory **13** (1975), 158–166.   MR 52#6381

[39] (with Okubo, Kazuyoshi) Constants related to operators of class C_ρ. Manuscripta Math. **16** (1975), no. 4, 385–394.   MR 51#13733

[40] (with Shintani, Toshitada) Best approximants in L^1 space. Z. Wahrscheinlichkeits-theorie und Verw. Gebiete **33** (1975/76), no. 1, 33–39.   MR 52#1860

[41] Lebesgue-type decomposition of positive operators. Acta Sci. Math. (Szeged) **38** (1976), no. 3-4, 253–260.   MR 56#16424

[42] Unitary dilation for a triple of commuting contractions. (Russian summary) Bull. Acad. Polon. Sci. Sér. Sci. Math. Astronom. Phys. **24** (1976), no. 10, 851–853. MR 55#3828

[43] (with Nishio, Katsuyoshi) Characterizations of operations derived from network connections. J. Math. Anal. Appl. **53** (1976), no. 3, 539–549. MR 53#5181

[44] (with Okubo, Kazuyoshi) Operator radii of commuting products. Proc. Amer. Math. Soc. **56** (1976), 203–210.   MR 53#8927

[45] ρ-contration and ρ-radius. (Japanese) Sûgaku **28** (1976), no. 2, 107–120. MR 57#17318

[46] (with Ceauşescu, Zoia; Foiaş, C.) On intertwining dilations. II. Acta Sci. Math. (Szeged) **39** (1977), no. 1-2, 3–14.   MR 56#1097b

[47] On the predual of H^∞. Special issue dedicated to Władysław Orlicz on the occasion of his seventy-fifth birthday. Comment. Math. Special Issue 1 (1978), 33–40.   MR 80c:46063

[48] Topics on operator inequalities. Hokkaido University, Research Institute of Applied Electricity, Division of Applied Mathematics, Sapporo, 1978. 44pp MR 58#2451

[49] Structure of quadratic inequalities. J. Math. Anal. Appl. **70** (1979), no. 1, 72–84. MR 80h:15009

[50] (with Suciu, Ion; Timotin, D.) Characterization of some Harnack parts of contractions. J. Operator Theory **2** (1979), no. 2, 233–245. MR 82a:47014

[51] Concavity of certain maps on positive definite matrices and applications to Hadamard products. Linear Algebra Appl. **26** (1979), 203–241. MR 80f:15023

[52] Generalized Schur complements. Linear Algebra Appl. **27** (1979), 173–186.
MR 80m:15005

[53] (with Fan, Ky) Pick-Julia theorems for operators. Math. Z. **168** (1979), no. 1,
23–24.   MR 80h:47018

[54] Linear operators on Kreĭn spaces. Hokkaido University, Research Institute of
Applied Electricity, Division of Applied Mathematics, Sapporo, 1979. 59pp
MR 81e:47032

[55] Inequalities for M-matrices. Linear and Multilinear Algebra **8** (1979/80), no. 4,
291–316.   MR 81m:15015

[56] Hua-Marcus inequalities. Linear and Multilinear Algebra **8** (1979/80), no. 4,
347–352.   MR 81h:15013

[57] (with Kubo, Fumio) Means of positive linear operators. Math. Ann. **246**
(1979/80), no. 3, 205–224.   MR 84d:47028

[58] (with van Hemmen, J. Leo) An inequality for trace ideals. Comm. Math. Phys. **76**
(1980), no. 2, 143–148.   MR 81j:47014

[59] Limit of iterates of cascade addition of matrices. Numer. Funct. Anal. Optim. **2**
(1980), no. 7-8, 579–589 (1981).   MR 82h:93004

[60] Means of positive linear operators. Proceedings of the Fourth Conference on Operator Theory (Timişoara/Herculane, 1979), pp. 8–22, Univ. Timişoara, Timişoara,
1980.

[61] (with Suciu, Ion; Timotin, D.) Characterization of some Harnack parts of contractions. Proceedings of the Fourth Conference on Operator Theory (Timişoara/
Herculane, 1979), pp. 23–25, Univ. Timişoara, Timişoara, 1980.

[62] Fixed points of certain maps on positive semidefinite operators. Functional analysis
and approximation (Oberwolfach, 1980), pp. 29–38, Internat. Ser. Numer. Math.,
60, Birkhäuser, Basel-Boston, Mass., 1981.   MR 84g:47016

[63] Inequalities for permanents. Hokkaido Math. J. **10** (1981), Special Issue, 18–36.
MR 83i:15010

[64] Majorization, doubly stochastic matrices and comparison of eigenvalues. Hokkaido
University, Research Institute of Applied Electricity, Division of Applied Mathematics, Sapporo, 1982. 87pp.   MR 84a:15019
Linear Algebra Appl. **118** (1989), 163–248.   MR 90g:15034

[65] An inequality between symmetric function means of positive operators. Acta Sci.
Math. (Szeged) **45** (1983), no. 1-4, 19–22.   MR 85b:47014

[66] On the arithmetic-geometric-harmonic-mean inequalities for positive definite
matrices. Linear Algebra Appl. **52/53** (1983), 31–37.   MR 84j:15016

[67] (with Zerner, Martin) Sur une valeur propre d'un opérateur. (English summary)
Comm. Math. Phys. **93** (1984), no. 1, 123–139.   MR 85h:47047

[68] Totally positive matrices. Hokkaido University, Sapporo, 1984. 50pp
MR 85i:15033
Linear Algebra Appl. **90** (1987), 165–219.   MR 88b:15023

[69] Approximation in trace norm by positive semidefinite matrices. Linear Algebra Appl. **71** (1985), 15–21.   MR 87b:15037

[70] (with Szymański, Wacław) Order structure and Lebesgue decomposition of positive definite operator functions. Indiana Univ. Math. J. **35** (1986), no. 1, 157–173. MR 87k:47050

[71] (with Choi, M. D.) Nonlinear completely positive maps. Aspects of positivity in functional analysis (Tübingen, 1985), 3–13, North-Holland Math. Stud., 122, North-Holland, Amsterdam-New York, 1986.   MR 88a:46059

[72] (with Bunce, John W.) The geometric mean, operator inequalities, and the Wheatstone bridge. Linear Algebra Appl. **97** (1987), 77–91. MR 89a:47022

[73] (with Horn, Roger A.; Johnson, Charles Royal) The singular values of a Hadamard product: a basic inequality. Linear and Multilinear Algebra **21** (1987), no. 4, 345–365.   MR 89g:15018

[74] (with Bhatia, Rajendra) Eigenvalue inequalities associated with the Cartesian decomposition. Linear and Multilinear Algebra **22** (1987), no. 2, 133–147. MR 89c:15022

[75] On some operator inequalities. Math. Ann. **279** (1987), no. 1, 157–159. MR 89c:47019

[76] Reproducing kernel spaces and quadratic inequalities. Hokkaido University, Research Institute of Applied Electricity, Division of Applied Mathematics, Sapporo, 1987. 96pp   MR 88d:46002

[77] Matrix quadratic equations. Hokkaido University, Research Institute of Applied Electricity, Division of Applied Mathematics, Sapporo, 1988. 97pp MR 89c:15019

[78] (with Kubo, Fumio) Some matrix inequalities in multiport network connections. The Gohberg anniversary collection, Vol. I (Calgary, AB, 1988), pp. 111–131, Operator theory: advances and applications, 40, Birkhäuser, Basel, 1989. MR 91d:15038

[79] (with Halperin, I.) Bibliography: series of vectors and Riemann sums. Hokkaido University, Research Institute of Applied Electricity, Division of Applied Mathematics, Sapporo, 1989. 42pp   MR 90k:00036

[80] (with Bunce, J.; Trapp, G. E.) An alternate variational characterization of matrix Riccati equation solutions. Circuits Systems Signal Process. **9** (1990), no. 2, 223–228.   MR 91c:15007

[81] (with Alzer, Horst; Nakamura, Yoshihiro) The inequalities of W. Sierpiński and Ky Fan. J. Math. Anal. Appl. **149** (1990), no. 2, 497–512. MR 91d:26020

[82] De Branges spaces and analytic operator functions. Hokkaido University, Research Institute of Applied Electricity, Division of Applied Mathematics, Sapporo, 1990. 84pp

[83] (with Kubo, Fumio) Inequalities among operator symmetric function means. Signal processing, scattering and operator theory, and numerical methods (Amsterdam, 1989), pp. 535–542, Progr. Systems Control Theory, 5, Birkhäuser Boston, Boston, MA, 1990.

[84] (with Okubo, Kazuyoshi) Induced norms of the Schur multiplier operator. Linear Algebra Appl. **147** (1991), 181–199.

[85] (with Nakamura, Yoshihiro) Some extremal problems related to majorization. Topics in matrix and operator theory (Rotterdam, 1989), pp. 83–92, Operator theory: advances and applications, 50, Birkhäuser, Basel, 1991.

[86] Completely positive matrices. Hokkaido University, Research Institute of Applied Electricity, Division of Applied Mathematics, Sapporo, 1991. 42pp

January, 1992

Operator Theory:
Advances and Applications, Vol. 62
© 1993 Birkhäuser Verlag Basel

ON CERTAIN (NEARLY) CONVEX JOINT NUMERICAL RANGES

Hari Bercovici

Dedicated to Professor T. Ando, on the occasion of his sixtieth anniversary.

Let H be a complex Hilbert space with scalar product denoted $\langle \cdot, \cdot \rangle$, and let $L(H)$ denote the algebra of bounded linear operators on H. For a family $\tau = \{T_i : i \in J\} \subset L(H)$ one defines the joint numerical range $W(\tau)$ by

$$W(\tau) = \{\{\langle T_i x, x \rangle : i \in J\} : x \in H, \|x\| = 1\}.$$

This set will be regarded as a subset of $\mathbf{C}^J$, and this space will be given the topology of pointwise (or componentwise) convergence. Cassier proved in [2] certain convexity theorems about the closure and interion or $W(\tau)$, assuming that the operators T_i belong to a commutative uniform subalgebra of $L(H)$. (Cassier considers countable families of operators.) We will demonstrate here that these convexity results lie at a more basic level, and do not require uniformity or commutativity. To do this we denote by $K(H)$ the ideal of $L(H)$ consisting of the compact operators, and denote by $\|T\|_e$ the distance of $T \in L(H)$ to $K(H)$. Our basic condition on τ is as follows; here I denotes the identity operator on H.

1. ASSUMPTION. *For every linear combination T of $\{T_i : i \in J\} \cup \{I\}$ we have* $\|T\|_e = \|T\|$.

Let S_e denote the convex set of essential states on $L(H)$. These are simply linear functionals $f \in L(H)^*$ such that $\|f\| = f(I) = 1$, and $f|K(H) = 0$. We introduce the auxilliary set

$$W_e(\tau) = \{\{f(T_i) : i \in J\} : f \in S_e\},$$

which may be called the essential numerical range of τ. We are now ready for the following result.

2. PROPOSITION. (i) $W_e(\tau) \subset \overline{W(\tau)}$.
   (ii) *If Assumption 1 is satisfied then* $W(\tau) \subset W_e(\tau)$.

Proof: Indeed, (i) follows immediately from a lemma of Glimm (cf. [4] or Lemma 11.2.1 of [3]) which implies that every $f \in S_e$ is a pointwise limit of functionals of the form

---

The author was supported in part by grants from the National Science Foundation.

$T \to \langle Tx, x \rangle, \|x\| = 1$. For (ii) we fix $x \in H, \|x\| = 1$, and suppose that Assumption 1 is satisfied. Then the map $T + K(H) \to \langle Tx, x \rangle$ is well defined on the linear subspace generated by $\{T_i + K(H) : i \in J\} \cup \{I + K(H)\}$. The Hahn-Banach theorem implies immediately the existence of $f \in S_e$ such that $\langle T_i x, x \rangle = f(T_i)$ for all $i \in J$. Q.E.D.

Since $W_e(\tau)$ is clearly convex, we have therefore the following version of Cassier's theorem 1 of [2].

3. COROLLARY. *If Assumption 1 is satisfied then $\overline{W(\tau)}$ is convex.*

We note without proof the following immediate consequence of Glimm's lemma quoted above.

4. LEMMA. *Assume that J is finite, $M \subset H$ is a finite set, $f \in S_e$, and $\varepsilon > 0$. There exists $x \in H$ such that $x \perp M$ and $|f(T_i) - \langle T_i x, x \rangle| < \varepsilon$ for all $i \in J$.*

Let C be a convex set in some linear space, and $x \in C$. We say that x is an internal point of C if for every $y \in C$ there exists $\varepsilon > 0$ such that $x + t(y - x) \in C$ for all $t \in (-\varepsilon, \varepsilon)$. We denote by $\text{int}(C)$ the set of internal points of C. We are now ready for our version of Theorem 2 of [2]. The proof has certain similarities with S. Brown's techniques from [1].

5. PROPOSITION. *Suppose that J is finite and Assumption 1 is satisfied. Then $\text{int}(\overline{W(\tau)}) \subset W(\tau)$.*

Proof: It clearly suffices to show that $0 \in W(\tau)$ if 0 is an internal point of the convex set $\overline{W(\tau)}$. Assume therefore that 0 is an internal point of $\overline{W(\tau)}$, and let X denote the linear space generated by $\overline{W(\tau)}$. It follows that $\{\lambda \in X : \max|\lambda_i| \leq r\} \subset \overline{W(\tau)}$ for some $r > 0$. Since $\overline{W_e(\tau)} = \overline{W(\tau)}$, we can find $f_0 \in S_e$ such that $|f_0(T_i)| < r/4$ for $i \in J$. Lemma 4 now implies that there exists $x_0 \in H, \|x_0\| = 1$, such that $\langle T_i x_0, x_0 \rangle < r/4$ for $i \in J$. Now, the vector $\{-4\langle T_i x_0, x_0 \rangle : i \in J\}$ belongs to X, and hence to $\overline{W(\tau)}$. We can therefore find $f_1 \in S_e$ such that

$$| - 4\langle T_i x_0, x_0 \rangle - f_1(T_i)| < r/4, \quad i \in J.$$

A further application of Lemma 4 yields a unit vector x_1 orthogonal onto x_0, $T_i x_0$, and $T_i^* x_0$, $i \in J$, such that

$$|4\langle T_i x_0, x_0 \rangle + \langle T_i x_1, x_1 \rangle| < r/4, \quad i \in J.$$

An inductive application of this argument yields an orthonormal sequence $x_0, x_1, x_2, \ldots$ such that $\langle T_i x_n, x_m \rangle = 0$ for all $i \in J$ and $m, n \geq 0$, and such that

$$|4^n \langle T_i x_0, x_0 \rangle + 4^{n-1}\langle T_i x_1, x_1 \rangle + \cdots + \langle T_i x_n, x_n \rangle| < r/4$$

for all $i \in J$ and $n \geq 0$. This last relation can be written as

$$|\langle T_i x_0, x_0 \rangle + \langle T_i(2^{-1}x_1), 2^{-1}x_1 \rangle + \cdots + \langle T_i(2^{-n}x_n), 2^{-n}x_n \rangle| < 4^{-n-1}r,$$

or, equivalently because of the orthogonality relations,

$$\left|\left\langle T_i \left(\sum_{k=0}^{n} 2^{-k} x_k \right), \sum_{k=0}^{n} 2^{-k} x_k \right\rangle\right| < 4^{-n-1}r$$

for $i \in J$ and $n \geq 0$. This clearly implies that $\langle T_i y, y \rangle = 0$, where $y = (2/\sqrt{3}) \sum_{k=0}^{\infty} 2^{-k} x_k$ is a unit vector. Thus 0 is in the numerical range of τ. Q.E.D.

REFERENCES

1. S. Brown, *Some invariant subspaces for subnormal operators*, Integral Equations Operator Theory 1(1978),310–333.

2. G. Cassier, *Image numérique simultanée d'une famille d'opérateurs sur l'espace de Hilbert*, C. R. Acad. Sci. Paris, Sér. I 305(1987), 681–684.

3. J. Dixmier, Les C*-algèbres et leurs représentations, Gauthier-Villars, Paris, 1966.

4. J. Glimm, *A Stone-Weierstrass theorem for C*-algebras*, Ann. of Math., 72(1960), 216–244.

Department of Mathematics
Indiana University
Bloomington, IN 47405
U.S.A.

AMS Subject Classification: A47A10

Operator Theory:
Advances and Applications, Vol. 62
© 1993 Birkhäuser Verlag Basel

The two-sided Nevanlinna-Pick problem in the Stieltjes class

Vladimir Bolotnikov

Dedicated to Professor T. Ando on the occasion of his sixtieth birthday

In this paper we consider the generalization of the classical interpolation problem in a special class of analytic matrix-valued functions. The method used to solve this problem is that of the fundamental matrix inequality, suitably adapted to the present situation.

1   Introduction

In this paper we consider the two-sided Nevanlinna–Pick problem in the Stieltjes class.

Definition 1.1 *A matrix-valued function $w(z)$ holomorphic in the complex plane with a cut along the semi-axis $[0, +\infty)$ is called a Stieltjes function if*

$$1)\frac{w(z) - w(z)^*}{z - \bar{z}} \geq 0, \quad \Im z \neq 0; \qquad 2)w(x) \geq 0, \ x < 0.$$

The class of all $\mathbb{C}^{m \times m}$ valued Stieltjes functions is denoted by S_m.

Note that according to the symmetry principle for every $w(z) \in S$

$$w(z) = w(\bar{z})^*. \tag{1.1}$$

In S_m we solve the following interpolation problem (IS):

given a set of matrices a_i, b_i, c_i, d_i, γ_i

$$a_i, c_i \in \mathbb{C}^{r_i \times m}; \ b_i, d_i \in \mathbb{C}^{s_i \times m}; \ \gamma_i \in \mathbb{C}^{r_i \times s_i} \ (r_i, s_i, m \in \mathbb{N})$$

and points $z_i \in \mathbb{C}^+$ $(i = 1, \ldots, n)$ find necessary and sufficient conditions which ensure the existence of a function $w(z)$ in S_m such that

$$a_i w(z_i) = c_i; \ b_i w(z_i)^* = d_i, \ a_i w'(z_i) b_i^* = \gamma_i, \ i = 1, \ldots, n \tag{1.2}$$

and describe the set of all solutions when these conditions prevail.

As a corollary we will obtain the solution of the interpolation problem (IN) in the Nevanlinna class $\mathbf{N}_m$ with the same interpolation conditions (1.2).

Definition 1.2 *The $\mathbb{C}^{m \times m}$-valued function $w(z)$ holomorphic in $\mathbb{C}^+$ is of the class $\mathbf{N}_m$ if $(w(z) - w(z)^*)/2i \geq 0$ for $Im\ z > 0$.*

The following theorem [12] establishes the connection between $\mathbf{N}_m$ and $\mathbf{S}_m$.

Theorem 1.3 *The function $w(z)$ belongs to S_m if and only if both $w(z)$ and $zw(z)$ belong to $\mathbf{N}_m$.*

As a corollary of the last theorem we obtain the Schwartz–Pick type inequalities for functions of the class S_m.

Theorem 1.4 *Let $w(z)$ be a function of the class S_m and let $z_1, ..., z_n$ be n points in $\mathbb{C}^+$. Then the following block matrices are nonnegative*

$$W_1 = \left(\frac{w(z_i) - w(z_j)^*}{z_i - \bar{z}_j} \right)^n_{i,j=1} \geq 0;$$

$$W_2 = \left(\frac{z_i w(z_i) - \bar{z}_j w(z_j)^*}{z_i - \bar{z}_j} \right)^n_{i,j=1} \geq 0;$$

The two-sided NP–problem can be set in other classes of analytic functions [1], [2], [5], [6], [9]. In the paper [11] was considered the nondegenerate case of the nontangential problem (IS). We also refer to the monographs [7], [10] where a number of approaches to solve some versions of the NP–problem have been developed. In the present paper a central role will be played by the method of the fundamental matrix inequality ([11],[12],[14],[15]).

The outline of the paper is as follows: in Section 2, following ideas of I. Kovalishina and V. Potapov, we establish the system of matrix inequalities for the problem (IS). The necessary condition of solvability of (IS) is the nonnegativity of special block-matrices K, K_p defined in (2.3)-(2.6) and their strict positivity is a sufficient one. The description of the set of all solutions to IS depends on whether K, K_p are degenerate or not.

In Section 3 we consider the case where K and K_p both are strictly positive. The set of all solutions is parametrized by a linear fractional tranformation of the form (3.25) with the resolvent matrix $\theta(z)$ of the class W_* (see Definition 3.1).

Section 4 deals with the degenerate case; we still have a description of all the solutions as a linear fractional transformation. As a corollary we obtain a criterion for uniqueness of the solution.

As for a number of classical interpolation problems problem (IS) can be solved recursively. This originates with the work of Schur [16]. A suitable version of the Schur algorithm will be given in Section 5.

2 The main matrix inequalities

In this section we characterize the solutions to (IS) in terms of a system of matrix inequalities. Descriptions of the set of solutions in terms of linear fractional transformations will be given in Sections 3 and 4.

Theorem 2.1 *Let w be a $\mathbf{C}^{m\times m}$-valued function analytic in $\mathbf{C}\setminus[0,+\infty)$. Then w is a solution to the interpolation problem (IS) if and only if it satisfies the system of inequalities*

$$
\left(
\begin{pmatrix} K_1 & K_2 \\ K_2^* & K_3 \end{pmatrix}
\quad
\begin{pmatrix} \psi_1(z) \\ \psi_2(z) \\ \frac{w(z)-w(z)^*}{z-\bar z} \end{pmatrix}
\right) \geq 0;
\tag{2.1}
$$

$$
\left(
\begin{pmatrix} K_{p1} & K_{p2} \\ K_{p2}^* & K_{p3} \end{pmatrix}
\quad
\begin{pmatrix} \psi_{p1}(z) \\ \psi_{p2}(z) \\ \frac{zw(z)-\bar z w(z)^*}{z-\bar z} \end{pmatrix}
\right) \geq 0
\tag{2.2}
$$

for $z \neq \bar z$, where K_l and K_{pl} $(l=1,2,3)$ are block-matrices defined by

$$
(K_1)_{ij} = \frac{c_i a_j^* - a_i c_j^*}{z_i - \bar z_j}; \quad (K_3)_{ij} = \frac{b_i d_j^* - d_i b_j^*}{z_j - \bar z_i};
\tag{2.3}
$$

$$
(K_2)_{ij} = \begin{cases} \dfrac{c_i b_j^* - a_i d_j^*}{z_i - \bar z_j} & \text{if } i \neq j \\ \gamma_i & \text{if } i = j \end{cases}
\tag{2.4}
$$

$$
(K_{p1})_{ij} = \frac{z_i c_i a_j^* - \bar z_j a_i c_j^*}{z_i - \bar z_j}; \quad (K_{p3})_{ij} = \frac{z_j b_i d_j^* - \bar z_i d_i b_j^*}{z_j - \bar z_i};
\tag{2.5}
$$

$$
(K_{p2})_{ij} = \begin{cases} \dfrac{z_i c_i b_j^* - \bar z_j a_i d_j^*}{z_i - \bar z_j} & \text{if } i \neq j \\ z_i \gamma_i + c_i b_i^* & \text{if } i=j \end{cases}
\tag{2.6}
$$

and $\psi_j(z)$, $\psi_{pj}(z)$ $(j=1,2)$ are the matrix-valued functions defined by

$$
(\psi_1(z))_i = \frac{a_i w(z)-c_i}{z-z_i}; \qquad (\psi_{p1}(z))_i = \frac{za_i w(z)-z_i c_i}{z-z_i};
$$
$$
(\psi_2(z))_i = \frac{b_i w(z)-d_i}{z-\bar z_i}; \qquad (\psi_{p2}(z))_i = \frac{zb_i w(z)-\bar z_i d_i}{z-\bar z_i}.
\tag{2.7}
$$

Proof: The necessity is proved as follows: one considers the system of Schwarz-Pick inequalities for the function w of the class S_m (Theorem 1.4) and multiplies them by the matrix

$$
T = \begin{pmatrix} \operatorname{diag}(a_i I_{r_i}) & 0 & 0 \\ 0 & \operatorname{diag}(b_i I_{s_i}) & 0 \\ 0 & 0 & I_m \end{pmatrix}
$$

on the left and by T^* on the right. Now we turn to proof of sufficiency. Suppose that the analytic matrix–valued function $w(z)$ satisfies the system of matrix inequalities (2.1), (2.2). Then, in particular,

$$\frac{w(z) - w(z)^*}{z - \bar{z}} \geq 0, \qquad \frac{zw(z) - \bar{z}w(z)^*}{z - \bar{z}} \geq 0.$$

By theorem 1.1, these inequalities imply that $w(z) \in S_m$. The positivity of the matrix function (2.1) implies the boundedness of the functions ψ_1, ψ_2 in compact neighbourhoods of the points z_i. Therefore, we obtain

$$\begin{aligned} a_i w(z_i) &= c_i; \\ b_i w(\bar{z}_i) &= d_i; \end{aligned} \tag{2.8}$$

Using the symmetry property (1.1) one can rewrite the last equality in the form

$$b_i w(z_i)^* = d_i. \tag{2.9}$$

To obtain conditions on the derivative of w, we note that (2.1) implies that, for $i = 1, \ldots, n$,

$$\begin{pmatrix} (K_1)_{ii} & (K_2)_{ii} & (\psi_1(z))_i \\ (K_2^*)_{ii} & (K_3)_{ii} & (\psi_2(z))_i \\ (\psi_1(z)^*)_i & (\psi_2(z)^*)_i & \frac{w(z)-w(z)^*}{z-\bar{z}} \end{pmatrix} \geq 0. \tag{2.10}$$

Multiplying (2.10) by the matrix

$$f = \begin{pmatrix} I_{r_i} & 0 & 0 \\ 0 & -I_{s_i} & b_i \end{pmatrix}$$

on the left, by f^* on the right we let $z \to z_i$. Taking into account (2.9) we obtain the inequality

$$\begin{pmatrix} (K_1)_{ii} & a_i w'(z_i) b_i^* - \gamma_i \\ b_i w'(z_i)^* z_i^* - \gamma_i^* & 0 \end{pmatrix} \geq 0$$

which implies

$$a_i w'(z_i) b_i^* = \gamma_i. \tag{2.11}$$

Equalities (2.8), (2.9), (2.11) mean that $w(z)$ is a solution of the problem (IS). The theorem is proved.

Definition 2.2 *The block matrices*

$$K = \begin{pmatrix} K_1 & K_2 \\ K_2^* & K_3 \end{pmatrix}, \quad K_p = \begin{pmatrix} K_{p1} & K_{p2} \\ K_{p2}^* & K_{p3} \end{pmatrix}$$

will be called the information blocks of the problem (IS).

It follows from (2.1), (2.2) that the nonnegativity of the information blocks is a necessary condition to ensure that (IS) has solution. As will be proved in Section 4 this condition is also sufficient.

Using the same methods, one can prove

Theorem 2.3 *Let $w(z)$ be a $\mathbb{C}^{m\times m}$ valued function analytic in $\mathbb{C}\setminus\mathbb{R}$. Then w is a solution to the problem* (IN) *if and only it it satisfies the inequality (2.1).*

3 Solutions to (IS): the nondegenerate case

In this section we suppose that the information blocks K, K_p are strictly positive and describe the set of all solutions to (IS). To begin with, we introduce some definitions.

Definition 3.1 *A matrix-valued function $\theta(z)$ meromorphic in $\mathbb{C}$ is of a class $\mathbf{W}_s$ if*

$$\theta(z)J\theta(z)^* = J \quad \text{for } z \in \mathbb{R}; \qquad \theta(z)J\theta(z)^* \geq J \quad \text{for } z \in \mathbb{C}^+; \tag{3.1}$$

$$\theta(x)J_\pi\theta(x)^* = J_\pi \quad \text{for } x < 0,$$

$$J = \begin{pmatrix} 0 & iI_m \\ -iI_m & 0 \end{pmatrix}, \quad J_\pi = \begin{pmatrix} 0 & I_m \\ I_m & 0 \end{pmatrix}. \tag{3.2}$$

and is of a class $\mathbf{W}$ if it satisfies only (3.1).

The following theorem [9] establishes the connection between $\mathbf{W}$ and $\mathbf{W}_s$.

Theorem 3.2 *The function $\theta(z)$ belongs to W_s if and only if*

$$\begin{array}{ll} \text{(i)} & \theta(z) \in \mathbf{W} \text{ and} \\ \text{(ii)} & \theta_p(z) = P(z)\theta(z)P^{-1}(z) \in W, \end{array} \tag{3.3}$$

where

$$P(z) = \begin{pmatrix} zI_m & 0 \\ 0 & I_m \end{pmatrix} \tag{3.4}$$

The following two lemmas which in fact are contained in [11] describe a number of functions of the classes W, W_s.

Lemma 3.3 *Let $G \in \mathbb{C}^{r\times m}$, $\Gamma \in \mathbb{C}^{m\times m}$ and let H be strictly positive matrix of $\mathbb{C}^{m\times m}$ which is a solution to the Lyapunov equation*

$$\Gamma H - H\Gamma^* = -iGJG^*.$$

Then the matrix-valued function

$$\hat{\theta}(z) = I + iG^*(zI - \Gamma^*)^{-1}H^{-1}GJ$$

is of the class **W** *and*

$$\hat{\theta}(z)J\hat{\theta}(\omega)^* - J = i(\bar{\omega} - z)G^*(zI - \Gamma^*)^{-1}H^{-1}(\bar{\omega}I - \Gamma)^{-1}G.$$

Lemma 3.4 *Let $G_j \in \mathbb{C}^{r \times m}$, $\Gamma \in \mathbb{C}^{m \times m}$ and let H_j $(j = 1, 2)$ be strictly positive matrices such that*

$$H_2 - \Gamma H_1 = G_2 G_1^*. \tag{3.5}$$

Then the matrix-valued function

$$\theta(z) = I + i \begin{pmatrix} G_1^* & G_1^*\Gamma^* \\ G_2^* & zG_2^* \end{pmatrix} \begin{pmatrix} (zI - \Gamma^*)^{-1}H_1^{-1}G_1 & 0 \\ 0 & (zI - \Gamma^*)^{-1}H_2^{-1}G_2 \end{pmatrix} J \tag{3.6}$$

is of the class W_s and

$$\theta(z)J\theta(\omega)^* - J = i(\bar{\omega} - z) \begin{pmatrix} G_1^* \\ G_2^* \end{pmatrix} (zI - \Gamma^*)^{-1}H_1^{-1}(\bar{\omega}I - \Gamma)^{-1}(G_1, G_2);$$

$$\theta_p(z)J\theta_p(\omega)^* - J = i(\bar{\omega} - z) \begin{pmatrix} G_1^*\Gamma^* \\ G_2^* \end{pmatrix} (zI - \Gamma^*)^{-1}H_2^{-1}(\bar{\omega}I - \Gamma)^{-1}(\Gamma G_1, G_2).$$

Note that under assumption (3.5) the matrix $\theta(z)$ admits the following representation which can be checked by a direct computation

$$\theta(z) = \{I + i \begin{pmatrix} G_1^* \\ G_2^* \end{pmatrix} (zI - \Gamma^*)^{-1}H_1^{-1}(G_1, G_2)J\} \begin{pmatrix} I & 0 \\ G_2^*H_2^{-1}G_2 & I \end{pmatrix}, \tag{3.7}$$

where the first factor is a function of the class W and the second one is a J-unitary matrix.

Returning to the main topic of this section we introduce the following matrices

$$A = \begin{pmatrix} a_1 \\ \vdots \\ a_n \end{pmatrix}, \ B = \begin{pmatrix} b_1 \\ \vdots \\ b_n \end{pmatrix}, \ C = \begin{pmatrix} c_1 \\ \vdots \\ c_n \end{pmatrix}, \ D = \begin{pmatrix} d_1 \\ \vdots \\ d_n \end{pmatrix}, \tag{3.8}$$

$$\Gamma = \begin{pmatrix} \text{diag}(z_i I_{r_i}) & 0 \\ 0 & \text{diag}(\bar{z}_i I_{s_i}) \end{pmatrix}; \ \Gamma(z) = (zI - \Gamma)^{-1} \tag{3.9}$$

and note the identity

$$K_p - \Gamma K = \begin{pmatrix} A \\ B \end{pmatrix} (C^*, D^*), \tag{3.10}$$

which follows from (2.3)-(2.7), (3.8), (3.9) and will play a central role in the sequel. Particularly, (3.10) means that the matrices

$$H_1 = K, \ H_2 = K_p, \ G_1 = \begin{pmatrix} C \\ D \end{pmatrix}, \ G_2 = \begin{pmatrix} A \\ B \end{pmatrix}, \ \Gamma$$

satisfy the conditions of Lemma 3.2. So the matrix–valued function

$$\theta(z) = I + i \begin{pmatrix} C^* & D^* & 0 & 0 \\ 0 & 0 & A^* & B^* \end{pmatrix} \begin{pmatrix} \Gamma(\bar{z})^* & \Gamma^*\Gamma(\bar{z})^* \\ \Gamma(\bar{z})^* & z\Gamma(\bar{z})^* \end{pmatrix} \begin{pmatrix} K^{-1} & 0 \\ 0 & K_p^{-1} \end{pmatrix} \begin{pmatrix} C & 0 \\ D & 0 \\ 0 & A \\ 0 & B \end{pmatrix} J$$

$$(3.11)$$

is of the class W, and

$$\theta(z)J\theta(\omega)^* - J = i(\bar{\omega} - z)JM^*\Gamma(\bar{z})^*K^{-1}\Gamma(\bar{\omega})MJ;$$

$$\theta_p J(z)\theta_p(\omega)^* - J = i(\bar{\omega} - z)JM_p^*\Gamma(\bar{z})^*K_p^{-1}\Gamma(\bar{\omega})M_pJ,$$

$$(3.12)$$

where

$$M = \begin{pmatrix} A & -C \\ B & -D \end{pmatrix}, \quad M_p = \begin{pmatrix} A & | -\Gamma\begin{pmatrix} C \\ D \end{pmatrix} \\ B & \end{pmatrix}.$$

$$(3.13)$$

In view of (3.7), $\theta(z)$ admits the representation

$$\theta(z) = \hat{\theta}(z) \begin{pmatrix} I & 0 \\ (A^*, B^*)K_p^{-1}\begin{pmatrix} A \\ B \end{pmatrix} & I \end{pmatrix},$$

$$(3.14)$$

where the function

$$\hat{\theta}(z) = I + iJM^*\Gamma(\bar{z})^*K^{-1}M$$

$$(3.15)$$

is, in view of (3.10) and Lemma 3.1, in the class $\mathbf{W}$.

Since $\theta(z)$, $\theta_p(z)$ are both J-unitary on the real axis then the following symmetry relations hold

$$\theta^{-1}(z) = J\theta(\bar{z})^*J, \qquad \theta_p^{-1}(z) = J\theta_p(\bar{z})^*J,$$

which together with (3.12) imply that

$$J - \theta(z)^{-*}J\theta^{-1}(z) = J(J - \theta(\bar{z})J\theta(\bar{z})^*)J =$$

$$i(\bar{z} - z)M^*\Gamma(z)^*K^{-1}\Gamma(z)M;$$

$$(3.16)$$

$$J - \theta_p(z)^{-*}J\theta_p^{-1}(z) = i(\bar{z} - z)M_p^*\Gamma(z)^*K_p^{-1}\Gamma(z)M_p.$$

$$(3.17)$$

Using (3.8), (3.9), (3.13) we rewrite the inequalities (2.1), (2.2) in the form

$$\begin{pmatrix} K & \Gamma(z)M\begin{pmatrix} w(z) \\ I \end{pmatrix} \\ \star & \frac{w(z) - w(z)^*}{z - \bar{z}} \end{pmatrix} \geq 0; \qquad \begin{pmatrix} K_p & \Gamma(z)M_p\begin{pmatrix} zw(z) \\ I \end{pmatrix} \\ \star & \frac{zw(z) - \bar{z}w(z)^*}{z - \bar{z}} \end{pmatrix} \geq 0.$$

$$(3.18)$$

Since $K > 0$, $K_p > 0$, inequalities (3.18) are equivalent to

$$(w(z)^*, I)\{\tfrac{J}{i(\bar{z} - z)} - M^*\Gamma(z)^*K^{-1}\Gamma(z)M\}\begin{pmatrix} w(z) \\ I \end{pmatrix} \geq 0;$$

$$(\bar{z}w(z)^*, I)\{\tfrac{J}{i(\bar{z} - z)} - M_p^*\Gamma(z)^*K_p^{-1}\Gamma(z)M_p\}\begin{pmatrix} zw(z) \\ I \end{pmatrix} \geq 0.$$

$$(3.19)$$

Using (3.16), (3.17) the inequalities (3.19) can be rewritten as

$$(w(z)^*, I)\frac{\theta(z)^{*-}J\theta^{-1}(z)}{i(\bar{z}-z)}\begin{pmatrix} w(z) \\ I \end{pmatrix} \geq 0; \tag{3.20}$$

$$(\bar{z}w(z)^*, I)\frac{\theta_p(z)^{*-}J\theta_p^{-1}(z)}{i(\bar{z}-z)}\begin{pmatrix} zw(z) \\ I \end{pmatrix} \geq 0. \tag{3.21}$$

The last inequality in view of (3.3), (3.4) can be rewritten as

$$(w(z)^*, I)\frac{\theta(z)^{*-}P(z)^*JP(z)\theta^{-1}(z)}{i(\bar{z}-z)}\begin{pmatrix} w(z) \\ I \end{pmatrix} \geq 0. \tag{3.22}$$

The description of all the solutions of the system of inequalities (3.20), (3.21) for the nontangential problem was described in [11]. Since the adaptation of this result to the two-sided problem is rather trivial, the proof of Theorem 3.6 is partially omitted. To formulate this theorem we first need some definitions.

Definition 3.5 A pair $\{p(z),\, q(z)\}$ of $\mathbb{C}^{m\times m}$ valued functions meromorphic in $\mathbb{C}\setminus[0,+\infty)$ is called a Stieltjes pair if

$$1)\ \ det(p(z)p^*(z) + q(z)q^*(z)) \not\equiv 0; \tag{3.23}$$

$$2)\frac{q^*(z)p(z) - p^*(z)q(z)}{z-\bar{z}} \geq 0,\ \ Imz \neq 0; \tag{3.24}$$

$$3)\frac{zq^*(z)p(z) - \bar{z}p^*(z)q(z)}{z-\bar{z}} \geq 0,\ \ Imz \neq 0.$$

We introduce an equivalence relation on the set of Stieltjes pairs $\bar{S}_m$: a pair $\{p, q\}$ is said to be equivalent to the pair $\{p_1, q_1\}$ if there exists a $\mathbb{C}^{m\times m}$-valued function $\Omega(z)\,(det\Omega(z) \not\equiv 0)$ meromorphic in $\mathbb{C}\setminus[0,\infty)$ such that $p_1(z) = p(z)\Omega(z),\ q_1(z) = q(z)\Omega(z)$.

Using the matrices A, B defined in (3.8) we introduce the following subset of $\bar{S}_m$:

$$\bar{S}_m^0 = \{\{p(z), q(z)\} \in \bar{S}_m : det(p(z)^*(A^*, B^*)\begin{pmatrix} A \\ B \end{pmatrix} p(z) + q(z)^*q(z) \not\equiv 0\}$$

and note that in the nontangential case $\bar{S}_m^0 = \bar{S}_m$.

Now we state the main result of this section.

Theorem 3.6 *Under the hypothesis $K > 0, K_p > 0$, let $\theta \in W_*$ be defined as in (3.11) and let $\theta = (\theta_{ij})$ be the block decomposition of θ into four $\mathbb{C}^{m\times m}$-valued functions. Then the solutions to the problem (IS) are parametrized by the linear-fractional transformation*

$$w(z) = (\theta_{11}(z)p(z) + \theta_{12}(z)q(z))(\theta_{21}(z)p(z) + \theta_{22}(z)q(z))^{-1} \tag{3.25}$$

where the equivalence classes in the set $\bar{S}_m^0$ of Stieltjes pairs $\{p(z), q(z)\}$ are the independent parameters of (3.25). More precisely: every solution w of the problem (IS) is of the form (3.25) for some Stieltjes pair $\{p(z), q(z)\} \in \bar{S}_m^0$; conversely, for every Stieltjes pair $\{p(z), q(z)\} \in \bar{S}_m^0$ the transformation (3.25) is well-defined $(det(\theta_{21}(z)p(z) + \theta_{22}(z)q(z)) \not\equiv 0)$ and the corresponding $w(z)$ defined by (3.25) is in S_m and is a solution of (IS), and finally, pairs $\{p(z), q(z)\}$, $\{p_1(z), q_1(z)\}$ are equivalent if and only if they lead under the transformation (3.25) to the same $w(z)$.

We first need the following lemma.

Lemma 3.7 *Let $\{p(z), q(z)\}$ be a Stieltjes pair. Then the linear fractional transformation (3.25) is well-defined*

$$det(\theta_{21}(z)p(z) + \theta_{22}q(z)) \not\equiv 0$$

if and only if $\{p(z), q(z)\}$ belongs to $\bar{S}_m^0$.

Proof: To begin with we compute the dual to (3.12) J-form of $\theta(z)$ using (3.13), (3.14) and identity (3.10)

$$\theta(\omega)^* J\theta(z) - J = i(\bar{\omega} - z)\begin{pmatrix} I & (A^*, B^*)K_p^{-1}\begin{pmatrix} A \\ B \end{pmatrix} \\ 0 & I \end{pmatrix} M^* K^{-1}\Gamma(\bar{\omega})K\Gamma(\bar{z})^* K^{-1}M \times$$

$$\times \begin{pmatrix} I & 0 \\ (A^*, B^*)K_p^{-1}\begin{pmatrix} A \\ B \end{pmatrix} & I \end{pmatrix}.$$

$$(3.26)$$

From (3.13)–(3.15) we obtain

$$v(z) = \theta_{21}p(z) + \theta_{22}(z)q(z) =$$

$$= q(z) + (A^*, B^*)\Gamma(\bar{z})^* K^{-1}M\begin{pmatrix} I & 0 \\ (A^*, B^*) K_p^{-1}\begin{pmatrix} A \\ B \end{pmatrix} & I \end{pmatrix}\begin{pmatrix} p(z) \\ q(z) \end{pmatrix}. \qquad (3.27)$$

Let a pair $\{p(z), q(z)\}$ be in $\bar{S}_m^0$. We introduce a pair

$$\begin{pmatrix} u(z) \\ v(z) \end{pmatrix} = \theta(z)\begin{pmatrix} p(z) \\ q(z) \end{pmatrix}$$

and show that $det v(z) \not\equiv 0$.

Indeed, suppose that the point $z \in \mathbb{C}^+$ and the nonzero vector $h \in \mathbb{C}^m$ are such that

$$v(z)h = 0. \qquad (3.28)$$

Since

$$h^*(p(z)^*, q(z)^*)\theta(z)^* J\theta(z) \begin{pmatrix} p(z) \\ q(z) \end{pmatrix} h = (h^* u(z)^*, 0) J \begin{pmatrix} u(z)h \\ 0 \end{pmatrix} = 0$$

then

$$h^*(p(z)^*, q(z)^*)\frac{J}{z - \bar{z}} \begin{pmatrix} p(z) \\ q(z) \end{pmatrix} h = h^*(p(z)^*, q(z)^*)\frac{J - \theta(z)^* J\theta(z)}{z - \bar{z}} \begin{pmatrix} p(z) \\ q(z) \end{pmatrix} h \geq 0.$$
$$\tag{3.29}$$

Substituting (3.26) to (3.29) and taking into account the strict positivity of the kernel $K^{-1}\Gamma(\bar{\omega})K\Gamma(\bar{z})^* K^{-1}$ we obtain the equality

$$M \begin{pmatrix} I & 0 \\ (A^*, B^*) K_p^{-1} \begin{pmatrix} A \\ B \end{pmatrix} & I \end{pmatrix} \begin{pmatrix} p(z) \\ q(z) \end{pmatrix} h = 0 \tag{3.30}$$

which both with (3.27), (3.28) implies

$$q(z)h = 0. \tag{3.31}$$

Substituting (3.13),(3.31) to (3.30) and using (3.10) we obtain the equality

$$\left\{ \begin{pmatrix} A \\ B \end{pmatrix} - \begin{pmatrix} C \\ D \end{pmatrix} (A^*, B^*) K_p^{-1} \begin{pmatrix} A \\ B \end{pmatrix} \right\} p(z)h = K\Gamma^* K_p^{-1} \begin{pmatrix} A \\ B \end{pmatrix} p(z)h = 0$$

which is equivalent to

$$\begin{pmatrix} A \\ B \end{pmatrix} p(z)h = 0. \tag{3.32}$$

It follows from (3.31), (3.32) that

$$det(p(z)^*(A^*, B^*) \begin{pmatrix} A \\ B \end{pmatrix} p(z) + q(z)^* q(z)) = 0. \tag{3.33}$$

Since $\{p(z), q(z)\}$ belongs to $\bar{S}_m^0$, the last equality can hold only for a set of isolated points. Consequently $detv(z) \neq 0$ everywhere outside this exceptional set.

Let conversely $\{p(z), q(z)\} \in \bar{S}_m \backslash \bar{S}_m^0$. Then (3.33) holds for all $z \in \mathbb{C}$ and for every z there exists the nonzero vector $h \in \mathbb{C}^m$ such that (3.31), (3.32) are valid. Substitute (3.31), (3.32) to (3.27) to obtain $v(z)h = 0$. Since z is arbitrary $detv(z) \equiv 0$. Lemma is proved. ∎

Reproducing the arguments of [11] for the nontangential problem and applying Lemma 3.7 we obtain assertions of Theorem 3.6.

As a corollary we obtain that the strict positivity of informative matrices K, K_p ensure the solvability of the problem (IS).

Definition 3.8 *The matrix of the linear fractional transformation describing all the solutions of the interpolation problem is called the resolvent matrix of this problem.*

We conclude this section with a description of all solutions to the problem (IN) in the Nevanlinna class N_m.

Definition 3.9 *A pair $\{p(z), q(z)\}$ of $\mathbb{C}^{m \times m}$ valued functions meromorphic in $\mathbb{C} \setminus \mathbb{R}$ is called a Nevanlinna pair if it satisfies the conditions (3.23), (3.24).*

On the set of Nevanlinna pairs $\bar{N}_m$ we introduce the same equivalence which was which was defined earlier for Stieltjes pairs and define a subset

$$
\begin{aligned}
\bar{N}_m^o &= \{\{p(z), q(z)\} \in \bar{N}_m : det(p(z)^*(A^*, B^*) \begin{pmatrix} A \\ B \end{pmatrix} p(z) + q(z)^* q(z)) \not\equiv 0\} = \\
&= \{\{p(z), q(z)\} \in \bar{N}_m : det((p(z)^*, 0)M^*M \begin{pmatrix} p(z) \\ 0 \end{pmatrix} + q(z)^* q(z)) \not\equiv 0\}
\end{aligned}
$$

which coincides with $\bar{N}_m$ in the nontangential case.

Theorem 3.10 *Let $M \in \mathbb{C}^{r \times m}$, $K > 0$ and Γ be elements of $\mathbb{C}^{m \times m}$ such that*

$$
\Gamma K - K \Gamma^* = -i M J M^*. \tag{3.34}
$$

Then the solutions to the inequality (3.18) are parametrized by a linear fractional transformation

$$
w(z) = (\hat{\theta}_{11}(z)p(z) + \hat{\theta}_{12}(z)q(z))(\hat{\theta}_{21}(z)p(z) + \hat{\theta}_{22}(z)q(z))^{-1} \tag{3.35}
$$

with the resolvent matrix $\hat{\theta}(z)$ defined in (3.15) and Nevanlinna pairs $\{p(z), q(z)\} \in \bar{N}_m^0$ in the role of parameters.

Remark 3.11 The resolvent matrix of the problem (IN) is defined up to J-unitary right factor. So in view of (3.14) the matrix $\theta(z)$ defined in (3.11) (the resolvent matrix of (IS)) is the resolvent matrix of (IN) too. The identity (3.36) for matrices K, Γ, M defined in (2.3), (2.4), (3.19), (3.13) evidently follows from (3.10).

Therefore theorems 2.2 and 3.9 imply

Theorem 3.11 *Under the hypothesis $K > 0$ the solutions to the problem (IN) are parametrized by the linear fractional transformation (3.35), (or (3.25)) with the resolvent matrix $\hat{\theta}(z)$ (or $\theta(z)$) defined in (3.15) ((3.11)) and parameters $\{p(z), q(z)\}$ varying in $\bar{N}_m^0$.*

4 The degenerate case

In this section we consider the case where $K \geq 0$, $K_p \geq 0$, K, K_p being possibly singular. The method of the previous section has to be suitable adapted and as we will see there is still a description of the set of all solutions of (IS) in terms of a linear fractional transformation. Let K, K_p be elements of $\mathbb{C}^{N \times N}$, $rank K = r$, $rank K_p = s$ and let $e_{i_1}, \ldots, e_{i_r}; e_{j_1}, \ldots, e_{j_s}$ be vectors from the standard basis in $\mathbb{C}^N$ such that

$$Lin(e_{i_k})_{k=1,r} \cap Ker K = \{0\};$$
$$Lin(e_{j_l})_{l=1,s} \cap Ker K_p = \{0\}. \tag{4.1}$$

Let $Q \in \mathbb{C}^{r \times N}$, $Q_p \in \mathbb{C}^{s \times N}$ be defined by

$$Q = \begin{pmatrix} e_{i_1} \\ \vdots \\ e_{i_r} \end{pmatrix}, \quad Q_p = \begin{pmatrix} e_{j_1} \\ \vdots \\ e_{j_s} \end{pmatrix}. \tag{4.2}$$

Then it holds that

$$QKQ^* > 0, \quad Q_p K_p Q_p^* > 0,$$

and

$$rank QKQ^* = rank K = r, \quad rank Q_p K_p Q_p^* = rank K_p = s.$$

Hence, the system of inequalities (3.18) is equivalent to the system

$$\begin{pmatrix} QKQ^* & Q\Gamma M \begin{pmatrix} w(z) \\ I \end{pmatrix} \\ \star & \frac{w(z) - w^*(z)}{z - \bar{z}} \end{pmatrix} \geq 0; \tag{4.3}$$

$$\begin{pmatrix} Q_p K_p Q_p{\star} & Q_p \Gamma M_p \begin{pmatrix} zw(z) \\ I \end{pmatrix} \\ \star & \frac{zw(z) - \bar{z}w^*(z)}{z - \bar{z}} \end{pmatrix} \geq 0; \tag{4.4}$$

$$P_{Ker K} \Gamma(z) M \begin{pmatrix} w(z) \\ I \end{pmatrix} = 0; \tag{4.5}$$

$$P_{Ker K_p} \Gamma(z) M_p \begin{pmatrix} zw(z) \\ I \end{pmatrix} = 0. \tag{4.6}$$

where $P_{Ker K}, P_{Ker K_p}$ are orthogonal projections on subspaces $Ker K$, $Ker K_p$ respectively.

As in the nondegenerate case we introduce subclasses

$$\bar{S}_m^0 = \{\{p(z), q(z)\} \in \bar{S}_m \ : \ det((p(z)^*, 0) M^* Q^* QM \begin{pmatrix} p(z) \\ 0 \end{pmatrix} + q(z)^* q(z) \not\equiv 0\} \subset \bar{S}_m;$$

$$\bar{N}_m^0 = \{\{p(z), q(z)\} \in \bar{N}_m \ : \ det((p(z)^*, 0) M^* Q^* QM \begin{pmatrix} p(z) \\ 0 \end{pmatrix} + q(z)^* q(z) \not\equiv 0\} \subset \bar{N}_m.$$

Theorem 4.1 *Let $K \geq 0$, $K_p \geq 0$, Q, Q_p be matrices defined in (4.2). Then the linear fractional transformation*

$$w(z) = (\theta_{11}(z)p(z) + \theta_{12}(z)q(z))(\theta_{21}(z)p(z) + \theta_{22}(z)q(z))^{-1} \tag{4.7}$$

with the resolvent matrix

$$\theta(z) = \hat{\theta}(z) \begin{pmatrix} I & 0 \\ F & I \end{pmatrix}, \tag{4.8}$$

where

$$\hat{\theta}(z) = I + iJM^*\Gamma(\bar{z})^*Q^*(QKQ^*)^{-1}QM, \tag{4.9}$$

$$F = (A^*, B^*)Q_p^*(Q_pK_pQ_p^*)^{-1}Q_p \begin{pmatrix} A \\ B \end{pmatrix}, \tag{4.10}$$

gives a parametrization of all the solutions to (IS) (or, equivalently, to the system (4.3)-(4.6)), when $\{p(z), q(z)\}$ varies in $\bar{S}_m^0$ and is of the form

$$p(z) = U \begin{pmatrix} \hat{p}(z) & & \\ & O_\mu & \\ & & I_\nu \end{pmatrix}, \quad q(z) = U \begin{pmatrix} \hat{q}(z) & & \\ & I_\mu & \\ & & O_\nu \end{pmatrix}, \tag{4.11}$$

with unitary U and Stieltjes pair $\{\hat{p}(z), \hat{q}(z)\} \in \bar{S}_{m-\mu-\nu}$ where

$$\mu = rank(P_{KerK_p} \begin{pmatrix} A \\ B \end{pmatrix}); \tag{4.12}$$

$$\nu = rank(P_{KerK} \begin{pmatrix} C \\ D \end{pmatrix}). \tag{4.13}$$

The proof is divided into a number of steps.

Step 1 All solutions $w(z)$ to the inequality (4.3) are parametrized by a linear fractional transformation (4.7) with the resolvent matrix $\theta(z)$ defined by (4.8) and $\{p(z), q(z)\}$ varying in $\bar{N}_m^0$.

Step 2 For all such $w(z)$ the conditions (4.5), (4.6) are equivalent to the following ones

$$P_{KerK} \begin{pmatrix} C \\ D \end{pmatrix} q(z) \equiv 0; \quad P_{KerK_p} \begin{pmatrix} A \\ B \end{pmatrix} p(z) \equiv 0. \tag{4.14}$$

Step 3 Let $w(z)$ of the form (4.7) with $\{p(z), q(z)\} \in \bar{N}_m^0$ satisfy (4.3)-(4.5). Then $w(z)$ satisfies (4.6) if and only if the Nevanlinna pair $\{p(z), q(z)\}$ belongs to $\bar{S}_m^0$.

Step 4 The Stieltjes pair $\{p(z), q(z)\}$ satisfies (4.14) if and only if it admits the representation (4.11).

Proof of Step 1 Since the matrix Γ is diagonal then

$$Q\Gamma = Q\Gamma Q^*Q. \tag{4.15}$$

From (3.26), (4.15) we obtain the identity

$$Q\Gamma Q^* Q K Q^* - Q K Q^* Q \Gamma Q^* = -iQMJM^*Q^*.$$

Therefore matrices QM, QKQ^*, ΓQ^* satisfy the conditions of Theorem 3.10 and all the solutions to the inequality (4.3) are parametrized by the linear fractional transformation (4.7) with the resolvent matrix $\hat{\theta}(z)$ defined in (4.9) and parameters $\{p(z), q(z)\} \in \bar{N}_m^0$. In view of Remark 3.1 the same description may be obtained by a linear fractional transformation with the same parameters and the resolvent matrix $\theta(z)$ defined by (4.8).

Proof of Step 2 From (4.7) we obtain

$$\begin{pmatrix} w(z) \\ I \end{pmatrix} = \theta(z) \begin{pmatrix} p(z) \\ q(z) \end{pmatrix} (\theta_{21}(z)p(z) + \theta_{22}(z)q(z))^{-1}. \tag{4.16}$$

Substituting (4.16) to (4.5), (4.6) and taking into account nondegeneracy of the matrix $(\theta_{21}(z)p(z) + \theta_{22}(z)q(z))$ we rewrite conditions (4.5), (4.6) in the equivalent form

$$P_{KerK}\Gamma(z)M\theta(z) \begin{pmatrix} p(z) \\ q(z) \end{pmatrix} = 0; \tag{4.17}$$

$$P_{KerK}\Gamma(z)M_p P(z)\theta(z) \begin{pmatrix} p(z) \\ q(z) \end{pmatrix} = 0. \tag{4.18}$$

Using (3.34), (4.8) we obtain

$$\Gamma(z)M\theta(z) = \{K\Gamma(\bar{z})^*Q^*(QKQ^*)^{-1}QM + \Gamma(z)(I - KQ^*(QKQ^*)^{-1}Q)M\} \begin{pmatrix} I & 0 \\ F & I \end{pmatrix}.$$

Substituting this equality to (4.17) we obtain

$$P_{KerK}\Gamma(z)(I - KQ^*(QKQ^*)^{-1}Q)M \begin{pmatrix} I & 0 \\ F & I \end{pmatrix} \begin{pmatrix} p(z) \\ q(z) \end{pmatrix} \equiv 0. \tag{4.19}$$

Since

$$I - KQ^*(QKQ^*)^{-1}Q = (I - KQ^*(QKQ^*)^{-1}Q)P_{KerK} \tag{4.20}$$

and

$$z\Gamma(z) = I + \Gamma\Gamma(z),$$

identity (4.19) can be rewritten as

$$\{I + P_{KerK}\Gamma\Gamma(z)(I - KQ^*(QKQ^*)^{-1}Q)\}P_{KerK}M \begin{pmatrix} I & 0 \\ F & I \end{pmatrix} \begin{pmatrix} p(z) \\ q(z) \end{pmatrix} \equiv 0$$

and thus

$$P_{KerK}M \begin{pmatrix} I & 0 \\ F & I \end{pmatrix} \begin{pmatrix} p(z) \\ q(z) \end{pmatrix} \equiv 0. \tag{4.21}$$

As to (4.18) we note that in view of (3.10) (3.34), (4.8)

$$\Gamma(z)M_pP(z)\theta(z) = \{\, K_p\Gamma(\bar{z})^*Q^*(qQKQ^*)^{-1}QM +$$

$$+\Gamma(z)\Gamma(I - KQ^*(QKQ^*)^{-1}Q)M + \begin{pmatrix} A & 0 \\ B & 0 \end{pmatrix} \} \begin{pmatrix} I & 0 \\ F & I \end{pmatrix} \tag{4.22}$$

Substituting this equality to (4.18) we obtain

$$P_{KerK}\{\Gamma(z)\Gamma(I - KQ^*(QKQ^*)^{-1}Q)M + \begin{pmatrix} A & 0 \\ B & 0 \end{pmatrix} \} \begin{pmatrix} I & 0 \\ F & I \end{pmatrix} \begin{pmatrix} p(z) \\ q(z) \end{pmatrix} \equiv 0$$

and, in view of (4.20), (4.21)

$$P_{Ker\,K_p} \begin{pmatrix} A \\ B \end{pmatrix} p(z) \equiv 0. \tag{4.23}$$

Substituting (3.13), (4.10), to (4.21) and using (3.10), (4.23) and an equality

$$I - K_pQ_p^*(Q_pK_pQ_p^*)^{-1}Q_p = (I - K_pQ_p^*(Q_pK_pQ_p^*)^{-1}Q_p)P_{KerK_p}$$

we obtain

$$\begin{aligned}
0 &= P_{KerK}M \begin{pmatrix} I & 0 \\ F & I \end{pmatrix} \begin{pmatrix} p(z) \\ q(z) \end{pmatrix} = \\
&= P_{KerK}\{(\begin{pmatrix} A \\ B \end{pmatrix} - (K_p - K\Gamma^*)Q_p^*(Q_pK_pQ_p^*)^{-1}Q_p \begin{pmatrix} A \\ B \end{pmatrix}) - \begin{pmatrix} C \\ D \end{pmatrix} q(z)\} = \\
&= P_{KerK}\{(I - K_pQ_p^*(Q_pK_pQ_p^*)^{-1}Q_p) \begin{pmatrix} A \\ B \end{pmatrix})p(z) - \begin{pmatrix} C \\ D \end{pmatrix} q(z)\} = \\
&= P_{KerK} \begin{pmatrix} C \\ D \end{pmatrix} q(z) \equiv 0.
\end{aligned}$$

Proof of Step 3 Substituting (4.16) to (4.4) we obtain the following inequality

$$\begin{pmatrix} Q_pK_pQ_p^* & Q_p\Gamma(z)M_pP(z)\theta(z) \begin{pmatrix} p(z) \\ q(z) \end{pmatrix} \\ \star & (p(z)^*, q(z)^*)\frac{\theta(z)^*P(z)^*JP(z)\theta(z)}{i(z-\bar{z})} \begin{pmatrix} p(z) \\ q(z) \end{pmatrix} \end{pmatrix} \geq 0. \tag{4.24}$$

Using (3.10), (4.8) we obtain

$$\theta(z)P(z)^*JP(z)\theta(z) = \begin{pmatrix} i(\bar{z} - z)F & i\bar{z}I \\ -izI & 0 \end{pmatrix} + R(z) + T(z) + T^*(z), \tag{4.25}$$

where

$$R(z) = i(\bar{z} - z)\{\, M^*Q^*(QKQ^*)^{-1}Q\Gamma(\bar{z})K_p\Gamma(\bar{z})^*Q^*(QKQ^*)^{-1}QM +$$

$$+ \begin{pmatrix} A^* & B^* \\ 0 & 0 \end{pmatrix} \Gamma(\bar{z})^*Q^*(QKQ^*)^{-1}QM + M^*Q^*(QKQ^*)^{-1}Q\Gamma(\bar{z}) \begin{pmatrix} A & 0 \\ B & 0 \end{pmatrix} \} \begin{pmatrix} I & 0 \\ F & I \end{pmatrix};$$

$$\tag{4.26}$$

$$T(z) = -iz \begin{pmatrix} I & F \\ 0 & I \end{pmatrix} M^*(I - Q^*(QKQ^*)^{-1}QK)\Gamma(\bar{z})^*Q^*(QKQ^*)^{-1}QM \begin{pmatrix} I & 0 \\ F & I \end{pmatrix}. \tag{4.27}$$

Since $Q_pK_pQ_p^* > 0$ then the inequality (4.24) is equivalent to

$$(p(z)^*, q(z)^*)\left\{ \frac{\theta(z)^*P(z)^*JP(z)\theta(z)}{i(\bar{z}-z)} - \right.$$
$$\left. -\theta(z)^*P(z)^*M_p^*\Gamma(z)^*Q_p^*(Q_pK_pQ_p^*)^{-1}Q_p\Gamma(z)M_pP(z)\theta(z) \right\} \begin{pmatrix} p(z) \\ q(z) \end{pmatrix} \geq 0. \tag{4.28}$$

Substituting (4.10), (4.21), (4.25)–(4.27) to (4.28) and taking into account (4.14) we obtain

$$\frac{(p(z)^*, q(z)^*)\begin{pmatrix} 0 & i\bar{z}I \\ -izI & 0 \end{pmatrix}\begin{pmatrix} p(z) \\ q(z) \end{pmatrix}}{i(\bar{z}-z)} \geq 0,$$

thus $\frac{\bar{z}q(z)^*p(z) - zp(z)^*q(z)}{\bar{z}-z} \geq 0$ and $\{p(z), q(z)\}$ is a Stieltjes pair. Therefore $\{p(z), q(z)\} \in \bar{S}_m \cap \bar{N}_m^0 = \bar{S}_m^0$.

Proof of Step 4

Lemma 4.2 *[8] Let $\{p(z), q(z)\}$ be in $\bar{S}_m$.*

1. *If $(0, I_\nu)p(z) \equiv 0$ then there exists a pair $\{p_1(z), q_1(z)\} \subset \bar{S}_{m-\nu}$ such that pairs $\{p(z), q(z)\}$ and $\{\begin{pmatrix} p_1(z) & 0 \\ 0 & O_\nu \end{pmatrix}, \begin{pmatrix} q_1(z) & 0 \\ 0 & I_\nu \end{pmatrix}\}$ are equivalent.*

2. *If $(0, I_\mu)q(z) \equiv 0$ then there exists a pair $\{p_2(z), q_2(z)\} \in S_{m-\mu}$ such that pairs $\{p(z), q(z)\}$ and $\{\begin{pmatrix} p_2(z) & 0 \\ 0 & I_\mu \end{pmatrix}, \begin{pmatrix} q_2(z) & 0 \\ 0 & O_\mu \end{pmatrix}\}$ are equivalent.*

Lemma 4.3 *The subspaces $Ran(P_{KerK}\begin{pmatrix} C \\ D \end{pmatrix})$ and $Ran(P_{KerK}\begin{pmatrix} A \\ B \end{pmatrix})$ are orthogonal.*

Proof: Let h be in $Ran(P_{KerK}\begin{pmatrix} C \\ D \end{pmatrix})$ and g be in $Ran(P_{KerK}\begin{pmatrix} A \\ B \end{pmatrix})$.

So $h = f\begin{pmatrix} C \\ D \end{pmatrix}$, $g = e\begin{pmatrix} A \\ B \end{pmatrix}$ for some nonzero vectors $f \in KerK$, $e \in KerK_p$. In view of (3.10)

$$hg^* = f\begin{pmatrix} C \\ D \end{pmatrix}(A^*, B^*)e^* = fK_pe^* - fK\Gamma^*e = 0.$$

Since vectors h, g are arbitrary ones, lemma is proved.

In view of Lemma 4.1 there exist unitary matrices V, U such that

$$VP_{KerK}\begin{pmatrix} C \\ D \end{pmatrix}U = \begin{pmatrix} 0 & T \\ 0 & 0 \end{pmatrix}; \tag{4.29}$$

$$V P_{KerK_p} \begin{pmatrix} A \\ B \end{pmatrix} U = \begin{pmatrix} 0 & W & O_{\mu \times \nu} \\ 0 & 0 & 0 \end{pmatrix}, \tag{4.30}$$

where $T \in \mathbb{C}^{\nu \times \nu}$, $W \in \mathbb{C}^{\mu \times \mu}$ are nondegenerate matrices. Substituting (4.29), (4.30) to (4.14) we obtain the equivalent conditions

$$(0, T) U^* q(z) \equiv 0; \qquad (0, W, O_{\mu \times \nu}) U^* p(z) \equiv 0.$$

Taking into account the nondegeneracy of T, W and applying twice lemma 4.1 to the Stieltjes pair $\{U^* p(z), U^* q(z)\}$ we obtain the equivalence of pairs

$$\{p(z), q(z)\} \text{ and } \left\{ U \begin{pmatrix} \hat{p}(z) & \\ & O_\mu \\ & & I_\nu \end{pmatrix}, U \begin{pmatrix} \hat{q}(z) & \\ & I_\mu \\ & & O_\nu \end{pmatrix} \right\}$$

for some unitary matrix U and a pair $\{\hat{p}(z), \hat{q}(z)\} \in \bar{S}_{m-\mu-\nu}$. This ends the proof of the Theorem. $\blacksquare$

Corollary 4.4 *Let*

$$rank(A, B) = m. \tag{4.31}$$

> *Then the problem (IS) is solvable if and only if matrices K, K_p are both nonnegative.*

Proof: The necessity follows from Theorem 2.1. The sufficiency follows from Theorem 4.1: in view of (4.31) $\bar{S}_m = \bar{S}_m^0$ and the set of elements of the form (4.11) is nonempty.

The following example shows that the nonnegativity of matrices K, K_p does not suffice the solvability of the problem (IS).

Example 4.6 Let $n = 1$, $m = 3$, $r = 2, s = 0$, $z_1 = i$,

$$a_1 = \begin{pmatrix} 1 & 0 & 0 \\ 0 & 1 & 0 \end{pmatrix}; \quad c_1 = \begin{pmatrix} i & 0 & 0 \\ 0 & 0 & 1 \end{pmatrix}.$$

Therefore,

$$K = \begin{pmatrix} 1 & 0 \\ 0 & 0 \end{pmatrix}, \quad K_p = \begin{pmatrix} 0 & 0 \\ 0 & 0 \end{pmatrix};$$

$$P_{KerK} c_1 = \begin{pmatrix} 0 & 0 & 0 \\ 0 & 0 & 1 \end{pmatrix}; \quad P_{KerK_p} a_1 = \begin{pmatrix} 1 & 0 & 0 \\ 0 & 1 & 0 \end{pmatrix}. \tag{4.32}$$

According to Theorem 4.1 the problem (IS) is solvable if and only if there exists a pair $\{p, q\} \in \bar{S}_m^0$ such that

$$P_{KerK_p} a_1 p(z) \equiv 0, \quad P_{KerK_p} c_1 q(z) \equiv 0. \tag{4.33}$$

Substituting (4.32) to (4.33) and using Lemma 4.2 we obtain that up to equivalence

$$p(z) = \begin{pmatrix} 0 & & \\ & 0 & \\ & & 1 \end{pmatrix}, \quad q(z) = \begin{pmatrix} 1 & & \\ & 1 & \\ & & 0 \end{pmatrix}. \tag{4.34}$$

But the pair $\{p, q\}$ of the form (4.34) does not belong to $\bar{S}_m^0$ since

$$det(p^* a^* a p + q^* q) = det \begin{pmatrix} 1 & & \\ & 1 & \\ & & 0 \end{pmatrix} = 0.$$

Note the following version of Theorem 4.1 for the problem (IN).

Theorem 4.5 *Let $K \geq 0$, Q be the matrix defined by (4.2). Then the linear fractional transformation (4.7) with the resolvent matrix $\hat{\theta}(z)U$, where $\hat{\theta}(z)$ is the matrix defined by (4.9) and U is a J-unitary matrix gives a parametrization of all solutions to (IN) when parameter $\{p(z), q(z)\}$ varies in $\bar{N}_m^0$ and is of the form*

$$p(z) = V \begin{pmatrix} \hat{p}(z) & 0 \\ 0 & O_\mu \end{pmatrix}, \quad q(z) = V \begin{pmatrix} \hat{q}(z) & 0 \\ 0 & I_\mu \end{pmatrix}$$

where $V \in \mathbb{C}^{m \times m}$ is a unitary matrix, $\{\hat{p}(z), \hat{q}(z)\} \in \bar{N}_{m-\mu}$, and

$$\mu = rank(P_{KerK} \begin{pmatrix} A - iC \\ B - iD \end{pmatrix}).$$

To conclude this section we note that adding of the real negative interpolation points $x_i < 0$, $i = 1, \ldots, k$ with the interpolation conditions

$$f_i \omega(x_i) = g_i, \quad f_i \omega'(x_i) f_i^* = \eta_i,$$

for matrices $f_i, g_i \in \mathbb{C}^{t_i \times m}$, $\eta_i \in \mathbb{C}^{t_i \times t_i}$ has no influence on the character of the previous results. Namely all previous theorems and formulas (only for the problem (IS), as Nevanlinna functions in general may be not defined on the real axis) are valid for

$$\tilde{K} = \begin{pmatrix} K & \begin{pmatrix} K_4 \\ K_5 \end{pmatrix} \\ (K_4^*, K_5^*) & K_x \end{pmatrix}, \quad \tilde{\Gamma} \begin{pmatrix} \Gamma & 0 \\ 0 & diag(x_i I_{t_i}) \end{pmatrix},$$

$$\tilde{K}_p = \Gamma K + \begin{pmatrix} A \\ B \\ F \end{pmatrix} (C^*, D^*, G^*);$$

$$\tilde{M} = \begin{pmatrix} A & -C \\ B & -D \\ F & -G \end{pmatrix}, \quad \tilde{M}_p = \left(\begin{pmatrix} A \\ B \\ F \end{pmatrix} | - \tilde{\Gamma} \begin{pmatrix} C \\ D \\ G \end{pmatrix} \right),$$

where K, Γ, A, B, C, D are matrices defined as above and

$$F = \begin{pmatrix} f_1 \\ \vdots \\ f_k \end{pmatrix}; \qquad G = \begin{pmatrix} g_1 \\ \vdots \\ g_k \end{pmatrix};$$

$$(K_x)_{ij} = \begin{cases} \eta_i & \text{for } i = j \\ \dfrac{g_i f_j^* - f_i g_i}{x_i - x_j} & \text{for } i \neq j \end{cases}$$

$$(K_4)_{ij} = \frac{c_i f_j^* - a_i g_j}{z_i - x_j}; \qquad (K_5)_{ij} = \frac{d_i f_j^* - b_i g_j}{\bar{z}_i - x_j}.$$

5 The Schur algorithm

The problem (IS) as a number of classical interpolation problems may be solved by a recursive algorithm [16]. Note that recursions of Schur to the tangential and the two-sided Nevanlinna-Pick problems were extended in [4], [6], [13].

The Schur algorithm to the problem (IS) is based on the following theorems.

Theorem 5.1 *Let us consider the following system of the matrix inequalities*

$$\begin{pmatrix} K & \Gamma(z)M \begin{pmatrix} p(z) \\ q(z) \end{pmatrix} \\ \star & \dfrac{q^*(z)p(z) - p^*(z)q(z)}{z - \bar{z}} \end{pmatrix} \geq 0,$$

$$\begin{pmatrix} K_p & \Gamma(z)M_p \begin{pmatrix} zp(z) \\ q(z) \end{pmatrix} \\ \star & \dfrac{zq^*(z)p(z) - \bar{z}p^*(z)q(z)}{z - \bar{z}} \end{pmatrix} \geq 0, \tag{5.1}$$

(the "projective" analogue of the system (2.7)). Then

1) all solutions of the system (5.1) are parametrized by the linear transformation

$$\begin{pmatrix} p(z) \\ q(z) \end{pmatrix} = \theta(z) \begin{pmatrix} p_1(z) \\ q_1(z) \end{pmatrix}, \tag{5.2}$$

where $\theta(z) \in W_s$ is a matrix defined in (3.11) and $\{p_1(z), q_1(z)\}$ varies in $\bar{S}$.

2) The pairs $\{p, q\}$ and $\{u, v\}$ are equivalent if and only if the corresponding parameters $\{p_1, q_1\}$ and $\{u_1, v_1\}$ are equivalent.

The proof of this theorem is similar to the proof of Theorem 3.1. Note that pairs $\{p(z), q(z)\}$ satisfying to (5.1) are the solutions of some interpolation problem in the class of Stieltjes pairs $\bar{S}$. Such interpolation problem in the set $\bar{N}$ of Nevalinna pairs was considered in [1]. According to Theorem 2.1 functions $w(z) \in S_m$ satisfying the conditions

$$a_1 w(z_1) = c_1, \quad b_1 w^*(z_1) = d_1, \quad a_1 w'(z_1) b_1^* = \gamma_1$$

are parametrized by the linear fractional transformation

$$w(z) = (\theta_{11}^{(1)}(z) p(z) + \theta_{12}^{(1)}(z) q(z))(\theta_{21}^{(1)}(z) p(z) + \theta_{22}^{(1)}(z) q(z))^{-1}, \tag{5.3}$$

where

$$\theta^{(1)}(z) = (\theta_{ik}^{(1)}(z))_{i,k=1,2} =$$

$$= I_{2m} + i \begin{pmatrix} c_1^* & d_1^* & 0 & 0 \\ 0 & 0 & a_1^* & b_1^* \end{pmatrix} \begin{pmatrix} \Gamma_1^*(\bar{z}) & \Gamma_1^* \Gamma_1^*(\bar{z}) \\ \Gamma_1^*(\bar{z}) & z\Gamma_1^*(\bar{z}) \end{pmatrix} \begin{pmatrix} k_1^{-1} & 0 \\ 0 & k_{p1}^{-1} \end{pmatrix} \begin{pmatrix} c_1 & 0 \\ d_1 & 0 \\ 0 & a_1 \\ 0 & b_1 \end{pmatrix} J; \tag{5.4}$$

$$\Gamma = \begin{pmatrix} z_1 I_{r_1} & 0 \\ 0 & \bar{z}_1 I_{s_1} \end{pmatrix}; \quad \Gamma_1(z) = (zI - \Gamma)^{-1};$$

$$k_1 = \begin{pmatrix} \dfrac{c_1 a_1^* - a_1 c_1^*}{z_1 - \bar{z}_1} & \gamma_1 \\ \gamma_1^* & \dfrac{b_1 d_1^* - d_1 b_1^*}{z_1 - \bar{z}_1} \end{pmatrix}; \tag{5.5}$$

$$k_{p1} = \begin{pmatrix} \dfrac{z_1 c_1 a_1^* - \bar{z}_1 a_1 c_1^*}{z_1 - \bar{z}_1} & \gamma_1 + c_1 b_1^* \\ \gamma_1^* + b_1 c_1^* & \dfrac{z_1 b_1 d_1^* - \bar{z}_1 d_1 b_1^*}{z_1 - \bar{z}_1} \end{pmatrix};$$

and $\{p(z), q(z)\}$ is an arbitrary Stieltjes pair such that

$$\det(p(z)^* (a_1^*, b_1^*) \begin{pmatrix} a_1 \\ b_1 \end{pmatrix} p(z) + q(z)^* q(z)) \not\equiv 0.$$

We establish the conditions on $\{p(z), q(z)\}$ which ensure $w(z)$ to be a solution of the "full" problem (IS). Let us define a new interpolation data $a_i^{(1)}, b_i^{(1)}, c_i^{(1)}, d_i^{(1)}, \gamma_i^{(1)}$ $(i = 2, \ldots, n)$ in the following way:

$$\begin{pmatrix} a_i^{(1)} \\ c_i^{(1)} \end{pmatrix} = \begin{pmatrix} a_i \\ c_i \end{pmatrix} - \begin{pmatrix} \dfrac{z_i c_i a_1^* - \bar{z}_1 a_i c_1^*}{z_i - \bar{z}_1} & \dfrac{z_i c_i b_1^* - \bar{z}_1 a_i d_1^*}{z_i - z_1} \\ \dfrac{c_i a_1^* - a_i c_1^*}{z_i - \bar{z}_1} & \dfrac{c_i b_1^* - a_i d_1^*}{z_i - z_1} \end{pmatrix} k_{p1}^{-1} \begin{pmatrix} a_1 \\ b_1 \end{pmatrix};$$

$$\begin{pmatrix} b_i^{(1)} \\ d_i^{(1)} \end{pmatrix} = \begin{pmatrix} b_i \\ d_i \end{pmatrix} - \begin{pmatrix} \dfrac{\bar{z}_i d_i a_1^* - \bar{z}_1 b_i c_1^*}{\bar{z}_i - \bar{z}_1} & \dfrac{z_1 b_i d_1^* - z_i d_i b_1^*}{z_1 - \bar{z}_i} \\ \dfrac{d_i a_1^* - b_i c_1^*}{\bar{z}_i - \bar{z}_1} & \dfrac{b_i d_1^* - d_i b_1^*}{z_1 - \bar{z}_i} \end{pmatrix} k_1^{-1} \begin{pmatrix} c_1 \\ d_1 \end{pmatrix}; \tag{5.6}$$

$$\gamma_i^{(1)} = \gamma_i - \left(\dfrac{c_i a_1^* - a_i c_1^*}{z_i - \bar{z}_1}, \dfrac{c_i b_1^* - a_i d_1^*}{z_i - z_1} \right) k_1^{-1} \begin{pmatrix} \dfrac{c_1 b_1^* - a_1 d_1^*}{z_i - z_1} \\ \dfrac{d_1 b_1^* - b_1 d_1^*}{\bar{z}_i - z_1} \end{pmatrix}.$$

After these data we construct as in (2.3)–(2.6) informative block-matrices $\tilde{K}$, $\tilde{K}_p$ and matrices

$$\tilde{A} = (a_i^{(1)}), \quad \tilde{B} = (b_i^{(1)}), \quad \tilde{C} = (c_i^{(1)}), \quad \tilde{D} = (d_i^{(1)}), \quad (i = 2, \ldots, n),$$

$$\tilde{M} = \begin{pmatrix} \tilde{A} & \tilde{C} \\ \tilde{B} & \tilde{D} \end{pmatrix}, \quad \tilde{M}_p = (\ \tilde{A} \atop \tilde{B}\ |\ -\Gamma \begin{pmatrix} \tilde{C} \\ \tilde{D} \end{pmatrix}), \tag{5.7}$$

where

$$\tilde{\Gamma} = \begin{pmatrix} \mathrm{diag}(z_i I r_i) & 0 \\ 0 & \mathrm{diag}(\bar{z}_i I s_i) \end{pmatrix}_{i=2,\ldots,n}.$$

Let us consider the following system of matrix inequalities

$$\begin{pmatrix} \tilde{K} & (zI - \tilde{\Gamma})^{-1}\tilde{M}\begin{pmatrix} p(z) \\ q(z) \end{pmatrix} \\ \star & \dfrac{q^*(z)p(z) - p^*(z)q(z)}{z - \bar{z}} \end{pmatrix} \geq 0;$$

$$\begin{pmatrix} \tilde{K}_p & (zI - \tilde{\Gamma})^{-1}\tilde{M}_p\begin{pmatrix} zp(z) \\ q(z) \end{pmatrix} \\ \star & \dfrac{zq^*(z)p(z) - \bar{z}p^*(z)q(z)}{z - \bar{z}} \end{pmatrix} \geq 0. \tag{5.8}$$

Theorem 5.2 *The matrix-valued function $w(z)$ of the form (5.3) is a solution of the problem (IS) if and only if the corresponding Stieltjes pair $\{p(z), q(z)\}$ satisfies the system (5.8).*

To prove this theorem it is sufficient to substitute (5.3), (5.4) to (3.18). Since the inequalities (5.8) preserve the structure of the inequalities (3.18), the recursion can be continued with successive use of Theorems 3.1, 3.2.

After n steps of the described algorithm we shall obtain Theorem 3.1, at the same time the resolvent matrix $\theta(z)$ of the problem (IS) will be represented in the form of Blaschke-Potapov product

$$\theta(z) = \sqcap_{l=1}^{n}\theta^{(l)}(z),$$

where

$$\theta^{(l)}(z) = I + i \begin{pmatrix} c_l^{(l-1)^*} & d_l^{(l-1)^*} & 0 & 0 \\ 0 & 0 & a_l^{(l-1)^*} & b_l^{(l-1)^*} \end{pmatrix} \begin{pmatrix} \Gamma_l(\bar{z})^* & \Gamma_l^*\Gamma_l(\bar{z})^* \\ \Gamma_l(\bar{z})^* & z\Gamma_l(\bar{z})^* \end{pmatrix} \times$$

$$\times \begin{pmatrix} k_l^{-1} & 0 \\ 0 & k_{pl}^{-1} \end{pmatrix} \begin{pmatrix} c_l^{(l-1)} & 0 \\ d_l^{(l-1)} & 0 \\ 0 & a_l^{(l-1)} \\ 0 & b_l^{(l-1)} \end{pmatrix} J$$

for

$$\Gamma_l = \begin{pmatrix} z_l I_{r_l} & 0 \\ 0 & \bar{z}_l I_{s_l} \end{pmatrix}, \quad \Gamma_l(z) = (zI - \Gamma_l)^{-1}$$

and matrices $a_i^{(1)}, b_i^{(1)}, c_i^{(1)}, d_i^{(1)}, \gamma_i^{(1)}$ defined after matrices $a_i^{(l-1)}, b_i^{(l-1)}, c_i^{(l-1)}, d_i^{(l-1)}$ by recursive relations similarly to (5.6).

References

1. D. Alpay, P. Bruinsma, A. Dijksma, H. de Snoo, "Interpolation problems, extensions of symmetric operators and reproducing kernel spaces I", Operator theory: Advances and Applications, Vol. 50 (1991), 35-82.

2. D. Alpay, V. Bolotnikov, "On a class of functions analytic in a half disk and an associated interpolation problem". To appear in J.M. Analysis and Appl.

3. D. Alpay, J. Ball, I. Gohberg, L. Rodman, "Two-sided residue interpolation in the Stieltjes class." In preparation.

4. D. Alpay, H. Dym, "On applications of reproducing kernel spaces to the Schur algorithm and rational J-unitary factorization", O.T. Ad. Appl. 18 (1986).

5. J.A. Ball, "Interpolation problems of Pick-Nevanlinna and Loewner types for meromorphic matrix function", Integral Equations Operator Theory, 6 (1983), 804-840.

6. J.A. Ball, "Nevanlinna-Pick interpolation: Generalizations and applications", Pitman research notes in mathematics 171, Longman, 1989, 51-94.

7. J. Ball, I. Gohberg, L. Rodman, Interpolation of Rational Matrix Functions, Operator THeory: Advances and Applications, 45 (1990)

8. V. Bolotnikov, "Bitangential problem of Nevanlinna-Pick in the Stieltjes class" Kharkov University, 1984 (Russian).

9. P. Bruinsma, "Degenerate interpolation problems for Nevanlinna pairs", Indag. Mathem. To be published.

10. H. Dym, J-contractive matrix functions, reproducing kernel Hilbert spaces and interpolation, Regional Conf. Series in Math. 71, Amer. Math. Soc., Providence, RI, 1989.

11. Y. Dyukarev, V. Katsnelson, "Multiplicative and additive classes of Stieltjes analytic matrix valued functions, and interpolation problems associated with them", Amer. Math. Soc. Transl. (2) Vol 131, 1986.

12. A. Efimov, V. Potapov, "J-expanding matrix functions and their role in the analytical theory of electrical circuits", Russian Math. Surveys 28 (1973).

13. I. Fedchina, "Descriptions of solutions of the tangential Nevanlinna-Pick problem", Doklady Akad. Nauk Arm. SSR, 60 (1975) 37-42 (Russian).

14. I. Kovalishina, "Analytic theory of a class of interpolation problems", Math. USSR. Izv. 22 (1984), 419-463.

15. I. Kovalishina, V. Potapov, "An indefinite metric in the Nevanlinna-Pick problem", Amer. Math. Soc. Transl. (2) (1988), 15-19.

16. I. Schur, "Uber Potenzreihen die im Innern des Einheitshreises beschränkt sind", Reine Ang. Math. 147 (1917).

Department of Mathematics and Computer Science
Ben Gurion University of the Negev
Beer-Sheva 84105, Israel.

Mathematics Subject Classification: 30E05

Operator Theory:
Advances and Applications, Vol. 62
© 1993 Birkhäuser Verlag Basel

STATE SPACE FORMULAS FOR COPRIME FACTORIZATIONS

P. A. Fuhrmann*[†]and R. Ober

Dedicated to Professor T. Ando on his sixtieth birthday

Abstract

In this paper we will give a uniform approach to the derivation of state space formulas of coprime factorizations, of different types, for rational matrix functions.

1   Introduction

The notion of coprimeness is as old as mathematics and goes back at least to the golden age of Greece, we refer to the Euclidean algorithm for the computation of the greatest common divisor of two integers.

Our interest in this paper lies in the representations of rational functions, i.e. quotients of coprime polynomials. By the Euclidean algorithm, or equivalently via ideal theory, coprimeness of two polynomials p, q is equivalent to the solvability of the Bezout equation

$$ap + bq = 1$$

over the ring of polynomials.

Changing our focus to the study of matrix rational functions, we use left and right matrix fractions of the form

$$G = ND^{-1} = \overline{D}^{-1}\overline{N}$$

with $N, \overline{N}, D, \overline{D}$ polynomial matrices. Such factorizations are called right and left coprime factorizations respectively if there exist polynomial matrix solution to the Bezout equations

$$XN + YD = I$$

and

$$\overline{NX} + \overline{DY} = I$$

respectively.

---

*Earl Katz Family Chair in Algebraic System Theory
[†]Partially supported by the Israeli Academy of Sciences

These polynomial coprime factorizations played an extremely important role in the development of algebraic system theory, and in particular in realization theory. In this connection we refer to Rosenbrock [1970], Fuhrmann [1976], Kailath [1980].

In a development parallel to system theory, operator theorists studied similar types of coprime factorizations, however over different rings (or rather algebras). The most prominent algebra in this connection is H^∞, the algebra of bounded analytic functions on the unit disc, or alternatively a half plane. In the wake of Beurling [1949] came the intensive study of shift operators. Cyclic vectors for the (right) shift operator were identified already by Beurling as outer functions. The next step was to determine the cyclic and noncyclic vectors of the backward shift. The noncyclic vectors of the backward shift are important inasmuch as they generalize the role of rational functions. The fundamental contribution in this connection is the work of Douglas, Shapiro and Shields [1971] and its generalization to the matrix case in Fuhrmann [1975]. The interesting point is that noncyclic vectors in H^2 are characterized in terms of special coprime factorizations over H^∞. We will refer to these factorizations as DSS (Douglas-Shapiro-Shields) factorizations.

As may be expected, the DSS factorization plays a central role in the development of infinite dimensional system theory. This is the theme of Fuhrmann [1981]. It is interesting to point out that the use of shift operators in infinite dimensional system theory predates their use in algebraic system theory, which was originated in Fuhrmann [1976].

Realization theory is but a tool in the development of control theory. Thus the real interest in the use of coprime factorizations is their application to the solution of control problems, in particular to the construction of stabilizing controllers. The cornerstone of this whole area is the Kucera-Youla parametrization of all stabilizing controllers which is based on coprime factorizations over H^∞. Pioneering works in this direction are Desoer et al. [1980], McFarlane and Glover [1989]. State space formulas for coprime factorizations were first developed by Khargonekar and Sontag [1982], Nett [1984]. The proof of coprimeness was done via explicit construction of doubly coprime factorizations. Specific choice was made for the solution of the Bezout equations, however no attempt was made to give an intrinsic characterization of the resulting doubly coprime factorization. We remedy this by showing that special choices lead to minimal McMillan degree doubly coprime factorizations. For the DSS factorization the state space formulas are due to Doyle [1984], and for the case of normalized coprime factorizations to Meyer and Franklin [1987], see also Vidyasagar [1985]. A polynomial approach to the derivation of normalized coprime factorizations was given in Fuhrmann and Ober [1992]. This method is powerful enough to lead to the unified derivation of state space formulas for various types of coprime factorizations, and this is the theme of this paper. For results concerning coprime factorization for nonlinear systems see e.g. Hammer [1985] and Verma [1988]

After some preliminary results on polynomial models we will present a unified approach to the derivation of state space formulas for coprime factorizations and normalized coprime factorizations for various classes of functions. Thus we will study the classes of all rational functions of given McMillan degree, the class of unstable ones, bounded real and positive real functions. Except for the case of unstructured coprime factorizations, all other coprime factorizations are naturally given in terms of special indefinite metrics. Applications of these coprime factorizations, as well as the methods used to obtain them, to other control problems will be given in a subsequent paper.

The first author would like to thank the Center for Engineering Mathematics at the University of Texas at Dallas for its hospitality and support during the work on this research. Both authors wish to thank Prof. D. Prätzel-Wolters and the Department of Mathematics at the University of Kaiserslautern for their hospitality and support during the final preparations of this manuscript.

2 General factorizations

In this section we are going to analyze general factorizations of proper rational functions such that the factors are stable rational functions. The precise definition is as follows.

Definition 2.1 *Let G be a proper rational matrix-valued function. Then the factorization*

1. *$G = NM^{-1}$ is called a* right *factorization (RF) of G, if N, M are stable rational functions and M is invertible with proper inverse.*
   If N, M are right coprime, *i.e. if there exist stable rational functions $\tilde{U}$, $\tilde{V}$ such that*

   $$M\tilde{V} - N\tilde{U} = I,$$

   then the factorization is called a right coprime factorization *(RCF).*

2. *$G = \tilde{M}^{-1}\tilde{N}$ is called a* left *factorization (LF) of G, if $\tilde{N}, \tilde{M}$ are stable rational functions and $\tilde{M}$ is invertible with proper inverse.*
   If $\tilde{N}$, $\tilde{M}$ are left coprime, *i.e. if there exist stable rational functions U, V such that*

   $$V\tilde{M} - U\tilde{N} = I,$$

   then the factorization is called a left coprime factorization *(LCF).*

It is a standard result (see e.g. Vidyasagar [1985]) that right (left) coprime factorizations are unique up to right (left) multiplication by a stable rational function with proper stable inverse.

We are in particular interested in factorizations such that the McMillan degree of $\begin{pmatrix} M \\ N \end{pmatrix}$ equals the McMillan degree of G. The following Lemma shows that the McMillan degree of the function $\begin{pmatrix} M \\ N \end{pmatrix}$ is always larger than the McMillan degree of the $G = NM^{-1}$.

Lemma 2.1 *Let $G = NM^{-1}$ be a not necessarily coprime right factorization of the proper rational function G with M, N proper. Then,*

1. *If* $\begin{pmatrix} M \\ N \end{pmatrix} \equiv \left(\begin{array}{c|c} A_1 & B_1 \\ \hline C_1 & D_1 \\ C_2 & D_2 \end{array} \right)$ *is a realization of* $\begin{pmatrix} M \\ N \end{pmatrix}$ *then*

$$G \equiv \left(\begin{array}{c|c} A_1 - B_1 D_1^{-1} C_1 & B_1 D_1^{-1} \\ \hline C_2 - D_2 D_1^{-1} C_1 & D_2 D_1^{-1} \end{array} \right)$$

is a realization of G.

2. *if we denote by* $\delta(F)$ *the McMillan degree of the proper rational function* F *then*

$$\delta(G) \leq \delta \begin{pmatrix} M \\ N \end{pmatrix}.$$

Proof: 1.) Note that, since $M = D_1 + C_1(sI - A_1)^{-1} B_1$, it follows that $M^{-1} = D_1^{-1} - D_1^{-1} C_1(sI - A_1 + B_1 D_1^{-1} C_1)^{-1} B_1 D_1^{-1}$. Therefore

$$G = NM^{-1} = [D_2 + C_2(sI - A_1)^{-1} B_1][D_1^{-1} - D_1^{-1} C_1(sI - A_1 + B_1 D_1^{-1} C_1)^{-1} B_1 D_1^{-1}]$$

$$\begin{aligned}
&= D_2 D_1^{-1} + C_2(sI - A_1)^{-1} B_1 D_1^{-1} \\
&\quad - D_2 D_1^{-1} C_1(sI - A_1 + B_1 D_1^{-1} C_1)^{-1} B_1 D_1^{-1} \\
&\quad - C_2(sI - A_1)^{-1} B_1 D_1^{-1} C_1(sI - A_1 + B_1 D_1^{-1} C_1)^{-1} B_1 D_1^{-1}
\end{aligned}$$

$$\begin{aligned}
&= D_2 D_1^{-1} - D_2 D_1^{-1} C_1(sI - A_1 + B_1 D_1^{-1} C_1)^{-1} B_1 D_1^{-1} \\
&\quad + C_2(sI - A_1)^{-1}[sI - A_1 + B_1 D_1^{-1} C_1 - B_1 D_1^{-1} C_1](sI - A_1 + B_1 D_1^{-1} C_1)^{-1} B_1 D_1^{-1}
\end{aligned}$$

$$= D_2 D_1^{-1} + (C_2 - D_2 D_1^{-1} C_1)(sI - A_1 + B_1 D_1^{-1} C_1)^{-1} B_1 D_1^{-1}.$$

2.) This follows immediately from part 1.)   □

The following proposition gives a method to obtain factorizations using polynomial matrices. It establishes the existence of a factorization $G = NM^{-1}$ such that the McMillan degree of $\begin{pmatrix} M \\ N \end{pmatrix}$ equals the McMillan degree of G. A key step in the proof of this proposition is the following result that follows from the realization theory via polynomial models. For information on polynomial system and realization theory see Fuhrmann [1981].

Theorem 2.1 *Let* $G = ND^{-1}$ *be a polynomial coprime factorization and let* (A, B, C) *be a minimal realization of* G. *Let* $G' = MD^{-1}$ *with* M *a polynomial matrix. Then* G' *has a realization* (A, B, C_0) *for some* C_0.

With the help of this theorem we can now prove the desired existence result of right factorizations with a given McMillan degree constraint.

Proposition 2.1 *Let* G *be a proper rational transfer function and let* $G = ED^{-1}$ *($G = \overline{D}^{-1}\overline{E}$) be a polynomial right (left) coprime factorization. Let* T *($\tilde{T}$) be a square stable polynomial matrix of the same dimensions as* D *($\overline{D}$), such that* $N := ET^{-1}$ *($\tilde{N} := \overline{T}^{-1}\overline{E}$) and* $M := DT^{-1}$ *($\tilde{M} := \overline{T}^{-1}\overline{D}$) are proper and* M *($\tilde{M}$) has a proper inverse, then*

$$\begin{pmatrix} M \\ N \end{pmatrix} := \begin{pmatrix} DT^{-1} \\ ET^{-1} \end{pmatrix} \quad (\begin{pmatrix} -\tilde{N} & \tilde{M} \end{pmatrix} := \begin{pmatrix} -\tilde{T}^{-1}\tilde{E} & \tilde{T}^{-1}\tilde{D} \end{pmatrix})$$

is a right (left) factorization of G $(-G)$ and the McMillan degree of $\begin{pmatrix} M \\ N \end{pmatrix}$ $(\begin{pmatrix} -\tilde{N} & \tilde{M} \end{pmatrix})$ equals the McMillan degree of G $(-G)$.

Proof: The construction implies that NM^{-1} is a right factorization of G.
Let $G \equiv (A, B, C, D)$ be a minimal realization of G. Since $G = ED^{-1}$ and $M^{-1} = TD^{-1}$, Theorem 2.1 implies that M^{-1} has a realization given by

$$M^{-1} \equiv \left(\begin{array}{c|c} A & B \\ \hline C_0 & M^{-1}(\infty) \end{array} \right),$$

for some C_0. Hence M has a realization given by

$$M \equiv \left(\begin{array}{c|c} A - BM(\infty)C_0 & BM(\infty) \\ \hline -M(\infty)C_0 & M(\infty) \end{array} \right).$$

Since $N = ET^{-1}$ it follows again from Theorem 2.1 that

$$N \equiv \left(\begin{array}{c|c} A - BM(\infty)C_0 & BM(\infty) \\ \hline C_1 & D_1 \end{array} \right),$$

for some C_1 and D_1 and therefore

$$\begin{pmatrix} M \\ N \end{pmatrix} \equiv \left(\begin{array}{c|c} A - BM(\infty)C_0 & BM(\infty) \\ \hline -M(\infty)C_0 & M(\infty) \\ C_1 & D_1 \end{array} \right).$$

This shows that there exists a right factorization whose state space realization has the same state-space as the realization of G. Therefore the McMillan degree of $\begin{pmatrix} M \\ N \end{pmatrix}$ is less than the McMillan degree of G and by Lemma 2.1 equal to the McMillan degree of G.
The statement concerning left factorizations is proved using the duality that $G = NM^{-1}$ is a right factorization if and only if $G^T = (M^Y)^{-1}N^T$ is a left factorization of G^T. $\qquad \square$

In the following theorem all right factorizations $G = NM^{-1}$ of a proper rational function G are characterized such that the McMillan degree of G equals the McMillan degree of $\begin{pmatrix} M \\ N \end{pmatrix}$. These factorizations are precisely those that can be obtained via the state feedback construction of Khargonekar and Sontag [1982] and later of Nett et. al. [1984]. Clearly this approach also provides a proof for the existence of right factorizations.

Theorem 2.2 *Let G be a proper rational function G and let $G \equiv \left(\begin{array}{c|c} A & B \\ \hline C & D \end{array} \right)$ be a minimal realization.*

Then, with M, N being proper rational matrices, $G = NM^{-1}$ is a, not necessarily coprime, right factorization of G such that the McMillan degree of $\begin{pmatrix} M \\ N \end{pmatrix}$ equals the McMillan degree of G if and only if there exists a state feedback F such that $A - BF$ is stable, and an invertible matrix D_1, s.t. $\begin{pmatrix} M \\ N \end{pmatrix}$ has a realization given by

$$\begin{pmatrix} M \\ N \end{pmatrix} \equiv \left(\begin{array}{c|c} A - BF & BD_1 \\ \hline -F & D_1 \\ C - DF & DD_1 \end{array} \right).$$

Proof: Let $G = NM^{-1}$ be a right factorization. Let

$$\begin{pmatrix} M \\ N \end{pmatrix} \equiv \left(\begin{array}{c|c} A_1 & B_1 \\ \hline C_1 & D_1 \\ C_2 & D_2 \end{array} \right)$$

be a minimal realization and assume that G and $\begin{pmatrix} M \\ N \end{pmatrix}$ have the same McMillan degree. Since by assumption M has a proper inverse, D_1 is necessarily invertible. The stability of M and N implies that A_1 is stable. By Lemma 2.1 we have that

$$G \equiv \left(\begin{array}{c|c} A_1 - B_1 D_1^{-1} C_1 & B_1 D_1^{-1} \\ \hline C_2 - D_2 D_1^{-1} C_1 & D_2 D_1^{-1} \end{array} \right).$$

Since G and $\begin{pmatrix} M \\ N \end{pmatrix}$ have the same McMillan degree this implies that this realization of G is also minimal. Hence we can assume without loss of generality that

$$\left(\begin{array}{c|c} A & B \\ \hline C & D \end{array} \right) = \left(\begin{array}{c|c} A_1 - B_1 D_1^{-1} C_1 & B_1 D_1^{-1} \\ \hline C_2 - D_2 D_1^{-1} C_1 & D_2 D_1^{-1} \end{array} \right).$$

From here we can see that we have

$$D_2 = DD_1,$$

$$B_1 = BD_1,$$

$$A_1 = A + BC_1,$$

$$C_2 = C + DC_1.$$

Since A_1 is stable and $G \equiv \left(\begin{array}{c|c} A & B \\ \hline C & D \end{array} \right)$ and is minimal, this shows that $F := -C_1$ is a stabilizing state feedback such that,

$$\begin{pmatrix} M \\ N \end{pmatrix} \equiv \left(\begin{array}{c|c} A - BF & BD_1 \\ \hline -F & D_1 \\ C - DF & DD_1 \end{array} \right).$$

Since M has a proper inverse by the assumption, this shows that $D_1 = M(\infty)$ is invertible. Conversely, let D_1 be invertible and let F be such that $A - BF$ is stable. Define $\begin{pmatrix} M \\ N \end{pmatrix}$ by

$$\begin{pmatrix} M \\ N \end{pmatrix} \equiv \left(\begin{array}{c|c} A - BF & BD_1 \\ \hline -F & D_1 \\ C - DF & DD_1 \end{array} \right).$$

Then clearly the McMillan degree of $\begin{pmatrix} M \\ N \end{pmatrix}$ is less than or equal to that of G since both have a realization on the same state-space. It can be verified easily that $G = NM^{-1}$. Hence by Lemma 2.1 $\begin{pmatrix} M \\ N \end{pmatrix}$ and G have the same McMillan degree.

$\square$

The following corollary summarizes the analogous results concerning left factorizations.

Corollary 2.1 *Let G be a proper rational function and let $G \equiv \left(\begin{array}{c|c} A & B \\ \hline C & D \end{array} \right)$ be a minimal realization. Then $G = \tilde{M}^{-1}\tilde{N}$ is a, not necessarily coprime, left factorization of G such that the McMillan degree of $\left(-\tilde{N} \quad \tilde{M} \right)$ equals the McMillan degree of G if and only if there exists a output injection H, such that $A - HC$ is stable, and an invertible matrix $\overline{D}_1$, s.t. $\left(-\tilde{N} \quad \tilde{M} \right)$ has a realization given by*

$$\left(-\tilde{N} \quad \tilde{M} \right) \equiv \left(\begin{array}{c|cc} A - HC & HD - B & -H \\ \hline \overline{D}_1 C & -\overline{D}_1 D & \overline{D}_1 \end{array} \right).$$

Proof: The result can be obtained from the previous theorem, by using the duality that $G = NM^{-1}$ is a right factorization if and only if $G^T = (M^T)^{-1}N^T$ is a left factorization. $\square$

In this theorem and corollary we examined right (left) factorizations. It was left open whether these factorizations are in fact coprime. In order to answer this question we will make use of so-called doubly coprime factorizations.

Definition 2.2 *The two proper stable rational block matrices*

$$\begin{pmatrix} M & U \\ N & V \end{pmatrix}, \quad \begin{pmatrix} \tilde{V} & -\tilde{U} \\ -\tilde{N} & \tilde{M} \end{pmatrix},$$

with M ($\tilde{M}$) having a proper inverse, form a doubly coprime factorization of the proper rational function G, if

$$\begin{pmatrix} \tilde{V} & -\tilde{U} \\ -\tilde{N} & \tilde{M} \end{pmatrix} \begin{pmatrix} M & U \\ N & V \end{pmatrix} = \begin{pmatrix} I & 0 \\ 0 & I \end{pmatrix}$$

and

$$G = NM^{-1} = \tilde{M}^{-1}\tilde{N}.$$

We are particularly interested in doubly coprime factorizations such that all three functions

$$G, \quad \begin{pmatrix} \tilde{V} & -\tilde{U} \\ -\tilde{N} & \tilde{M} \end{pmatrix}, \quad \begin{pmatrix} M & U \\ N & V \end{pmatrix}$$

have the same McMillan degree.

Before we can derive state-space realizations of the doubly coprime factorizations we need to state the following lemma, which is a key step in the proof of the subsequent corollary. It is a consequence of standard arguments in realization theory.

Lemma 2.2 *Let $G = [G_1, G_2]$ be a proper rational function. Assume that G_1 has McMillan degree n, then the McMillan degree of G is n if and only if for the Hankel operators H_{G_1} and H_{G_2} we have*

$$range(H_{G_2}) \subseteq range(H_{G_1}).$$

If this is the case and $G_1 \equiv \left(\begin{array}{c|c} A_1 & B_1 \\ \hline C_1 & D_1 \end{array} \right)$ is a minimal state space realization, then G_2 has a state space realization of the form

$$G_2 \equiv \left(\begin{array}{c|c} A_1 & L \\ \hline C_1 & D_2 \end{array} \right)$$

for some L and D_2.

The following corollary answers two questions. Given a rational function of McMillan degree n, all doubly coprime factors are characterized that have McMillan degree n. As a consequence of the construction, we see that a right factorization $G = NM^{-1}$ such that $\begin{pmatrix} M \\ N \end{pmatrix}$ has McMillan degree n is necessarily right coprime.

Corollary 2.2 *Let G be a proper rational function of McMillan degree n. All proper, doubly coprime factorizations of G such that*

$$\begin{pmatrix} M & U \\ N & V \end{pmatrix}, \quad \begin{pmatrix} \tilde{V} & -\tilde{U} \\ -\tilde{N} & \tilde{M} \end{pmatrix},$$

have McMillan degree n are given by

$$\begin{pmatrix} M & U \\ N & V \end{pmatrix} \equiv \left(\begin{array}{c|cc} A - BF & BD_1 & BD_2 + H\bar{D}_1^{-1} \\ \hline -F & D_1 & D_2 \\ C - DF & DD_1 & \bar{D}_1^{-1} + DD_2 \end{array} \right)$$

and

$$\begin{pmatrix} \tilde{V} & -\tilde{U} \\ -\tilde{N} & \tilde{M} \end{pmatrix} \equiv \left(\begin{array}{c|cc} A - HC & HD - B & -H \\ \hline -D_1^{-1}F - D_1^{-1}D_2\bar{D}_1C & D_1^{-1} + D_1^{-1}D_2\bar{D}_1D & -D_1^{-1}D_2\bar{D}_1 \\ \bar{D}_1C & -\bar{D}_1D & \bar{D}_1 \end{array} \right),$$

where $G \equiv \left(\begin{array}{c|c} A & B \\ \hline C & D \end{array} \right)$ *is a minimal realization and* F (H) *is such that* $A - BF$ $(A - HC)$ *is stable* $D_1, \overline{D}_1$ *are invertible and* D_2 *is arbitrary.*

Proof: Let $\left(\begin{array}{cc} M & U \\ N & V \end{array} \right)$, $\left(\begin{array}{cc} \tilde{V} & -\tilde{U} \\ -\tilde{N} & \tilde{M} \end{array} \right)$ be doubly coprime factors of McMillan degree n.

Then $G = NM^{-1}$ is a right factorization such that $\left(\begin{array}{c} M \\ N \end{array} \right)$ has McMillan degree n. Let

$$\left(\begin{array}{c} M \\ N \end{array} \right) \equiv \left(\begin{array}{c|c} A - BF & BD_1 \\ \hline -F & D_1 \\ C - DF & DD_1 \end{array} \right)$$

be the minimal realization of $\left(\begin{array}{c} M \\ N \end{array} \right)$ of Theorem 2.2, where F is a stabilizing state feedback and D_1 is invertible. By assumption U, V are such that the McMillan degree of $\left(\begin{array}{c} M \\ N \end{array} \right)$ is equal to that of $\left(\begin{array}{cc} M & U \\ N & V \end{array} \right)$. By Lemma 2.2 thus has a $\left(\begin{array}{c} U \\ V \end{array} \right)$ has a realization

$$\left(\begin{array}{c} U \\ V \end{array} \right) \equiv \left(\begin{array}{c|c} A - BF & L \\ \hline -F & D_2 \\ C - DF & D_3 \end{array} \right)$$

for some $L : \mathcal{K}^m \to X$, $D_2 : \mathcal{K}^m \to \mathcal{K}^p$ and $D_3 : \mathcal{K}^m \to \mathcal{K}^m$ and therefore,

$$\left(\begin{array}{cc} M & U \\ N & V \end{array} \right) \equiv \left(\begin{array}{c|cc} A - BF & BD_1 & L \\ \hline -F & D_1 & D_2 \\ C - DF & DD_1 & D_3 \end{array} \right).$$

Similarly, the other factor has a state-space realization

$$\left(\begin{array}{cc} \tilde{V} & -\tilde{U} \\ -\tilde{N} & \tilde{M} \end{array} \right) \equiv \left(\begin{array}{c|cc} A - HC & HD - B & -H \\ \hline \overline{L} & \overline{D}_3 & \overline{D}_2 \\ \overline{D}_1 C & -\overline{D}_1 D & \overline{D}_1 \end{array} \right),$$

for some $\overline{L} : X \to \mathcal{K}^m$, $\overline{D}_2 : \mathcal{K}^m \to \mathcal{K}^m$ and $\overline{D}_3 : \mathcal{K}^p \to \mathcal{K}^m$ and where H is a stabilizing output injection and $\overline{D}_1$ is invertible.

We first consider the feedthrough terms. Since we have a doubly coprime factorization we need to have that,

$$\left(\begin{array}{cc} I & 0 \\ 0 & I \end{array} \right) = \left(\begin{array}{cc} \overline{D}_3 & \overline{D}_2 \\ -\overline{D}_1 D & \overline{D}_1 \end{array} \right) \left(\begin{array}{cc} D_1 & D_2 \\ DD_1 & D_3 \end{array} \right)$$

$$= \left(\begin{array}{cc} \overline{D}_3 D_1 + \overline{D}_2 DD_1 & \overline{D}_3 D_2 + \overline{D}_2 D_3 \\ -\overline{D}_1 DD_1 + \overline{D}_1 DD_1 & -\overline{D}_1 DD_2 + \overline{D}_1 D_3 \end{array} \right)$$

$$= \left(\begin{array}{cc} \overline{D}_3 D_1 + \overline{D}_2 DD_1 & \overline{D}_3 D_2 + \overline{D}_2 D_3 \\ 0 & -\overline{D}_1 DD_2 + \overline{D}_1 D_3 \end{array} \right).$$

Solving these equations we obtain after some calculations that

$$\overline{D}_3 = D_1^{-1} + D_1^{-1} D_2 \overline{D}_1 D,$$

$$D_3 = \overline{D}_1^{-1} + D D_2,$$

$$\overline{D}_2 = -D_1^{-1} D_2 \overline{D}_1.$$

Hence, we necessarily have that

$$\begin{pmatrix} M & U \\ N & V \end{pmatrix} \equiv \left(\begin{array}{c|cc} A - BF & BD_1 & L \\ \hline -F & D_1 & D_2 \\ C - DF & DD_1 & \overline{D}_1^{-1} + DD_2 \end{array} \right)$$

and

$$\begin{pmatrix} \tilde{V} & -\tilde{U} \\ -\tilde{N} & \tilde{M} \end{pmatrix} \equiv \left(\begin{array}{c|cc} A - HC & HD - B & -H \\ \hline \overline{L} & D_1^{-1} + D_1^{-1} D_2 \overline{D}_1 D & -D_1^{-1} D_2 \overline{D}_1 \\ \overline{D}_1 C & -\overline{D}_1 D & \overline{D}_1 \end{array} \right).$$

To determine L and $\overline{L}$, we calculate the state-space realizations of the cascaded system

$$\begin{pmatrix} I & 0 \\ 0 & I \end{pmatrix} = \begin{pmatrix} \tilde{V} & -\tilde{U} \\ -\tilde{N} & \tilde{M} \end{pmatrix} \begin{pmatrix} M & U \\ N & V \end{pmatrix} \equiv$$

$$\left(\begin{array}{cc|cc} A - HC & \begin{bmatrix} HD - B & -H \end{bmatrix} \begin{bmatrix} -F \\ C - DF \end{bmatrix} & \begin{bmatrix} HD - B & -H \end{bmatrix} \begin{bmatrix} D_1 & D_2 \\ DD_1 & \overline{D}_1^{-1} + DD_2 \end{bmatrix} \\ 0 & A - BF & BD_1 L \\ \hline \overline{L} & \begin{bmatrix} D_1^{-1} + D_1^{-1} D_2 \overline{D}_1 D & -D_1^{-1} D_2 \overline{D}_1 \end{bmatrix} \begin{bmatrix} -F \\ C - DF \end{bmatrix} & I & 0 \\ \overline{D}_1 C & \begin{bmatrix} -\overline{D}_1 D & \overline{D}_1 \end{bmatrix} \begin{bmatrix} -F \\ C - DF \end{bmatrix} & 0 & I \end{array} \right)$$

$$= \left(\begin{array}{cc|cc} A - HC & BF - HC & -BD_1 & -BD_2 - H\overline{D}_1^{-1} \\ 0 & A - BF & BD_1 & L \\ \hline \overline{L} & -D_1^{-1}F - D_1^{-1}D_2\overline{D}_1 C & I & 0 \\ \overline{D}_1 C & \overline{D}_1 C & 0 & I \end{array} \right).$$

A state-space transformation by $\begin{pmatrix} I & -I \\ 0 & I \end{pmatrix}$ gives

$$\left(\begin{array}{cc|cc} A - HC & 0 & 0 & -BD_2 - H\overline{D}_1^{-1} + L \\ 0 & A - BF & BD_1 & L \\ \hline \overline{L} & -D_1^{-1}F - D_1^{-1}D_2\overline{D}_1 C - \overline{L} & I & 0 \\ \overline{D}_1 C & 0 & 0 & I \end{array} \right).$$

Consider the $(2,2)$ subsystem

$$I \equiv \left(\begin{array}{c|c} A - HC & -BD_2 - H\overline{D}_1^{-1} + L \\ \hline \overline{D}_1 C & I \end{array} \right)$$

and the $(1,1)$ subsystem

$$I \equiv \left(\begin{array}{c|c} A - BF & BD_1 \\ \hline -D_1^{-1}F - D_1^{-1}D_2\overline{D}_1C - \overline{L} & I \end{array} \right).$$

Since $\overline{D}_1$ and D_1 are invertible, the first system is observable and the second system is reachable. Hence these two systems are I if and only if

$$-BD_2 - H\overline{D}_1^{-1} + L = 0$$

$$-D_1^{-1}F - D_1^{-1}D_2\overline{D}_1C - \overline{L} = 0,$$

or if and only if

$$L = BD_2 + H\overline{D}_1^{-1}$$

and

$$\overline{L} = -D_1^{-1}F - D_1^{-1}D_2\overline{D}_1C.$$

Conversely, let D_2 be arbitrary and let $\begin{pmatrix} M & U \\ N & V \end{pmatrix}$ and $\begin{pmatrix} \tilde{V} & -\tilde{U} \\ -\tilde{N} & \tilde{M} \end{pmatrix}$ be defined through the state-space realizations in the statement of the corollary. Then it can be checked in a straightforward way that

$$\begin{pmatrix} \tilde{V} & -\tilde{U} \\ -\tilde{N} & \tilde{M} \end{pmatrix} \begin{pmatrix} M & U \\ N & V \end{pmatrix} = \begin{pmatrix} I & 0 \\ 0 & I \end{pmatrix}.$$

By construction, the McMillan degrees of these two functions are less than or equal to that of G. Lemma 2.1 then implies that the McMillan degrees of all three functions are the same.
$\square$

The following corollary states that McMillan degree n factors of rational functions of McMillan degree n are necessarily coprime.

Corollary 2.3 *Let $G = NM^{-1}$ be a, not necessarily coprime, right factorization of G. If the McMillan degree of $\begin{pmatrix} M \\ N \end{pmatrix}$ equals the McMillan degree of G then N an M are right coprime.*
Similarly, let $G = \tilde{M}^{-1}\tilde{N}$ be a, not necessarily coprime, left factorization of G. If the McMillan degree of $\begin{pmatrix} -\tilde{N} & \tilde{M} \end{pmatrix}$ equals the McMillan degree of G then $\tilde{N}$ an $\tilde{M}$ are left coprime.

Proof: This follows immediately from the previous corollary where solutions to the Bezout equations were constructed.
$\square$

3   Antistable functions

One of the main purposes of this paper is to derive state-space realizations of factorizations that are normalized in certain ways. The first class of systems for which we are going to consider normalized factorizations is the class of antistable functions. By an antistable function we mean a function whose poles are in the open right half plane. Here the factorization is normalized so that the denominator M is inner, i.e. $M^*M = I$. This type of factorization has been introduced by Douglas, Shapiro and Shields [1971] for scalar functions and by Fuhrmann [1981] for matrix-valued functions. It is therefore referred to as the Douglas-Shapiro-Shields factorization (DDS). This factorization is amongst other applications particularly important in the theory of Hankel operators. State space formulae for DSS-factorizations appear in the control literature, see e.g. Doyle [1984].

Let G be a proper antistable function, i.e. all poles of G are in the open right half plane. A right (left) coprime factorization $G = NM^{-1}$ $(G = \tilde{M}^{-1}\tilde{N})$ is called a right (left) *Douglas-Shapiro-Shields (DSS)* factorization if $M^*M = I$ $(\tilde{M}\tilde{M}^* = I.)$

The existence of a DSS factorization is guaranteed by the following proposition.

Proposition 3.1 *Let G be an antistable proper rational function of McMillan degree n. Then there exists a right (left) factorization*

$$G = NM^{-1} \quad (G = \tilde{M}^{-1}\tilde{N})$$

with N, M proper and $M^*M = I$ $(\tilde{M}^*\tilde{M} = I)$. Moreover, M and $\tilde{M}$ have McMillan degree n. The right (left) factorization with this property is unique up to right (left) multiplication by a unitary constant matrix. All such factorizations are coprime and such that $\begin{pmatrix} M \\ N \end{pmatrix}$ ($\begin{pmatrix} -\tilde{N} & \tilde{M} \end{pmatrix}$) has McMillan degree n.

Proof: Let $G = ED^{-1}$ be a right polynomial coprime factorization. By assumption D is antistable. Let T be a square stable spectral factor of D^*D, i.e. $D^*D = T^*T$. Then

$$\begin{pmatrix} M \\ N \end{pmatrix} := \begin{pmatrix} DT^{-1} \\ ET^{-1} \end{pmatrix}$$

defines by Proposition 2.1 right factors of G, i.e. $G = NM^{-1}$ with N, M and M^{-1} proper and $\begin{pmatrix} M \\ N \end{pmatrix}$ has the same McMillan degree as G. Hence the factorization is coprime by Corollary 2.3. Clearly, $M^*M = I$. Since D is antistable and T is stable, there are no pole-zero cancellations and therefore M is of McMillan degree n. Let $G = NM^{-1} = N_1 M_1^{-1}$ be two DSS factorizations of G. Since both are coprime factorizations, there exists a stable function Q with proper stable inverse that relates the two factorizations (see e.g. Vidyasagar [1985]). In particular, $M = M_1 Q$. Since $I = M^*M = Q^*M_1^*M_1 Q = Q^*Q$, this shows that $Q = Q^{-*}$. Since Q is stable with proper stable inverse, this implies that Q must be a constant unitary matrix.

The statement concerning left factorizations follows analogously. $\Box$

The following Lemma will be important in the proofs of the main theorems of this and subsequent sections.

Lemma 3.1 *Let $U(s)$ be a proper rational function such that*

$$U(s) = J_1 U(s)^{-*} J_2, \quad s \in \mathcal{C} \setminus \sigma(U),$$

where J_i, $i = 1, 2$, are constant matrices such that $J_i = J_i^{-1} = J_i^*$, $i = 1, 2$. Let $U \equiv \left(\begin{array}{c|c} A & B \\ \hline C & D \end{array} \right)$ be a minimal realization of $U(s)$. Then,

1. $D = J_1 D^{-*} J_2$.

2. *$U(s)$ has another minimal realization of the form*

$$U \equiv \left(\begin{array}{c|c} A & B \\ \hline C & D \end{array} \right) \equiv \left(\begin{array}{c|c} -A^* + C^* J_1 D J_2 B^* & C^* J_1 D \\ \hline D J_2 B^* & D \end{array} \right),$$

   and there exists a unique non-singular state-space transformation Y between these two realizations such that $Y = Y^*$, and

   $$Y A = (-A^* + C^* J_1 D J_2 B^*) Y,$$

   $$Y B = C^* J_1 D,$$

   $$C = D J_2 B^* Y.$$

Proof: Note that

$$U^{-1} \equiv \left(\begin{array}{c|c} A - B D^{-1} C & B D^{-1} \\ \hline -D^{-1} C & D^{-1} \end{array} \right)$$

and

$$U^{-*} \equiv \left(\begin{array}{c|c} -A^* + C^* D^{-*} B^* & C^* D^{-*} \\ \hline D^{-*} B^* & D^{-*} \end{array} \right)$$

and therefore

$$U = J_1 U^{-*} J_2 \equiv \left(\begin{array}{c|c} -A^* + C^* D^{-*} B^* & C^* D^{-*} J_2 \\ \hline J_1 D^{-*} B^* & J_1 D^{-*} J_2 \end{array} \right).$$

Since $U = J_1 U^{-*} J_2$ we clearly have that $D = J_1 D^{-*} J_2$. Since $U \equiv \left(\begin{array}{c|c} A & B \\ \hline C & D \end{array} \right)$ is a minimal realization there exists a unique non-singular state-space transformation T, between this and the second realization of U, i.e.

$$C = J_1 D^{-*} B^* T,$$

$$B = T^{-1} C^* D^{-*} J_2,$$

$$A = T^{-1}(-A^* + C^* D^{-*} B^*) T.$$

Dualizing these equations, we obtain,

$$C^* = T^* B D^{-1} J_1,$$

$$B^* = J_2 D^{-1} C T^{-*},$$

$$A^* = T^*(-A + BD^{-1}C)T^{-*}.$$

Solving for A, B and C we have,

$$C = DJ_2 B^* T^* = J_1 D^{-*} B^* T^*$$

$$B = T^{-*} C^* J_1 D = T^{-*} C^* D^{-*} J_2,$$

$$A = -T^{-*} A^* T^* + BD^{-1}C = -T^{-*} A T^* + T^{-*} C^* D^{-*} J_2 D^{-1} J_1 D^{-*} B^* T^*$$

$$= T^{-*}(-A^* + C^* D^{-*} B^*)T^*.$$

But this shows that T^* is also a state-space transformation between the two systems. The uniqueness of the transformation therefore implies that $T^* = T$. The form of the realization of U given in the statement, follows immediately from the above identities. $\qquad\square$

Proposition 3.1 showed that a right (left) DSS-factorization is unique up to right multiplication by a constant unitary matrix. The factorization is such that the McMillan degree of $\begin{pmatrix} M \\ N \end{pmatrix}$ equals the McMillan degree of G. In the following theorem all doubly-coprime factorizations are characterized which are such that $\begin{pmatrix} M \\ N \end{pmatrix}$ forms a DSS-factorization of G.

Theorem 3.1 *Let G be an antistable proper rational transfer function of McMillan degree n with minimal state-space realization (A, B, C, D). Then all doubly coprime factorizations, such that $G = NM^{-1}$, with $MM^* = I$, and $G = \tilde{M}^{-1}\tilde{N}$ with $\tilde{M}\tilde{M}^* = I$, and the McMillan degrees of $\begin{pmatrix} M & U \\ N & V \end{pmatrix}$ and $\begin{pmatrix} \tilde{V} & -\tilde{U} \\ -\tilde{N} & \tilde{M} \end{pmatrix}$ are n, are given by*

$$\begin{pmatrix} M & U \\ N & V \end{pmatrix} \equiv \left(\begin{array}{c|cc} A - BB^*Y & BD_1 & BD_2 + ZC^*\overline{D}_1^* \\ \hline -B^*Y & D_1 & D_2 \\ C - DB^*Y & DD_1 & \overline{D}_1^* + DD_2 \end{array} \right),$$

$$\begin{pmatrix} \tilde{V} & -\tilde{U} \\ -\tilde{N} & \tilde{M} \end{pmatrix} \equiv \left(\begin{array}{c|cc} A - ZC^*C & ZC^*D - B & -ZC^* \\ -D_1^*B^*Y - D_1^*D_2\overline{D}_1C & D_1^* + D_1^*D_2\overline{D}_1D & -D_1^*D_2\overline{D}_1 \\ \overline{D}_1C & -\overline{D}_1D & \overline{D}_1 \end{array} \right),$$

where D_1 is arbitrary such that $D_1 = D_1^{-*}$, $\overline{D}_1$ is arbitrary such that $\overline{D}_1 = \overline{D}_1^{-*}$, D_2 arbitrary, and Y and Z are the unique positive definite solutions to the Riccati equations

$$AY + YA^* - YBB^*Y = 0,$$

$$A^*Z + ZA - ZC^*CZ = 0.$$

Proof: Let $G = NM^{-1}$ be a DSS factorization such that $\begin{pmatrix} M \\ N \end{pmatrix}$ has McMillan degree n. This exists by Proposition 3.1. By Theorem 2.2 any right factorization $G = NM^{-1}$ such $\begin{pmatrix} M \\ N \end{pmatrix}$ has McMillan degree n is of the form

$$\begin{pmatrix} M \\ N \end{pmatrix} \equiv \left(\begin{array}{c|c} A - BF & BD_1 \\ \hline -F & D_1 \\ C - DF & DD_1 \end{array} \right),$$

where F is a stabilizing feedback and $D_1 = M(\infty)$ is invertible. Since M is such that $M^*M = I$, we have that

$$M = M^{-*}.$$

Since M has McMillan degree n, the realization

$$M \equiv \left(\begin{array}{c|c} A - BF & BD_1 \\ \hline -F & D_1 \end{array} \right)$$

is minimal. Lemma 3.1 now implies that there exists a unique non-singular state-space transformation $\hat{Y} = \hat{Y}^*$ such that

$$D_1 = D_1^{-*},$$

$$\hat{Y} B D_1 = -F^* D_1,$$

$$-F = D_1 D_1^* B^* \hat{Y} = B^* \hat{Y},$$

$$\hat{Y}(A - BF) = (-A^* + F^*B^* - F^*D_1 D_1^* B^*)\hat{Y} = -A^* \hat{Y}.$$

Using that $F = -B^* \hat{Y}$ we can rewrite the equation

$$\hat{Y}(A - BF) = -A^* \hat{Y}$$

as

$$A^* \hat{Y} + \hat{Y} A + \hat{Y} BB^* \hat{Y} = 0.$$

Setting $Y := -\hat{Y}$, this equation is equivalent to the more conventional equation,

$$A^* Y + YA - YBB^* Y = 0.$$

Since Y is invertible, this Riccati equation is equivalent to the Lyapunov equation,

$$Y^{-1} A^* + AY^{-1} - BB^* = 0,$$

which shows that Y^{-1} and therefore Y is positive definite. Since A^* is antistable, Y^{-1} is the unique positive definite solution of this equation. Hence Y is the unique positive definite solution to the Riccati equation. A state space realization of $\begin{pmatrix} M \\ N \end{pmatrix}$ is given by

$$\begin{pmatrix} M \\ N \end{pmatrix} \equiv \left(\begin{array}{c|c} A - BB^*Y & BD_1 \\ \hline -B^*Y & D_1 \\ C - DB^*Y & DD_1 \end{array} \right),$$

where D_1 is arbitrary such that $D_1 = D_1^{-*}$.

The expressions for the doubly coprime factors now follow from Corollary 2.2.

An analogous argument or the duality consideration that $G = NM^{-1}$ is a right factorization if and only if $G^T = (M^T)^{-1}N^T$, shows that a state space realization of $\begin{bmatrix} -\tilde{N} & \tilde{M} \end{bmatrix}$ is given by

$$\begin{bmatrix} -\tilde{N} & \tilde{M} \end{bmatrix} \equiv \left(\begin{array}{c|cc} A - ZC^*C & ZC^*D - B & -ZC^* \\ \hline \overline{D}_1 C & -\overline{D}_1 D & \overline{D}_1 \end{array} \right),$$

where Z is the unique positive solution of the Riccati equation

$$AZ + ZA^* - ZC^*CZ = 0,$$

and $\overline{D}_1$ is arbitrary such that $\overline{D}_1 = \overline{D}_1^{-*}$. The remaining part of the argument is analogous to the above derivation.

Conversely, let Y be the unique positive definite solution to the Riccati equation

$$A^*Y + YA - YBB^*Y = 0.$$

Constructing the DSS factorization of G as in Proposition 3.1, proceeding as above and using the uniqueness of the solution Y, shows that

$$\begin{pmatrix} M \\ N \end{pmatrix} \equiv \left(\begin{array}{c|c} A - BB^*Y & BD_1 \\ \hline -B^*Y & D_1 \\ C - DB^*Y & DD_1 \end{array} \right),$$

gives a realization of the DSS factors of G. Hence $F = B^*Y$ is a stabilizing feedback and the state-space construction gives indeed the required factorizations.

$\square$

The expressions for the doubly-coprime factorizations can be simplified if we are only interested in a particular factorization and not in all of them. The choice $D_1 = I$, $\overline{D}_1^* = I$ and $D_2 = 0$ would lead to such a simplification.

As part of the proof of the theorem we have also shown the well-known result that a certain degenerate Riccati equation has a stabilizing solution.

Corollary 3.1 *Let (A, B, C, D) be an antistable continuous-time minimal system. Then there exists a unique positive definite solution Y (Z) of the Riccati equation*

$$AY + YA^* - YBB^*Y = 0 \quad (A^*Z + ZA - ZC^*CZ = 0).$$

This solution is such that $A - BB^*Y$ $(A - ZC^*C)$ is stable, i.e. all eigenvalues of $A - BB^*Y$ $(A - ZC^*C)$ are in the open left half plane.

Proof: This statement was proved as part of the proof of the theorem. $\square$

4 Minimal systems

Normalized coprime factorizations proved to be powerful methods in control theoretic problems, especially in the area of robust control (see e.g. Vidyasagar [1985], McFarlane and Glover [1989], Fuhrmann and Ober [1992]). For their relevance in parametrization problems, see Ober and McFarlane [1989].

Let

$$J_L = \begin{pmatrix} I & 0 \\ 0 & I \end{pmatrix}.$$

A right (coprime) factorization $G = NM^{-1}$ of G is called a J_L-RF (J_L-RCF) of G if

$$\begin{pmatrix} M^* & N^* \end{pmatrix} \begin{pmatrix} I & 0 \\ 0 & I \end{pmatrix} \begin{pmatrix} M \\ N \end{pmatrix} = M^*M + N^*N = I.$$

Similarly a left coprime factorization $G = \overline{M}^{-1}\overline{N}$ the transfer function G is called J_L-LF (a J_L-LCF) of G if

$$\overline{NN}^* + \overline{MM}^* = I.$$

The following proposition guarantees the existence of such factorizations. The existence and uniqueness result is due to Vidyasagar (see e.g. Vidyasagar [1985]).

Proposition 4.1 *Let G be a proper rational function of McMillan degree n. Then there exists a J_L-right (left) factorization with proper functions M, N,*

$$G = NM^{-1} \quad (G = \tilde{M}^{-1}\tilde{N}).$$

This factorization is right (left) coprime and is unique up to right (left) multiplication by a constant unitary matrix. All such factorizations are such that $\begin{pmatrix} M \\ N \end{pmatrix}$ $((\begin{matrix} -\tilde{N} & \tilde{M} \end{matrix}))$ is of McMillan degree n.

Proof: Let $G = ED^{-1}$ be a right polynomial coprime factorization. Let T be a square stable spectral factor of $E^*E + D^*D$, i.e. $T^*T = E^*E + D^*D$. Then

$$\begin{pmatrix} M \\ N \end{pmatrix} := \begin{pmatrix} DT^{-1} \\ ET^{-1} \end{pmatrix}$$

defines by Proposition 2.1 right factors of G, i.e. $G = NM^{-1}$ with N, M and M^{-1} proper and $\begin{pmatrix} M \\ N \end{pmatrix}$ of McMillan degree n. This also implies by Corollary 2.3 that the factorization is coprime. Let now $G = N_1M_1^{-1}$ be another J_L-right coprime factorization. Then there exists a stable Q with proper stable inverse (Vidyasagar [1985]) such that $M = M_1Q$ and $N = N_1Q$. But $I = M^*M + N^*N = Q^*M_1^*M_1Q + Q^*N_1^*N_1Q = Q^*Q$, which shows that $Q = Q^{-*}$, which implies that Q is a constant matrix.

The statement concerning left factorizations follows analogously. $\qquad\square$

Note that for J_L-factorizations $G = NM^{-1} = \tilde{M}^{-1}\tilde{N}$ we have that

$$\begin{pmatrix} M & -\tilde{N}^* \\ N & \tilde{M}^* \end{pmatrix}^* \begin{pmatrix} M & -\tilde{N}^* \\ N & \tilde{M}^* \end{pmatrix} = \begin{pmatrix} I & 0 \\ 0 & I \end{pmatrix}.$$

The following Lemma will be useful in the proof of the subsequent theorem.

Lemma 4.1 *Consider*

$$\begin{pmatrix} D_1 & -D^*\overline{D}_1^* \\ DD_1 & \overline{D}_1^* \end{pmatrix} = \begin{pmatrix} I & -D^* \\ D & I \end{pmatrix} \begin{pmatrix} D_1 & 0 \\ 0 & \overline{D}_1^* \end{pmatrix},$$

with D_1 and $\overline{D}_1$ invertible. Then

1. $$\begin{pmatrix} D_1 & -D^*\overline{D}_1^* \\ DD_1 & \overline{D}_1^* \end{pmatrix}^{-1}$$

$$= \begin{pmatrix} D_1^{-1} & 0 \\ 0 & \overline{D}_1^{-*} \end{pmatrix} \begin{pmatrix} I & D^* \\ -D & I \end{pmatrix} \begin{pmatrix} (I+D^*D)^{-1} & 0 \\ 0 & (I+DD^*)^{-1} \end{pmatrix}.$$

2. *If*

$$\begin{pmatrix} D_1 & -D^*\overline{D}_1^* \\ DD_1 & \overline{D}_1^* \end{pmatrix} = \begin{pmatrix} D_1 & -D^*\overline{D}_1^* \\ DD_1 & \overline{D}_1^* \end{pmatrix}^{-*},$$

then

$$D_1 D_1^* = (I + D^*D)^{-1},$$

$$\overline{D}_1^*\overline{D}_1 = (I + DD^*)^{-1}.$$

Proof: The statements are checked in a straightforward way.   □

The following theorem characterizes all doubly-coprime factorizations such that $G = NM^{-1}$ is a J_L right coprime factorization. The state-space formulae for J_L- factorizations are not new. They are due to Meyer and Franklin [1987] for the strictly proper case and due to Vidyasagar [1988] for the general case. Our proof is however new in that the formulae are derived in a systematic way starting from the known existence of such factorizations.

Theorem 4.1 *Let (A, B, C, D) be a minimal realization of the proper rational function G of McMillan degree n.*
Then all doubly coprime factorizations of G, such that

1. *$G = NM^{-1}$ is a J_L-RF and $G = \tilde{M}^{-1}\tilde{N}$ is a J_L-LF,*

2. $$\begin{pmatrix} M & U \\ N & V \end{pmatrix}, \quad \begin{pmatrix} \tilde{V} & -\tilde{U} \\ -\tilde{N} & \tilde{M} \end{pmatrix},$$
are of McMillan degree n,

are given by

$$\begin{pmatrix} M & U \\ N & V \end{pmatrix} \equiv \left(\begin{array}{c|cc} A - BF & BD_1 & BD_2 + H\overline{D}_1^{-1} \\ \hline -F & D_1 & D_2 \\ C - DF & DD_1 & \overline{D}_1^{-1} + DD_2 \end{array} \right),$$

$$\begin{pmatrix} \tilde{V} & -\tilde{U} \\ -\tilde{N} & \tilde{M} \end{pmatrix} \equiv \left(\begin{array}{c|ccc} A - HC & HD - B & -H \\ \hline -D_1^{-1}F - D_1^{-1}D_2\overline{D}_1C & D_1^{-1} + D_1^{-1}D_2\overline{D}_1D & -D_1^{-1}D_2\overline{D}_1 \\ \overline{D}_1C & -\overline{D}_1D & \overline{D}_1 \end{array} \right),$$

where

- D_1 *is such that* $D_1 D_1^* = (I + D^*D)^{-1}$.
- $\overline{D}_1$ *is such that* $\overline{D}_1^* \overline{D}_1 = (I + DD^*)^{-1}$.
- D_2 *is arbitrary* .
- Y *and* Z *are solutions of the Riccati equations*

$$0 = (A - B(I + D^*D)^{-1}D^*C)^*Y + Y(A - B(I + D^*D)^{-1}D^*C)$$

$$-YB(I + D^*D)^{-1}B^*Y + C^*(I + DD^*)^{-1}C,$$

$$0 = (A - B(I + D^*D)^{-1}D^*C)Z + Z(A - B(I + D^*D)^{-1}D^*C)^*$$

$$-ZC^*(I + DD^*)^{-1}CZ + B(I + D^*D)^{-1}B^*,$$

such that $A - BF$ *and* $A - HC$ *are stable, where*

$$F = (I + D^*D)^{-1}D^*C + (I + D^*D)^{-1}B^*Y,$$

$$H = BD^*(I + DD^*)^{-1} + ZC^*(I + DD^*)^{-1}.$$

Proof: Let G have a minimal realization (A, B, C, D). Let $G = NM^{-1}$ be a right factorization G such that

$$\begin{pmatrix} M^* & N^* \end{pmatrix} J_L \begin{pmatrix} M \\ N \end{pmatrix} = I.$$

Such a factorization exists by Proposition 4.1 and has the same McMillan degree as G. By Theorem 2.2 any right factorization $\begin{pmatrix} M \\ N \end{pmatrix}$ such that the McMillan degree of $\begin{pmatrix} M \\ N \end{pmatrix}$ is the same as that of G, has a state space realization of the form,

$$\begin{pmatrix} M \\ N \end{pmatrix} \equiv \left(\begin{array}{c|c} A - BF & BD_1 \\ \hline -F & D_1 \\ C - DF & DD_1 \end{array} \right),$$

where F is a stabilizing state feedback and D_1 is invertible. Similarly, a left factorization $G = \tilde{M}^{-1}\tilde{N}$ such that

$$\begin{pmatrix} -\tilde{N} & \tilde{M} \end{pmatrix} J_L \begin{pmatrix} -\tilde{N}^* \\ \tilde{M}^* \end{pmatrix} = I,$$

exists and is of McMillan degree n and has a state-space representation of the form

$$\begin{pmatrix} -\tilde{N} & \tilde{M} \end{pmatrix} \equiv \left(\begin{array}{c|cc} A - HC & HD - B & -H \\ \hline \overline{D}_1 C & -\overline{D}_1 D & \overline{D}_1 \end{array} \right),$$

where H is a stabilizing output injection. Since $\begin{pmatrix} M \\ N \end{pmatrix}$ is stable and of McMillan degree n

and $\begin{pmatrix} -\tilde{N}^* \\ \tilde{M}^* \end{pmatrix}$ is antistable and also of McMillan degree n, the function

$$\begin{pmatrix} M & -\tilde{N}^* \\ N & \tilde{M}^* \end{pmatrix} \equiv \left(\begin{array}{cc|cc} A - BF & 0 & BD_1 & 0 \\ 0 & -A^* + C^*H^* & 0 & C^*\overline{D}_1^* \\ \hline -F & B^* - D^*H^* & D_1 & -D^*\overline{D}_1^* \\ C - DF & H^* & DD_1 & \overline{D}_1^* \end{array} \right)$$

$$=: \left(\begin{array}{c|c} \mathcal{A} & \mathcal{B} \\ \hline \mathcal{C} & \mathcal{D} \end{array} \right)$$

has McMillan degree $2n$. Note that since

$$\begin{pmatrix} M & -\tilde{N}^* \\ N & \tilde{M}^* \end{pmatrix}^{-*} = \begin{pmatrix} M & -\tilde{N}^* \\ N & \tilde{M}^* \end{pmatrix},$$

we have by Lemma 3.1 that $\begin{pmatrix} M & -\tilde{N}^* \\ N & \tilde{M}^* \end{pmatrix}$ has two equivalent realizations

$$\left(\begin{array}{c|c} \mathcal{A} & \mathcal{B} \\ \hline \mathcal{C} & \mathcal{D} \end{array} \right)$$

and

$$\left(\begin{array}{c|c} -\mathcal{A}^* + \mathcal{C}^*\mathcal{D}\mathcal{B}^* & \mathcal{C}^*\mathcal{D} \\ \hline \mathcal{D}\mathcal{B}^* & \mathcal{D} \end{array} \right).$$

Since $\mathcal{D} = \mathcal{D}^{-1}$, Lemma 4.1 implies that $\mathcal{D}$ and $\overline{D}_1$ are such that

$$D_1 D_1^* = (I + D^*D)^{-1},$$

$$\overline{D}_1^* \overline{D}_1 = (I + DD^*)^{-1}.$$

We need to compute $-\mathcal{A}^* + \mathcal{C}^*\mathcal{D}\mathcal{B}^*$,

$$-\mathcal{A}^* + \mathcal{C}^*\mathcal{D}\mathcal{B}^*$$

$$= \begin{pmatrix} -A^* + F^*B^* & 0 \\ 0 & A - HC \end{pmatrix}$$

$$+ \begin{pmatrix} -F^* & C^* - F^*D^* \\ B - HD & H \end{pmatrix} \begin{pmatrix} I & -D^* \\ D & I \end{pmatrix} \begin{pmatrix} D_1 & 0 \\ 0 & \overline{D}_1^* \end{pmatrix} \begin{pmatrix} D_1^*B^* & 0 \\ 0 & \overline{D}_1 C \end{pmatrix}$$

$$= \begin{pmatrix} -A^* + F^*B^* & 0 \\ 0 & A - HC \end{pmatrix}$$

$$+ \begin{pmatrix} -F^*(I + D^*D) + C^*D & C^* \\ B & -BD^* + H(I + DD^*) \end{pmatrix} \begin{pmatrix} D_1 D_1^* & 0 \\ 0 & \overline{D_1^* D_1} \end{pmatrix} \begin{pmatrix} B^* & 0 \\ 0 & C \end{pmatrix}$$

$$= \begin{pmatrix} -A^* + C^*DD_1D_1^*B^* & C^*\overline{D_1^*D_1}C \\ BD_1D_1^*B^* & A - BD^*\overline{D_1^*D_1}C \end{pmatrix}.$$

Since $\left(\begin{array}{c|c} A & B \\ \hline C & D \end{array} \right)$ and $\left(\begin{array}{c|c} -A^* + C^*DB^* & C^*D \\ \hline DB^* & D \end{array} \right)$ are both minimal realizations of $\begin{pmatrix} M & -\tilde{N}^* \\ N & \tilde{M}^* \end{pmatrix}$,

there exists a unique non-singular state-space transformation $Y = \begin{pmatrix} Y_{11} & Y_{12} \\ Y_{21} & Y_{22} \end{pmatrix}$, which by

Lemma 3.1 is such that $\begin{pmatrix} Y_{11} & Y_{12} \\ Y_{21} & Y_{22} \end{pmatrix} = \begin{pmatrix} Y_{11}^* & Y_{21}^* \\ Y_{12}^* & Y_{22}^* \end{pmatrix}$ and such that

$$C = DB^*Y,$$

$$YB = C^*D,$$

$$YA = (-A^* + C^*DB^*)Y.$$

More explicitly, we have writing the equation $YB = C^*D$ componentwise,

$$\begin{pmatrix} Y_{11} & Y_{12} \\ Y_{12}^* & Y_{22} \end{pmatrix} \begin{pmatrix} BD_1 & 0 \\ 0 & C^*\overline{D_1^*} \end{pmatrix}$$

$$= \begin{pmatrix} -F^* & C^* - F^*D^* \\ B - HD & H \end{pmatrix} \begin{pmatrix} I & -D^* \\ D & I \end{pmatrix} \begin{pmatrix} D_1 & 0 \\ 0 & \overline{D_1^*} \end{pmatrix}$$

and therefore

$$\begin{pmatrix} Y_{11} & Y_{12} \\ Y_{12}^* & Y_{22} \end{pmatrix} \begin{pmatrix} B & 0 \\ 0 & C^* \end{pmatrix} = \begin{pmatrix} -F^*(I + D^*D) + C^*D & C^* \\ B & -BD^* + H(I + DD^*) \end{pmatrix}.$$

Hence, we have that

$$-F^*(I + D^*D) + C^*D = Y_{11}B,$$

$$C^* = Y_{12}C^*,$$

$$H(I + DD^*) - BD^* = Y_{22}C^*,$$

$$B = Y_{12}^*B,$$

which shows that

$$F = (I + D^*D)^{-1}D^*C - (I + D^*D)^{-1}B^*Y_{11},$$

$$H = BD^*(I + DD^*)^{-1} + Y_{22}C^*(I + DD^*)^{-1}.$$

Writing $Y\mathcal{A} = (-\mathcal{A}^* + \mathcal{C}^*\mathcal{D}\mathcal{B}^*)Y$, componentwise, we have for the $(1,1)$ entry,

$$Y_{11}(A - BF) = (-A^* + C^*DD_1D_1^*B^*)Y_{11} + C^*\overline{D}_1^*\overline{D}_1CY_{12}^*,$$

and using the above identities, this gives,

$$0 = (A^* - C^*DD_1D_1^*B^*)Y_{11} + Y_{11}(A - B[(I + D^*D)^{-1}D^*C - (I + D^*D)^{-1}B^*Y_{11}])$$

$$-C^*\overline{D}_1^*\overline{D}_1C$$

$$= (A^* - C^*D(I + D^*D)^{-1}B^*)Y_{11} + Y_{11}(A - B(I + D^*D)^{-1}D^*C)+$$

$$Y_{11}B(I + D^*D)^{-1}B^*Y_{11} - C^*(I + DD^*)^{-1}C.$$

Setting $Y := -Y_{11}$ we obtain the Riccati equation

$$0 = (A - B(I + D^*D)^{-1}D^*C)^*Y + Y(A - B(I + D^*D)^{-1}D^*C)$$

$$-YB(I + D^*D)^{-1}B^*Y + C^*(I + DD^*)^{-1}C.$$

Moreover, with $F = (I + D^*D)^{-1}D^*C + (I + D^*D)^{-1}B^*Y$ we have that $A - BF$ is stable. Evaluating the $(2,2)$ entry we obtain,

$$Y_{22}(-A^* + C^*H^*) = BD_1D_1^*B^*Y_{12} + (A - BD^*\overline{D}_1^*\overline{D}_1C)Y_{22},$$

or,

$$0 = (A - BD^*(I + DD^*)^{-1}C)Z + Z(A^* - C^*(I + DD^*)^{-1}DB^*)$$

$$-ZC^*(I + DD^*)^{-1}CZ + B(I + D^*D)^{-1}B^*,$$

where we have set $Z := Y_{22}$. Note that $A - HC$ is stable with $H = BD^*(I + DD^*)^{-1} + ZC^*(I + DD^*)^{-1}$.
It can be verified in a straightforward but tedious way that if state space representations are given as in the statement of the theorem that the transfer functions of these representations give doubly coprime factorizations with the required properties.

□

In the proof of the theorem we also established the well-known result that the algebraic Riccati equation has a stabilizing solution.

Corollary 4.1 *Let (A, B, C, D) be a minimal continuous-time system. Then there exist hermitian solutions Y and Z of the Riccati equations*

$$0 = (A - B(I + D^*D)^{-1}D^*)^*Y + Y(A - B(I + D^*D)^{-1}D^*C)$$

$$-YB(I + D^*D)^{-1}B^*Y + C^*(I + DD^*)^{-1}C,$$

respectively,

$$0 = (A - B(I + D^*D)^{-1}D^*C)Z + Z(A - B(I + D^*D)^{-1}D^*C)^*$$

$$-ZC^*(I + DD^*)^{-1}CZ + B(I + D^*D)^{-1}B^*,$$

such that $A - BF$ and $A - HC$ are stable, where

$$F = (I + D^*D)^{-1}D^*C + (I + D^*D)^{-1}B^*Y$$

$$H = BD^*(I + DD^*)^{-1} + ZC^*(I + DD^*)^{-1}.$$

Proof: This statement was proved as part of the proof of the theorem. $\square$

5 Bounded-real functions

In this section we are going to consider stable rational functions that are bounded in magnitude by 1 in the right half plane. We are going to define factorizations for this class of functions and are going to give the corresponding doubly coprime factorizations.

Definition 5.1 *A proper stable rational function is called* bounded real, *if*

$$I - G^*(i\omega)G(i\omega) > 0,$$

for all $\omega \in \Re \cup \{\pm\infty\}$. Let

$$J_B = \begin{pmatrix} I & 0 \\ 0 & -I \end{pmatrix}.$$

A right (coprime) factorization of $G = NM^{-1}$ is called a J_B-normalized right (coprime) factorization ($J_B - RF$ respectively $J_B - RCF$) of G if

$$\begin{pmatrix} M^* & N^* \end{pmatrix} J_B \begin{pmatrix} M \\ N \end{pmatrix} = I.$$

Similarly, a left (coprime) factorization is called a J_B-normalized left (coprime) factorization ($J_B - LF$ respectively $J_B - LCF$) of G if

$$-\begin{pmatrix} \tilde{N} & \tilde{M} \end{pmatrix} J_B \begin{pmatrix} \tilde{N}^* \\ \tilde{M}^* \end{pmatrix} = I.$$

The existence and uniqueness of such factorizations is established in the following proposition.

Proposition 5.1 *Let G be a proper rational bounded-real function of McMillan degree n. Then there exists a J_B-right (left) factorization with proper functions M, N,*

$$G = NM^{-1} \quad (G = \tilde{M}^{-1}\tilde{N}).$$

This factorization is right (left) coprime and is unique up to right (left) multiplication by a constant unitary matrix. All such factorizations are such that $\begin{pmatrix} M \\ N \end{pmatrix}$ $((-\tilde{N} \quad \tilde{M}))$ *is of McMillan degree n.*

Proof: Let $G = ED^{-1}$ be a right polynomial coprime factorization. Let T be a square stable spectral factor of $D^*D - E^*E$, i.e. $T^*T = D^*D - E^*E$. Then

$$\begin{pmatrix} M \\ N \end{pmatrix} := \begin{pmatrix} DT^{-1} \\ ET^{-1} \end{pmatrix}$$

defines by Proposition 2.1 right factors of G, i.e. $G = NM^{-1}$ with N, M and M^{-1} proper and $\begin{pmatrix} M \\ N \end{pmatrix}$ of McMillan degree n. This also implies by Corollary 2.3 that the factorization is coprime. The remaining parts of the proof follow in the standard way. $\quad\square$

Some properties of J_B-factorizations are summarized in the following lemma.

Lemma 5.1 *Let* $G = NM^{-1}$ *be a* J_B-RF *and* $G = \tilde{M}^{-1}\tilde{N}$ *a* J_B-LF *of the stable bounded-real function G. Then*

$$1. \qquad \begin{pmatrix} M^* & N^* \\ \tilde{N} & \tilde{M} \end{pmatrix} J_B \begin{pmatrix} M & \tilde{N}^* \\ N & \tilde{M}^* \end{pmatrix} = \begin{pmatrix} I & 0 \\ 0 & -I \end{pmatrix},$$

$$2. \qquad \begin{pmatrix} M & \tilde{N}^* \\ N & \tilde{M}^* \end{pmatrix} = J_B \begin{pmatrix} M & \tilde{N}^* \\ N & \tilde{M}^* \end{pmatrix}^{-*} J_B.$$

Proof: The statements are easily verified. $\quad\square$

The following Lemma will be useful in the proof the subsequent theorem.

Lemma 5.2 *Consider*

$$\begin{pmatrix} D_1 & D^*\overline{D}_1^* \\ DD_1 & \overline{D}_1^* \end{pmatrix} = \begin{pmatrix} I & D^* \\ D & I \end{pmatrix} \begin{pmatrix} D_1 & 0 \\ 0 & \overline{D}_1^* \end{pmatrix},$$

with D_1 *and* $\overline{D}_1$ *invertible and D such that* $I - DD^* > 0$. *Then*

$$1. \qquad \begin{pmatrix} D_1 & D^*\overline{D}_1^* \\ DD_1 & \overline{D}_1^* \end{pmatrix}^{-1}$$

$$= \begin{pmatrix} D_1^{-1} & 0 \\ 0 & \overline{D}_1^{-*} \end{pmatrix} \begin{pmatrix} I & -D^* \\ -D & I \end{pmatrix} \begin{pmatrix} (I - D^*D)^{-1} & 0 \\ 0 & (I - DD^*)^{-1} \end{pmatrix}.$$

$$2. \text{ If }$$

$$\begin{pmatrix} D_1 & D^*\overline{D}_1^* \\ DD_1 & \overline{D}_1^* \end{pmatrix} = J_B \begin{pmatrix} D_1 & D^*\overline{D}_1^* \\ DD_1 & \overline{D}_1^* \end{pmatrix}^{-*} J_B,$$

then

$$D_1 D_1^* = (I - D^*D)^{-1},$$

$$\overline{D}_1^* \overline{D}_1 = (I - DD^*)^{-1}.$$

Proof: The statements are checked in a straightforward way. $\qquad\square$

We are now in a position to characterize all doubly coprime factorizations so that $G = NM^{-1}(= \tilde{M}^{-1}\tilde{N})$ is a J_B-RF (J_B-LF).

Theorem 5.1 *Let (A, B, C, D) be a minimal realization of the proper bounded-real rational function G of McMillan degree n.*
Then all doubly coprime factorizations of G, such that

1. $G = NM^{-1}$ *is a J_B-RF and $G = \tilde{M}^{-1}\tilde{N}$ is a J_B-LF,*

2. $\begin{pmatrix} M & U \\ N & V \end{pmatrix}, \quad \begin{pmatrix} \tilde{V} & -\tilde{U} \\ -\tilde{N} & \tilde{M} \end{pmatrix},$
   are of McMillan degree n,

are given by

$$\begin{pmatrix} M & U \\ N & V \end{pmatrix} \equiv \left(\begin{array}{c|cc} A - BF & BD_1 & BD_2 + H\overline{D}_1^{-1} \\ \hline -F & D_1 & D_2 \\ C - DF & DD_1 & \overline{D}_1^{-1} + DD_2 \end{array} \right),$$

$$\begin{pmatrix} \tilde{V} & -\tilde{U} \\ -\tilde{N} & \tilde{M} \end{pmatrix} \equiv \left(\begin{array}{c|cc} A - HC & HD - B & -H \\ \hline -D_1^{-1}F - D_1^{-1}D_2\overline{D}_1 C & D_1^{-1} + D_1^{-1}D_2\overline{D}_1 D & -D_1^{-1}D_2\overline{D}_1 \\ \overline{D}_1 C & -\overline{D}_1 D & \overline{D}_1 \end{array} \right),$$

where

- D_1 *is such that $D_1 D_1^* = (I - D^*D)^{-1}$.*
- $\overline{D}_1$ *is such that $\overline{D}_1^* \overline{D}_1 = (I - DD^*)^{-1}$.*
- D_2 *is arbitrary .*
- Y *and Z are solutions of the Riccati equations*

$$0 = (A + B(I - D^*D)^{-1}D^*C)^*Y + Y(A + B(I - D^*D)^{-1}D^*C)$$

$$+YB(I - D^*D)^{-1}B^*Y + C^*(I - DD^*)^{-1}C,$$

$$0 = (A + B(I - D^*D)^{-1}D^*C)Z + Z(A + B(I - D^*D)^{-1}D^*C)^*$$

$$+ZC^*(I - DD^*)^{-1}CZ + B(I - D^*D)^{-1}B^*,$$

such that $A - BF$ and $A - HC$ are stable, where

$$F = -(I - D^*D)^{-1}D^*C - (I - D^*D)^{-1}B^*Y,$$

$$H = -BD^*(I - DD^*)^{-1} - ZC^*(I - DD^*)^{-1}.$$

Proof: Let G have a minimal realization (A, B, C, D). Let $G = NM^{-1}$ be a right factorization G such that

$$\left(\begin{array}{cc} M^* & N^* \end{array} \right) J_B \left(\begin{array}{c} M \\ N \end{array} \right) = I.$$

Such a factorization exists by Proposition 4.1 and has the same McMillan degree as G. By Theorem 2.2 any right factorization $\left(\begin{array}{c} M \\ N \end{array} \right)$ such that the McMillan degree of $\left(\begin{array}{c} M \\ N \end{array} \right)$ is the same as that of G has a state space realization of the form,

$$\left(\begin{array}{c} M \\ N \end{array} \right) \equiv \left(\begin{array}{c|c} A - BF & BD_1 \\ \hline -F & D_1 \\ C - DF & DD_1 \end{array} \right),$$

where F is a stabilizing state feedback and D_1 is invertible. Similarly, a left factorization $G = \tilde{M}^{-1}\tilde{N}$ such that

$$\left(\begin{array}{cc} \tilde{N} & \tilde{M} \end{array} \right) J_B \left(\begin{array}{c} \tilde{N}^* \\ \tilde{M}^* \end{array} \right) = -I,$$

exists and is of McMillan degree n and has a state-space representation of the form

$$\left(\begin{array}{cc} \tilde{N} & \tilde{M} \end{array} \right) \equiv \left(\begin{array}{c|cc} A - HC & B - HD & -H \\ \hline \overline{D_1}C & \overline{D_1}D & \overline{D_1} \end{array} \right),$$

where H is a stabilizing output injection. Since $\left(\begin{array}{c} M \\ N \end{array} \right)$ is stable and of McMillan degree n and $\left(\begin{array}{c} \tilde{N}^* \\ \tilde{M}^* \end{array} \right)$ is antistable and also of McMillan degree n, the function

$$\left(\begin{array}{cc} M & \tilde{N}^* \\ N & \tilde{M}^* \end{array} \right) \equiv \left(\begin{array}{cc|cc} A - BF & 0 & BD_1 & 0 \\ 0 & -A^* + C^*H^* & 0 & C^*\overline{D_1^*} \\ \hline -F & D^*H^* - B^* & D_1 & D^*\overline{D_1^*} \\ C - DF & H^* & DD_1 & \overline{D_1^*} \end{array} \right)$$

$$=: \left(\begin{array}{c|c} \mathcal{A} & \mathcal{B} \\ \hline \mathcal{C} & \mathcal{D} \end{array} \right)$$

has McMillan degree $2n$. Note that since

$$\left(\begin{array}{cc} M & \tilde{N}^* \\ N & \tilde{M}^* \end{array} \right) = J_B \left(\begin{array}{cc} M & \tilde{N}^* \\ N & \tilde{M}^* \end{array} \right)^{-*} J_B,$$

we have by Lemma 3.1 that $\left(\begin{array}{cc} M & \tilde{N}^* \\ N & \tilde{M}^* \end{array} \right)$ has the two equivalent realizations

$$\left(\begin{array}{c|c} \mathcal{A} & \mathcal{B} \\ \hline \mathcal{C} & \mathcal{D} \end{array}\right)$$

and

$$\left(\begin{array}{c|c} -\mathcal{A}^* + \mathcal{C}^* J_B \mathcal{D} J_B \mathcal{B}^* & \mathcal{C}^* J_B \mathcal{D} \\ \hline \mathcal{D} J_B \mathcal{B}^* & \mathcal{D} \end{array}\right).$$

Since $\mathcal{D} = J_B D^{-1} J_B$, Lemma 4.1 implies that $\mathcal{D}$ and $\overline{D}_1$ are such that

$$D_1 D_1^* = (I - D^* D)^{-1},$$

$$\overline{D}_1^* \overline{D}_1 = (I - DD^*)^{-1}.$$

We need to compute $-\mathcal{A}^* + \mathcal{C}^* J_B \mathcal{D} J_B \mathcal{B}^*$,

$$-\mathcal{A}^* + \mathcal{C}^* J_B \mathcal{D} J_B \mathcal{B}^*$$

$$= \left(\begin{array}{cc} -A^* + F^* B^* & 0 \\ 0 & A - HC \end{array}\right)$$

$$+ \left(\begin{array}{cc} -F^* & C^* - F^* D^* \\ HD - B & H \end{array}\right) \left(\begin{array}{cc} I & -D^* \\ -D & I \end{array}\right) \left(\begin{array}{cc} D_1 & 0 \\ 0 & \overline{D}_1^* \end{array}\right) \left(\begin{array}{cc} D_1^* B^* & 0 \\ 0 & \overline{D}_1 C \end{array}\right)$$

$$= \left(\begin{array}{cc} -A^* + F^* B^* & 0 \\ 0 & A - HC \end{array}\right)$$

$$+ \left(\begin{array}{cc} -F^*(I - D^* D) - C^* D & C^* \\ -B & BD^* + H(I - DD^*) \end{array}\right) \left(\begin{array}{cc} D_1 D_1^* & 0 \\ 0 & \overline{D}_1^* \overline{D}_1 \end{array}\right) \left(\begin{array}{cc} B^* & 0 \\ 0 & C \end{array}\right)$$

$$= \left(\begin{array}{cc} -A^* - C^* D D_1 D_1^* B^* & C^* \overline{D}_1^* \overline{D}_1 C \\ -B D_1 D_1^* B^* & A + BD^* \overline{D}_1^* \overline{D}_1 C \end{array}\right).$$

Since $\left(\begin{array}{c|c} \mathcal{A} & \mathcal{B} \\ \hline \mathcal{C} & \mathcal{D} \end{array}\right)$ and $\left(\begin{array}{c|c} -\mathcal{A}^* + \mathcal{C}^* J_B \mathcal{D} J_B \mathcal{B}^* & \mathcal{C}^* J_B \mathcal{D} \\ \hline \mathcal{D} J_B \mathcal{B}^* & \mathcal{D} \end{array}\right)$ are both minimal realizations of $\left(\begin{array}{cc} M & \tilde{N}^* \\ N & \tilde{M}^* \end{array}\right)$, there exists a unique non-singular state-space transformation $Y = \left(\begin{array}{cc} Y_{11} & Y_{12} \\ Y_{21} & Y_{22} \end{array}\right)$, which by Lemma 3.1 is such that $\left(\begin{array}{cc} Y_{11} & Y_{12} \\ Y_{21} & Y_{22} \end{array}\right) = \left(\begin{array}{cc} Y_{11}^* & Y_{21}^* \\ Y_{12}^* & Y_{22}^* \end{array}\right)$, and such that

$$\mathcal{C} = \mathcal{D} J_B \mathcal{B}^* Y,$$

$$Y \mathcal{B} = \mathcal{C}^* J_B \mathcal{D},$$

$$Y \mathcal{A} = (-\mathcal{A}^* + \mathcal{C}^* J_B \mathcal{D} J_B \mathcal{B}^*) Y.$$

More explicitly, we have writing the equation $Y \mathcal{B} = \mathcal{C}^* J_B \mathcal{D}$ componentwise,

$$\begin{pmatrix} Y_{11} & Y_{12} \\ Y_{12}^* & Y_{22} \end{pmatrix} \begin{pmatrix} BD_1 & 0 \\ 0 & C^*\overline{D}_1^* \end{pmatrix}$$

$$= \begin{pmatrix} -F^* & C^* - F^*D^* \\ HD - B & H \end{pmatrix} \begin{pmatrix} I & D^* \\ -D & -I \end{pmatrix} \begin{pmatrix} D_1 & 0 \\ 0 & \overline{D}_1^* \end{pmatrix}$$

and therefore

$$\begin{pmatrix} Y_{11} & Y_{12} \\ Y_{12}^* & Y_{22} \end{pmatrix} \begin{pmatrix} B & 0 \\ 0 & C^* \end{pmatrix} = \begin{pmatrix} -F^*(I - D^*D) - C^*D & -C^* \\ -B & -BD^* - H(I - DD^*) \end{pmatrix}.$$

Hence, we have that

$$-F^*(I - D^*D) - C^*D = Y_{11}B,$$

$$-C^* = Y_{12}C^*,$$

$$-H(I - DD^*) - BD^* = Y_{22}C^*,$$

$$-B = Y_{12}^*B,$$

which shows that

$$F = -(I - D^*D)^{-1}D^*C - (I - D^*D)^{-1}B^*Y_{11},$$

$$H = -BD^*(I - DD^*)^{-1} - Y_{22}C^*(I - DD^*)^{-1}.$$

Writing $Y\mathcal{A} = (-\mathcal{A}^* + \mathcal{C}^*J_B\mathcal{D}J_B\mathcal{B}^*)Y$, componentwise, we have for the $(1,1)$ entry,

$$Y_{11}(A - BF) = (-A^* - C^*DD_1D_1^*B^*)Y_{11} + C^*\overline{D}_1^*\overline{D}_1CY_{12}^*,$$

and using the above identities, this gives,

$$0 = (A^* + C^*DD_1D_1^*B^*)Y_{11} + Y_{11}(A + B[(I - D^*D)^{-1}D^*C + (I - D^*D)^{-1}B^*Y_{11}])$$

$$+ C^*\overline{D}_1^*\overline{D}_1C$$

$$= (A^* + C^*D(I - D^*D)^{-1}B^*)Y + Y(A + B(I - D^*D)^{-1}D^*C) +$$

$$YB(I - D^*D)^{-1}B^*Y + C^*(I - DD^*)^{-1}C,$$

where we have set $Y := Y_{11}$. Evaluating the $(2,2)$ entry we obtain,

$$Y_{22}(-A^* + C^*H^*) = -BD_1D_1^*B^*Y_{12} + (A + BD^*\overline{D}_1^*\overline{D}_1C)Y_{22},$$

or,

$$0 = (A + BD^*(I - DD^*)^{-1}C)Z + Z(A^* + C^*(I - DD^*)^{-1}DB^*)$$

$$+ ZC^*(I - DD^*)^{-1}CZ + B(I - D^*D)^{-1}B^*,$$

where we have set $Z := Y_{22}$. Note that $A - BF$ and $A - HC$ are stable with F and H as above.

It can be verified in a straightforward but tedious way that if state space representations are given as in the statement of the theorem that the transfer functions of these representations give doubly coprime factorizations with the required properties.

$\square$

In the proof of the theorem we also gave a proof of the well-known result that the so called bounded-real Riccati equation has a stabilizing solution.

Corollary 5.1 *Let (A, B, C, D) be a minimal realization of a bounded-real rational function. Then there exist hermitian solutions Y respectively Z to the Riccati equations*

$$0 = (A + B(I - D^*D)^{-1}D^*B^*C)^* + Y(A + B(I - D^*D)^{-1}D^*C)+$$

$$YB(I - D^*D)^{-1}B^*Y + C^*(I - DD^*)^{-1}C,$$

respectively,

$$0 = (A + B(I - DD^*)^{-1}D^*C)Z + Z(A + B(I - DD^*)^{-1}D^*C)^*$$

$$+ZC^*(I - DD^*)^{-1}CZ + B(I - D^*D)^{-1}B^*,$$

such that $A - BF$ and $A - HC$ respectively are stable, where

$$F = -(I - D^*D)^{-1}D^*C - (I - D^*D)^{-1}B^*Y,$$

$$H = -BD^*(I - DD^*)^{-1} - ZC^*(I - DD^*)^{-1}.$$

Proof: This statement was proved as part of the proof of the theorem. $\square$

6 Positive-real functions

Positive real functions are of importance in many areas of system and control theory, e.g. in stochastic system theory or in adaptive control. We are now going to define what we mean by positive-real functions and by the J_P-factorization of such functions.

Definition 6.1 *A proper square stable rational function is called* positive-real, *if*

$$G(i\omega) + G^*(i\omega) > 0,$$

for all $\omega \in \Re \cup \{\pm\infty\}$. Let

$$J_P = \begin{pmatrix} 0 & I \\ I & 0 \end{pmatrix}.$$

A right (coprime) factorization of $G = NM^{-1}$ is called a J_P-normalized right (coprime) factorization $(J_P - RF$ respectively $J_P - RCF)$ of G if

$$\left(\begin{array}{cc} M^* & N^* \end{array} \right) J_P \left(\begin{array}{c} M \\ N \end{array} \right) = I.$$

Similarly, a left (coprime) factorization is called a J_P-normalized left (coprime) factorization $(J_P - LF$ respectively $J_P - LCF)$ of G if

$$\left(\begin{array}{cc} \tilde{N} & \tilde{M} \end{array} \right) J_P \left(\begin{array}{c} \tilde{N}^* \\ \tilde{M}^* \end{array} \right) = I.$$

The following proposition establishes the existence of J_P-factorizations.

Proposition 6.1 *Let G be a proper rational positive-real function of McMillan degree n. Then there exists a J_P-right (left) factorization with proper functions M, N,*

$$G = NM^{-1} \quad (G = \tilde{M}^{-1}\tilde{N}).$$

This factorization is right (left) coprime and is unique up to right (left) multiplication by a constant unitary matrix. All such factorizations are such that $\left(\begin{array}{c} M \\ N \end{array} \right)$ $\left(\left(\begin{array}{cc} -\tilde{N} & \tilde{M} \end{array} \right) \right)$ is of McMillan degree n.

Proof: Let $G = ED^{-1}$ be a right polynomial coprime factorization. Let T be a square stable spectral factor of $E^*D + D^*E$, i.e. $T^*T = E^*D + D^*E$ and proceed as previously.   □

Some properties of J_P-factorizations are summarized in the following lemma.

Lemma 6.1 *Let $G = NM^{-1}$ be a J_P-RF and $G = \tilde{M}^{-1}\tilde{N}$ a J_P-LF of the positive-real function G. Then*

1. $$\left(\begin{array}{cc} M & -\tilde{M}^* \\ N & \tilde{N}^* \end{array} \right)^* J_P \left(\begin{array}{cc} M & -\tilde{M}^* \\ N & \tilde{N}^* \end{array} \right) = \left(\begin{array}{cc} I & 0 \\ 0 & -I \end{array} \right),$$

2. $$\left(\begin{array}{cc} M & -\tilde{M}^* \\ N & \tilde{N}^* \end{array} \right) = J_P \left(\begin{array}{cc} M & -\tilde{M}^* \\ N & \tilde{N}^* \end{array} \right)^{-*} J_B.$$

Proof: The statements are easily verified.   □

A few useful identities are given in the following Lemma.

Lemma 6.2 *Consider*

$$\left(\begin{array}{cc} D_1 & -\overline{D_1^*} \\ DD_1 & D^*\overline{D_1^*} \end{array} \right) = \left(\begin{array}{cc} I & -I \\ D & D^* \end{array} \right) \left(\begin{array}{cc} D_1 & 0 \\ 0 & \overline{D_1^*} \end{array} \right),$$

with D_1 and $\overline{D}_1$ invertible and D is square such that $D + D^* > 0$. Then

1.
$$\begin{pmatrix} D_1 & -\overline{D}_1^* \\ DD_1 & D^*\overline{D}_1^* \end{pmatrix}^{-1}$$

$$= \begin{pmatrix} D_1^{-1} & 0 \\ 0 & \overline{D}_1^{-*} \end{pmatrix} \begin{pmatrix} D^* & I \\ -D & I \end{pmatrix} \begin{pmatrix} (D+D^*)^{-1} & 0 \\ 0 & (D+D^*)^{-1} \end{pmatrix}.$$

2. If

$$\begin{pmatrix} D_1 & -\overline{D}_1^* \\ DD_1 & D^*\overline{D}_1^* \end{pmatrix} = J_P \begin{pmatrix} D_1 & -\overline{D}_1^* \\ DD_1 & D^*\overline{D}_1^* \end{pmatrix}^{-*} J_B,$$

then

$$D_1 D_1^* = (D+D^*)^{-1},$$

$$\overline{D}_1^*\overline{D}_1 = (D+D^*)^{-1}.$$

Proof: The statements are checked in a straightforward way. $\qquad\square$

We can now characterize doubly coprime factorizations for positive-real functions.

Theorem 6.1 *Let* (A, B, C, D) *be a minimal realization of the proper positive-real rational function* G *of McMillan degree* n.
Then all doubly coprime factorizations of G*, such that*

1. $G = NM^{-1}$ *is a* J_P-*RF and* $G = \tilde{M}^{-1}\tilde{N}$ *is a* J_B-*LF,*

2. $\begin{pmatrix} M & U \\ N & V \end{pmatrix}$, $\begin{pmatrix} \tilde{V} & -\tilde{U} \\ -\tilde{N} & \tilde{M} \end{pmatrix}$,

   are of McMillan degree n.

are given by

$$\begin{pmatrix} M & U \\ N & V \end{pmatrix} \equiv \left(\begin{array}{c|cc} A - BF & BD_1 & BD_2 + H\overline{D}_1^{-1} \\ \hline -F & D_1 & D_2 \\ C - DF & DD_1 & \overline{D}_1^{-1} + DD_2 \end{array} \right),$$

$$\begin{pmatrix} \tilde{V} & -\tilde{U} \\ -\tilde{N} & \tilde{M} \end{pmatrix} \equiv \left(\begin{array}{c|cc} A - HC & HD - B & -H \\ \hline -D_1^{-1}F - D_1^{-1}D_2\overline{D}_1 C & D_1^{-1} + D_1^{-1}D_2\overline{D}_1 D & -D_1^{-1}D_2\overline{D}_1 \\ \overline{D}_1 C & -\overline{D}_1 D & \overline{D}_1 \end{array} \right),$$

where

- D_1 *is such that* $D_1 D_1^* = (D+D^*)^{-1}$.
- $\overline{D}_1$ *is such that* $\overline{D}_1^*\overline{D}_1 = (D+D^*)^{-1}$.
- D_2 *is arbitrary* .
- Y *and* Z *are solutions of the Riccati equations*

$$0 = (A - B(D + D^*)^{-1}C)^*Y + Y(A - B(D + D^*)^{-1}C)$$

$$+YB(D + D^*)^{-1}B^*Y + C^*(D + D^*)^{-1}C,$$

$$0 = (A - B(D + D^*)^{-1}C)Z + Z(A - B(D + D^*)^{-1}C)$$

$$+ZC^*(D + D^*)^{-1}CZ + B(D + D^*)^{-1}B^*,$$

such that $A - BF$ and $A - HC$ are stable, with

$$F = (D + D^*)^{-1}C - (D + D^*)^{-1}B^*Y,$$

$$H = B(D + D^*)^{-1} - ZC^*(D + D^*)^{-1}.$$

Proof: Let G have a minimal realization (A, B, C, D). Let $G = NM^{-1}$ be a right factorization G such that

$$\begin{pmatrix} M^* & N^* \end{pmatrix} J_P \begin{pmatrix} M \\ N \end{pmatrix} = I.$$

Such a factorization exists by Proposition 4.1 and has the same McMillan degree as G. By Theorem 2.2 any right factorization $\begin{pmatrix} M \\ N \end{pmatrix}$ such that the McMillan degree of $\begin{pmatrix} M \\ N \end{pmatrix}$ is the same as that of G has a state space realization of the form,

$$\begin{pmatrix} M \\ N \end{pmatrix} \equiv \left(\begin{array}{c|c} A - BF & BD_1 \\ \hline -F & D_1 \\ C - DF & DD_1 \end{array} \right),$$

where F is a stabilizing state feedback and D_1 is invertible. Similarly, a left factorization $G = \tilde{M}^{-1}\tilde{N}$ such that

$$\begin{pmatrix} -\tilde{M} & \tilde{N} \end{pmatrix} J_P \begin{pmatrix} -\tilde{M}^* \\ \tilde{N}^* \end{pmatrix} = -I,$$

exists and is of McMillan degree n and has a state-space representation of the form

$$\begin{pmatrix} -\tilde{M} & \tilde{N} \end{pmatrix} \equiv \left(\begin{array}{c|cc} A - HC & H & B - HD \\ \hline \overline{D}_1 C & -\overline{D}_1 & \overline{D}_1 D \end{array} \right),$$

where H is a stabilizing output injection. Since $\begin{pmatrix} M \\ N \end{pmatrix}$ is stable and of McMillan degree n and $\begin{pmatrix} -\tilde{M}^* \\ \tilde{N}^* \end{pmatrix}$ is antistable and also of McMillan degree n, the function

$$\begin{pmatrix} M & -\tilde{M}^* \\ N & \tilde{N}^* \end{pmatrix} \equiv \left(\begin{array}{cc|cc} A - BF & 0 & BD_1 & 0 \\ 0 & -A^* + C^*H^* & 0 & C^*\overline{D}_1^* \\ \hline -F & -H^* & D_1 & -\overline{D}_1^* \\ C - DF & D^*H^* - B^* & DD_1 & D^*\overline{D}_1^* \end{array} \right)$$

$$=: \left(\begin{array}{c|c} \mathcal{A} & \mathcal{B} \\ \hline \mathcal{C} & \mathcal{D} \end{array} \right)$$

has McMillan degree $2n$. Note that since

$$\left(\begin{array}{cc} M & -\tilde{M}^* \\ N & \tilde{N}^* \end{array} \right) = J_P \left(\begin{array}{cc} M & -\tilde{M}^* \\ N & \tilde{N}^* \end{array} \right)^{-*} J_B,$$

we have by Lemma 3.1 that $\left(\begin{array}{cc} M & -\tilde{M}^* \\ N & \tilde{N}^* \end{array} \right)$ has the two equivalent realizations

$$\left(\begin{array}{c|c} \mathcal{A} & \mathcal{B} \\ \hline \mathcal{C} & \mathcal{D} \end{array} \right)$$

and

$$\left(\begin{array}{c|c} -\mathcal{A}^* + \mathcal{C}^* J_P \mathcal{D} J_B \mathcal{B}^* & \mathcal{C}^* J_P \mathcal{D} \\ \hline \mathcal{D} J_B \mathcal{B}^* & \mathcal{D} \end{array} \right).$$

Since $\mathcal{D} = J_p \mathcal{D}^{-*} J_B$, Lemma 4.1 implies that $\mathcal{D}$ and $\overline{D}_1$ are such that

$$D_1 D_1^* = (D + D^*)^{-1},$$

$$\overline{D}_1^* \overline{D}_1 = (D + D^*)^{-1}.$$

We need to compute $-\mathcal{A}^* + \mathcal{C}^* J_P \mathcal{D} J_B \mathcal{B}^*$,

$$-\mathcal{A}^* + \mathcal{C}^* J_P \mathcal{D} J_B \mathcal{B}^*$$

$$= \left(\begin{array}{cc} -A^* + F^* B^* & 0 \\ 0 & A - HC \end{array} \right)$$

$$+ \left(\begin{array}{cc} -F^* & C^* - F^* D^* \\ -H & HD - B \end{array} \right) J_P \left(\begin{array}{cc} I & -I \\ D & D^* \end{array} \right) \left(\begin{array}{cc} D_1 & 0 \\ 0 & \overline{D}_1^* \end{array} \right) J_B \left(\begin{array}{cc} D_1^* B^* & 0 \\ 0 & \overline{D}_1 C \end{array} \right)$$

$$= \left(\begin{array}{cc} -A^* + F^* B^* & 0 \\ 0 & A - HC \end{array} \right)$$

$$+ \left(\begin{array}{cc} -F^* & C^* - F^* D^* \\ -H & HD - B \end{array} \right) \left(\begin{array}{cc} D & -D^* \\ I & I \end{array} \right) \left(\begin{array}{cc} D_1 D_1^* & 0 \\ 0 & \overline{D}_1^* \overline{D}_1 \end{array} \right) \left(\begin{array}{cc} B^* & 0 \\ 0 & C \end{array} \right)$$

$$= \left(\begin{array}{cc} -A^* + F^* B^* & 0 \\ 0 & A - HC \end{array} \right)$$

$$+ \left(\begin{array}{cc} -F^*(D + D^*) + C^* & C^* \\ -B & H(D + D^*) - B \end{array} \right) \left(\begin{array}{cc} D_1 D_1^* & 0 \\ 0 & \overline{D}_1^* \overline{D}_1 \end{array} \right) \left(\begin{array}{cc} B^* & 0 \\ 0 & C \end{array} \right)$$

$$= \left(\begin{array}{cc} -A^* + C^* D_1 D_1^* B^* & C^* \overline{D}_1^* \overline{D}_1 C \\ -B D_1 D_1^* B^* & A - B \overline{D}_1^* \overline{D}_1 C \end{array} \right).$$

Since $\left(\begin{array}{c|c} \mathcal{A} & \mathcal{B} \\ \hline \mathcal{C} & \mathcal{D} \end{array}\right)$ and $\left(\begin{array}{c|c} -\mathcal{A}^* + \mathcal{C}^* J_P \mathcal{D} J_B \mathcal{B}^* & \mathcal{C}^* J_P \mathcal{D} \\ \hline \mathcal{D} J_B \mathcal{B}^* & \mathcal{D} \end{array}\right)$ are both minimal realizations of $\left(\begin{array}{cc} M & -\tilde{M}^* \\ N & \tilde{N}^* \end{array}\right)$, there exists a unique non-singular state-space transformation $Y = \left(\begin{array}{cc} Y_{11} & Y_{12} \\ Y_{21} & Y_{22} \end{array}\right)$, which by Lemma 3.1 is such that $\left(\begin{array}{cc} Y_{11} & Y_{12} \\ Y_{21} & Y_{22} \end{array}\right) = \left(\begin{array}{cc} Y_{11}^* & Y_{21}^* \\ Y_{12}^* & Y_{22}^* \end{array}\right)$, and such that

$$\mathcal{C} = \mathcal{D} J_B \mathcal{B}^* Y,$$

$$Y\mathcal{B} = \mathcal{C}^* J_P \mathcal{D},$$

$$Y\mathcal{A} = (-\mathcal{A}^* + \mathcal{C}^* J_P \mathcal{D} J_B \mathcal{B}^*)Y.$$

More explicitly, we have writing the equation $Y\mathcal{B} = \mathcal{C}^* J_P \mathcal{D}$ componentwise,

$$\left(\begin{array}{cc} Y_{11} & Y_{12} \\ Y_{12}^* & Y_{22} \end{array}\right)\left(\begin{array}{cc} BD_1 & 0 \\ 0 & C^* \overline{D}_1^* \end{array}\right)$$

$$= \left(\begin{array}{cc} -F^* & C^* - F^* D^* \\ -H & HD - B \end{array}\right)\left(\begin{array}{cc} 0 & I \\ I & 0 \end{array}\right)\left(\begin{array}{cc} I & -I \\ D & D^* \end{array}\right)\left(\begin{array}{cc} D_1 & 0 \\ 0 & \overline{D}_1^* \end{array}\right)$$

and therefore

$$\left(\begin{array}{cc} Y_{11} & Y_{12} \\ Y_{12}^* & Y_{22} \end{array}\right)\left(\begin{array}{cc} B & 0 \\ 0 & C^* \end{array}\right) = \left(\begin{array}{cc} -F^*(D + D^*) + C^* & -C^* \\ -B & -H(D + D^*) + B \end{array}\right).$$

Hence, we have that

$$-F^*(D + D^*) + C^* = Y_{11}B,$$

$$-C^* = Y_{12}C^*,$$

$$-H(D + D^*) + B = Y_{22}C^*,$$

$$-B = Y_{12}^* B,$$

which shows that

$$F = (D + D^*)^{-1}C - (D + D^*)^{-1}B^* Y_{11},$$

$$H = B(D + D^*)^{-1} - Y_{22}C^*(D + D^*)^{-1}.$$

Writing $Y\mathcal{A} = (-\mathcal{A}^* + \mathcal{C}^* J_P \mathcal{D} J_B \mathcal{B}^*)Y$, componentwise, we have for the $(1,1)$ entry,

$$Y_{11}(A - BF) = (-A^* + C^* D_1 D_1^* B^*)Y_{11} + C^* \overline{D}_1^* \overline{D}_1 C Y_{12}^*,$$

and using the above identities, this gives,

$$0 = (A^* - C^* D_1 D_1^* B^*)Y_{11} + Y_{11}(A - B[(D + D^*)^{-1}C + (D + D^*)^{-1}B^* Y_{11}])$$

$$+ C^* \overline{D}_1^* \overline{D}_1 C$$

$$= (A^* - C^*(D + D^*)^{-1}B^*)Y + Y(A - B(D + D^*)^{-1}C)+$$

$$YB(D + D^*)^{-1}B^*Y + C^*(D + D^*)^{-1}C,$$

where we have set $Y := Y_{11}$. Evaluating the $(2,2)$ entry we obtain,

$$Y_{22}(-A^* + C^*H^*) = -BD_1D_1^*B^*Y_{12} + (A - B\overline{D}_1^*\overline{D}_1C)Y_{22},$$

or,

$$0 = (A - B(D + D^*)^{-1}C)Z + Z(A^* - C^*(D + D^*)^{-1}B^*)$$

$$+ZC^*(D + D^*)^{-1}CZ + B(D + D^*)^{-1}B^*.$$

The remaining part of the proof is analogous to the equivalent steps in the previous theorems.
$\square$

In the proof of the theorem we also established the well-known result that positive-real Riccati equations have a stabilizing solution.

Corollary 6.1 *Let (A, B, C, D) be a minimal realization of a positive-real rational function. Then there exist hermitian solutions Y respectively Z to the Riccati equations*

$$0 = (A^* - C^*(D + D^*)^{-1}B^*)Y + Y(A - B(D + D^*)^{-1}C)+$$

$$YB(D + D^*)^{-1}B^*Y + C^*(D + D^*)^{-1}C.$$

respectively,

$$0 = (A - B(D + D^*)^{-1}C)Z + Z(A^* - C^*(D + D^*)^{-1}B^*)$$

$$+ZC^*(D + D^*)^{-1}Z + B(D + D^*)^{-1}B^*.$$

such that $A - BF$ and $A - HC$ are stable, where

$$F = (D + D^*)^{-1}C - (D + D^*)^{-1},$$

$$H = B(D + D^*)^{-1} - ZC^*(D + D^*)^{-1}.$$

Proof: This statement was proved as part of the proof of the theorem. $\square$

7   REFERENCES

[1949] A. Beurling. "On two problems concerning linear transformations in Hilbert Space". Acta Math., 81, 239-255.

[1980] Desoer, C.A., R.W. Liu, J. Murray and R. Saeks. "Feedback system design: the fractional representation approach", IEEE TAC, 25, 399-412.

[1971] R.G. Douglas, H.S. Shapiro and A.L Shields. "Cyclic vectors and invariant subspaces for the backward shift." Ann. Inst. Fourier, Grenoble, 20,1, 37-76.

[1984] J.C. Doyle, " Lecture Notes in Advances in Multivariable Control", ONR/Honeywell Workshop, Minneapolis, MN.

[1975] P. A. Fuhrmann, "On Hankel operator ranges, meromorphic pseudo-continuation and factorization of operator valued analytic functions", *J. Lond. Math. Soc.*, (2) 13, 323-327.

[1976] P. A. Fuhrmann, "Algebraic system theory: An analyst's point of view", *J. Franklin Inst.*, 301, 521-540.

[1979] P. A. Fuhrmann, "Linear feedback via polynomial models", *Int. J. Contr.* 30, 363-377.

[1981] P. A. Fuhrmann, *Linear Systems and Operators in Hilbert Space*, McGraw-Hill, New York.

[1985] P. A. Fuhrmann, "The algebraic Riccati equation - a polynomial approach", *Syst. and Contr. Lett.*, 369-376.

[1991] P. A. Fuhrmann, " A polynomial approach to Hankel norm and balanced approximation", *Lin. Alg. Appl.*, 146, 133-220.

[1992] P. A. Fuhrmann and R. J. Ober, "A functional approach to LQG balancing", to appear in International Journal of Control.

[1985] J. Hammer, "Nonlinear systems, stabilization, and coprimeness", *International Journal of Control*, Vol. 42, pp. 1 - 20.

[1978] M. L. J. Hautus and M. Heymann, "Linear feedback-an algebraic approach", *SIAM J. Control* 16, 83-105.

[1980] T. Kailath, *Linear systems*, Prentice Hall, Englewood Cliffs, N.J.

[1982] P. Khargonekar and E. Sontag. "On the relation between stable matrix factorizations and regulable realizations of linear systems over rings", IEEE TAC, 27, 627-638.

[1989] D. McFarlane and K. Glover, "Robust controller design using normalized coprime factor plant descriptions", *Lecture Notes in Control and Information Sciences*, vol. 10, Springer Verlag.

[1987] D. Meyer and G. Franklin, "A connection between normalized coprime factorizations and linear quadratic regulator theory", *IEEE Trans. on Auto. Contr.* 32, 227-228.

[1975] A. S. Morse, "System invariants under feedback and cascade control", *Lecture Notes in Economics and Mathematical Systems*, vol. 131 (Proc. Symp. Udine), Springer Verlag.

[1984] C.N. Nett, C.A. Jacobson and M.J. Balas. " A connection between state-space and doubly coprime fractional representations". IEEE TAC, 29, 831-832.

[1989] R.J. Ober, D.C. McFarlane. "Balanced canonical forms: a normalized coprime factor approach." Linear Algebra and its Applications, 122-124: 23-640.

[1970] H. H. Rosenbrock, *State Space and Multivariable Theory*, J.Wiley, New York.

[1988] M.S. Verma, "Coprime Fractional Representations and Stability of Nonlinear Feedback Systems", International Journal of Control, 48, 897–918.

[1985] M. Vidyasagar, *Control System Synthesis: A Coprime Factorization Approach*, M.I.T. Press, Cambridge MA.

[1988] M. Vidyasagar. "Normalized coprime factorizations for non strictly proper systems". Automatica, 85-94.

[1976] D. C. Youla, J. J. Bongiorno and H. A. Jabr, "Modern Wiener-Hopf design of optimal controllers, Pt. 2 The multivariable case", IEEE Transactions on Automatic Control, 21, 319-338.

P. A. Fuhrmann
Department of Mathematics
Ben-Gurion University of the Negev
84105 Beer Sheva, Israel

R. Ober
Center for Engineering Mathematics
Programs in Mathematical Sciences
The University of Texas at Dallas
Richardson, Texas 75083 – 06688 , USA

MSC: 15A54 47A68 47N70 93B36

Operator Theory:
Advances and Applications, Vol. 62
© 1993 Birkhäuser Verlag Basel

GENERALIZATION OF HEINZ-KATO THEOREM
VIA FURUTA INEQUALITY

Takayuki Furuta

Dedicated to Professor Tsuyoshi Ando at his sixtieth birthday with respect and affection

A capital letter means a bounded linear operator on a complex Hilbert space H. An operator T is said to be positive if $(Tx, x) \geq 0$ for all $x \in H$.

We recall the following famous Heinz-Kato theorem [5][7]. If A and B are positive operators such that $\|Tx\| \leq \|Ax\|$ and $\|T^*y\| \leq \|By\|$ for all $x, y \in H$, then the following inequality holds; $|(Tx, y)| \leq \|A^\alpha x\|\|B^{1-\alpha}y\|$ for all $x, y \in H$ and for any $0 \leq \alpha \leq 1$.

Also we recall the following famous Löwner-Heinz theorem [5][8]. If $A \geq B \geq 0$, then $A^\alpha \geq B^\alpha$ for each $\alpha \in [0, 1]$. There are a lot of proofs of this famous theorem, among them, especially an elegant proof of this theorem is given in [9].

We have Furuta inequality [2] as some extension of this Löwner-Heinz theorem as follows.   If $A \geq B \geq 0$, then for each $r \geq 0$ (i) $(B^r A^p B^r)^{1/q} \geq B^{(p+2r)/q}$ and

(ii) $A^{(p+2r)/q} \geq (A^r B^p A^r)^{1/q}$ hold for each p and q such that $p \geq 0$ and $q \geq 1$ with $(1 + 2r)q \geq p + 2r$.

When we put $r = 0$ in (i) or (ii) in the Furuta inequality stated above, we have the famous Löwner-Heinz theorem.

Alternative proofs of the Furuta inequality are given in [1][3][6] and an elementary proof is shown in [4].

In this paper, firstly we shall show a generalization of the Heinz-Kato theorem as an application of the Furuta inequality and secondary we shall show that this generalization is equivalent to the Furuta inequality.

§1. GENERALIZATION OF HEINZ-KATO THEOREM

Theorem 1. *Let T be an operator on a Hilbert space H. If A and B are positive operators such that $\|Tx\| \le \|Ax\|$ and $\|T^*y\| \le \|By\|$ for all $x, y \in H$. Then for each $r \ge 0$ and $s \ge 0$, the following inequality holds for all $x, y \in H$;*

$$(1) \quad |(T|T|^{(1+2r)\alpha+(1+2s)\beta-1}x, y)|^2$$

$$\le ((|T|^{2r}A^{2p}|T|^{2r})^{(1+2r)\alpha/(p+2r)}x, x)((|T^*|^{2s}B^{2q}|T^*|^{2s})^{(1+2s)\beta/(q+2s)}y, y)$$

for any $p \ge 1$, $q \ge 1$ and $\alpha, \beta \in [0,1]$ such that $(1+2r)\alpha + (1+2s)\beta \ge 1$.

In the case $\alpha > 0$ and $\beta > 0$, the equality in (1) holds for some x and y iff the following (i),(ii) and (iii) hold together for some x and y;

(i) $|T|^{2(1+2r)\alpha}x$ and $|T|^{(1+2r)\alpha+(1+2s)\beta-1}T^*y$ are linearly dependent,

(ii) $|T|^{2(1+2r)\alpha}x = (|T|^{2r}A^{2p}|T|^{2r})^{(1+2r)\alpha/(p+2r)}x$,

(iii) $|T^*|^{2(1+2s)\beta}y = (|T^*|^{2s}B^{2q}|T^*|^{2s})^{(1+2s)\beta/(q+2s)}y$.

Remark 1. We remark that the condition $(1+2r)\alpha + (1+2s)\beta \ge 1$ in Theorem 1 is unnecessary if T is positive operator or T is invertible operator and this is easily seen in the proof of Theorem 1.

Corollary 1. *Let T be an operator on a Hilbert space H. If A and B are positive operators such that $\|Tx\| \le \|Ax\|$ and $\|T^*y\| \le \|By\|$ for all $x, y \in H$. Then the following inequality holds for all $x, y \in H$;*

$$(2) \qquad\qquad |(T|T|^{\alpha+\beta-1}x, y)|^2 \le \|A^\alpha x\|^2 \|B^\beta y\|^2$$

for any α and β such that $\alpha, \beta \in [0,1]$ and $\alpha + \beta \ge 1$.

In the case $\alpha > 0$ and $\beta > 0$, the equality in (2) holds for some x and y iff $|T|^{2\alpha}x$ and $|T|^{\alpha+\beta-1}T^*y$ are linearly dependent, $|T|^{2\alpha}x = A^{2\alpha}x$ and $|T^*|^{2\beta}y = B^{2\beta}y$ hold some x and y together.

Remark 2. In Corollary 1, we put $\alpha + \beta = 1$. Then (2) easily ensures

$$|(Tx, y)| \le \|A^\alpha x\| \|B^{1-\alpha}y\| \quad \text{for all } x, y \in H \text{ and for any } 0 \le \alpha \le 1$$

whenever $\|Tx\| \le \|Ax\|$ and $\|T^*y\| \le \|By\|$ for all $x, y \in H$ and for any positive operators A and B. This is just the Heinz-Kato theorem, so that Corollary 1 can be considered as a generalization of the Heinz-Kato theorem.

Corollary 2. *Let T be an operator on a Hilbert space H. If A and B are positive operators such that $\|Tx\| \leq \|Ax\|$ and $\|T^*y\| \leq \|By\|$ for all $x, y \in H$. Then for each $r \geq 0$, the following inequality holds for all $x, y \in H$;*

$$(3) \qquad |(T|T|^{2r+\beta}x, y)|^2 \leq ((|T|^{2r}A^{2p}|T|^{2r})^{(1+2r)/(p+2r)}x, x)\|B^\beta y\|^2$$

for any p and β such that $p \geq 1$ and $\beta \in [0, 1]$.

In the case $\beta > 0$, the equality in (3) holds for some x and y iff $|T|^{2(1+2r)}x$ and $|T|^{2r+\beta}T^*y$ are linearly dependent, $|T|^{2(1+2r)}x = (|T|^{2r}A^{2p}|T|^{2r})^{(1+2r)/(p+2r)}x$ and $|T^*|^{2\beta}y = B^{2\beta}y$ hold for some x and y together.

Also for each $s \geq 0$, the following inequality holds for all $x, y \in H$;

$$(4) \qquad |(T|T|^{\alpha+2s}x, y)|^2 \leq \|A^\alpha x\|^2((|T^*|^{2s}B^{2q}|T^*|^{2s})^{(1+2s)/(q+2s)}y, y)$$

for any q and α such that $q \geq 1$ and $\alpha \in [0, 1]$.

In the case $\alpha > 0$, the equality in (4) holds for some x and y iff $|T|^{2\alpha}x$ and $|T|^{\alpha+2s}T^*y$ are linearly dependent, $|T^*|^{2(1+2s)}y = (|T^*|^{2s}B^{2q}|T^*|^{2s})^{(1+2s)/(q+2s)}y$ and $|T|^{2\alpha}x = A^{2\alpha}x$ hold for some x and y together.

Corollary 3. *Let T be an operator on a Hilbert space H. If A and B are positive operators such that $\|Tx\| \leq \|Ax\|$ and $\|T^*y\| \leq \|By\|$ for all $x, y \in H$. Then for each $r \geq 0$ and $s \geq 0$, the following inequality holds for all $x, y \in H$;*

$$(5) \qquad |(T|T|^{2r+2s+1}x, y)|^2$$

$$\leq ((|T|^{2r}A^{2p}|T|^{2r})^{(1+2r)/(p+2r)}x, x)((|T^*|^{2s}B^{2q}|T^*|^{2s})^{(1+2s)/(q+2s)}y, y)$$

for any p and q such that $p \geq 1$ and $q \geq 1$.

The equality in (5) holds some x and y iff $|T|^{2(1+2r)}x$ and $|T|^{2r+2s+1}T^*y$ are linearly dependent,

$$|T|^{2(1+2r)}x = (|T|^{2r}A^{2p}|T|^{2r})^{(1+2r)/(p+2r)}x \text{ and}$$

$$|T^*|^{2(1+2s)}y = (|T^*|^{2s}B^{2q}|T^*|^{2s})^{(1+2s)/(q+2s)}y \text{ hold for some } x \text{ and } y \text{ together.}$$

Proof of Theorem 1.

In order to give a proof of Theorem 1, we need the following Furuta inequality.

Theorem A. (Furuta inequality) *If $A \geq B \geq 0$, then for each $r \geq 0$*

$$(i) \qquad (B^r A^p B^r)^{(1+2r)\theta/(p+2r)} \geq B^{(1+2r)\theta}$$

and

(ii)
$$A^{(1+2r)\theta} \geq (A^r B^p A^r)^{(1+2r)\theta/(p+2r)}$$

hold for any $p \geq 1$ and $\theta \in [0,1]$.

(i)   **Proof of the inequality (1)**

First of all, the two hypotheses $\|Tx\| \leq \|Ax\|$ and $\|T^*y\| \leq \|By\|$ for all $x, y \in H$ are equivalent to the following (6) and (7) respectively,

(6)
$$|T|^2 \leq A^2$$

and

(7)
$$|T^*|^2 \leq B^2.$$

Let $T = U|T|$ be the polar decomposition of an operator T, where U is partial isometry and $|T| = (T^*T)^{1/2}$ and $N(|T|) = N(U)$ where $N(X)$ denotes the kernel of an operator X. We put $a = (1 + 2r)\alpha$ and $b = (1 + 2s)\beta$.

In the case $0 < b$ and $1 \leq a + b$. Then for all $x, y \in H$ we have

(8)
$$\begin{aligned}
|(T|T|^{a+b-1}x, y)|^2 &= |(U|T|^{a+b}x, y)|^2 \\
&= |(|T|^a x, |T|^b U^* y)|^2 \\
&\leq \||T|^a x\|^2 \||T|^b U^* y\|^2 \\
&= (|T|^{2a}x, x)(U|T|^{2b}U^* y, y) \\
&= (|T|^{2a}x, x)(|T^*|^{2b}y, y)
\end{aligned}$$

since $|T^*|^{2b} = U|T|^{2b}U^*$ holds for any $b > 0$ in general.

Applying (i) of Theorem A to (6), then for each $r \geq 0$ we have the following (9);

(9)
$$(|T|^{2(1+2r)\alpha}x, x) \leq ((|T|^{2r}A^{2p}|T|^{2r})^{(1+2r)\alpha/(p+2r)}x, x)$$

for any $p \geq 1$ and $\alpha \in [0,1]$.

Also applying (i) of Theorem A to (7), then for each $s \geq 0$ we have the following (10);

(10)
$$(|T^*|^{2(1+2s)\beta}y, y) \leq ((|T^*|^{2s}B^{2q}|T^*|^{2s})^{(1+2s)\beta/(q+2s)}y, y)$$

for any $q \geq 1$ and $\beta \in [0,1]$. Therefore for each $r \geq 0$ and $s \geq 0$, we have the following inequality (1) for all $x, y \in H$ by (8),(9) and (10)

$$|(T|T|^{(1+2r)\alpha+(1+2s)\beta-1}x,y)|^2$$

$$\leq ((|T|^{2r}A^{2p}|T|^{2r})^{(1+2r)\alpha/(p+2r)}x,x)((|T^*|^{2s}B^{2q}|T^*|^{2s})^{(1+2s)\beta/(q+2s)}y,y)$$

for any $p \geq 1, q \geq 1$ and $\alpha, \beta \in [0,1]$ such that $(1+2r)\alpha + (1+2s)\beta \geq 1$.

In the case $b = 0$, the result is obvious and we omit to describe it in detail.

Hence the proof of (1) is complete.

(ii) **Scrutiny of the equality in the inequality** (1). Next we shall scrutinize the equality case in (1). First of all we cite the following obvious results.

For positive operator S, $(Sx,x) = 0$ holds for some vector x iff $Sx = 0$ holds

for some vector x. $(\star)$

For positive operator S, $N(S^q) = N(S)$ holds for any positive number q,

where $N(X)$ denotes the kernel of an operator X. $(\star\star)$

In the case $a = (1 + 2r)\alpha > 0$ and $b = (1 + 2s)\beta > 0$, the equality in (8) holds iff $|T|^a x$ and $|T|^b U^* y$ are linearly dependent, that is, $|T|^{2a}x$ and $|T|^{a+b-1}|T|U^*y$ are linearly dependent by $(\star\star)$, namely, $|T|^{2a}x$ and $|T|^{a+b-1}T^*y$ are linearly dependent, that is, we have a condition (i) in theorem 1.

The equality in (9) holds for some vector x iff $(|T|^{2r}A^{2p}|T|^{2r})^{(1+2r)\alpha/(p+2r)}x = |T|^{2(1+2r)\alpha}x$ holds for some vector x by $(\star)$ and also the equality in (10) holds for some vector y iff $(|T^*|^{2s}B^{2q}|T^*|^{2s})^{(1+2s)\beta/(q+2s)}y = |T^*|^{2(1+2s)\beta}y$ holds for some vector y by $(\star)$.

In the case $\alpha > 0$ and $\beta > 0$, the equality in (1) holds iff the equalities in (8),(9) and (10) hold together, that is, the conditions (i), (ii) and (iii) in Theorem 1 hold together by the results obtained above. Therefore the proof of the equality in (1) is complete.

Hence the proof of Theorem 1 is complete.

Remark 3.   Equivalent conditions . In the special case $\alpha = 0$, scrutiny of the equality in (1) is obvious. In the case $\beta = 0$, scrutiny of the equality in (1) is also obvious. In the equality in (1) of Theorem 1 in the case $\alpha > 0$ and $\beta > 0$, a condition (i) $|T|^{2(1+2r)\alpha}x$ and $|T|^{(1+2r)\alpha+(1+2s)\beta-1}T^*y$ are linearly dependent is equivalent to that $T|T|^{(1+2r)\alpha+(1+2s)\beta-1}x$ and $|T^*|^{2(1+2s)\beta}y$ are linearly dependent. We recall that $a = (1+2r)\alpha$ and $b = (1+2s)\beta$ in the proof of the inequality (1). In fact the former condition is equivalent to that $|T|^a x$ and $|T|^b U^* y$ are linearly dependent as stated in the proof of the equality in (1) and this condition is equivalent to that $U|T|^{a+b} x$ and $U|T|^{2b} U^* y$ are linearly dependent by $(\star\star)$ and $N(|T|) = N(U)$,that is, $T|T|^{(1+2r)\alpha+(1+2s)\beta-1} x$ and $|T^*|^{2(1+2s)\beta} y$ are linearly dependent.

Proof of Corollary 1. Put $r = s = 0$ in Theorem 1.

Proof of Corollary 2. Put $s = 0$ and $\alpha = 1$ in Theorem 1. Then we have (3). Also put $r = 0$ and $\beta = 1$ in theorem 1. Then we have (4).

Proof of Corollary 3. Put $\alpha = \beta = 1$ in Theorem 1.

§2. EQUIVALENCE RELATION BETWEEN THEOREM 1 AND THEOREM A

In §1, Theorem 1 is proved by Theorem A which is an extension of this Löwner-Heinz theorem. In this section, conversely we shall show that Theorem A can be derived from Theorem 1 as follows.

Theorem 1 $\implies$ Theorem A.

In (1) of theorem 1, we put $T = B,\, \alpha = \beta,\; r = s$ and also we put $y = x$. Then the hypothesis $\|Tx\| \le \|Ax\|$ is equivalent to $B^2 \le A^2$. By Theorem 1 and Remark 1, the following inequality holds for each $r \ge 0$ and $\alpha \in [0, 1]$

$$|(B^{2(1+2r)\alpha} x, x)|^2 \le ((B^{2r} A^{2p} B^{2r})^{(1+2r)\alpha/(p+2r)} x, x)(B^{2(1+2r)\alpha} x, x) \text{ for any } p \ge 1,$$

that is, for each $r \ge 0$ and $\alpha \in [0,1]$,

(11) $|(B^{2(1+2r)\alpha} x, x)| \le ((B^{2r} A^{2p} B^{2r})^{(1+2r)\alpha/(p+2r)} x, x)$ for any $p \ge 1$.

Then we have for each $r \ge 0$ and $\alpha \in [0,1]$,

(12) $B^{2(1+2r)\alpha} \le (B^{2r} A^{2p} B^{2r})^{(1+2r)\alpha/(p+2r)}$ for any $p \ge 1$

holds under the hypothesis $B^2 \le A^2$ and the inequality (12) is equivalent to (i) of Theorem A which is also equivalent to (ii) of Theorem A.

Hence we have just proved that Theorem 1 and Theorem A are mutually equivalent.

References

[1] M.Fujii: Furuta's inequality and its mean theoretic approach,

J. Operator Theory, **23**(1990), 67-72.

[2] T.Furuta: $A \geq B \geq 0$ assures $(B^r A^p B^r)^{1/q} \geq B^{(p+2r)/q}$ for $r \geq 0$, $p \geq 0$, $q \geq 1$ with $(1 + 2r)q \geq p + 2r$.

Proc. Amer. Math. Soc., **101**(1987), 85-88.

[3] T.Furuta: A proof via operator means of an order preserving inequality,

Linear Alg. and Its Appl., **113**(1989), 129-130.

[4] T.Furuta: Elementary proof of an order preserving inequality,

Proc. Japan Acad., **65**(1989), 126.

[5] E.Heinz, Beiträge zur Strüngstheorie der Spektralzerlegung,

Math. Ann., **123** (1951), 415-438.

[6] E.Kamei, A satellite to Furuta's inequality,

Math. Japon, **33**(1988), 883-886.

[7] T.Kato, Notes on some inequalities for linear operators,

Math. Ann., **125** (1952), 208-212.

[8] K.Löwner, Über monotone Matrixfunktion,

Math. Z., **38** (1934), 177-216.

[9] G.K.Pedersen, Some operator monotone functions,

Proc, Amer. Math. Soc., **36** (1972), 309-310.

Department of Applied Mathematics,

Faculty of Science,

Science University of Tokyo

1-3 Kagurazaka, Shinjuku-ku

Tokyo, 162

Japan

AMS Subject Classification: Primary 47A30, Secondary 47A63

Operator Theory:
Advances and Applications, Vol. 62
© 1993 Birkhäuser Verlag Basel

THE BAND METHOD FOR BORDERED ALGEBRAS

I. Gohberg, M. A. Kaashoek, and H. J. Woerdeman[*]

Dedicated to Professor T. Ando on the occasion of his sixtieth birthday

The band method for positive definite extension problems of non-band type is specified for the bordered case. A fractional type description of all positive extensions is new. We illustrate the result for Fredholm integral operators. The notion of Schur complement plays a crucial role.

§1. INTRODUCTION

In the paper [GKW4] a new set of extension problems of non-band type was solved, and the results were illustrated on concrete examples of the following kind. Let $\Delta = [0, 1] \times [0, 1]$, and consider the subsets

$$\mathcal{D} = \{(t, s) \in \Delta \mid t - \tfrac{1}{4} < s < t + \tfrac{1}{4} \text{ or } s < \tfrac{1}{4} \text{ or } t < \tfrac{1}{4}\},$$
$$\Delta_1 = \{(t, s) \in \Delta \mid t < s \text{ and } (t, s) \notin \mathcal{D}\},$$
$$\Delta_2 = \{(t, s) \in \Delta \mid t < s \text{ and } (t, s) \in \mathcal{D}\},$$
$$\Delta_3 = \{(t, s) \in \Delta \mid t > s \text{ and } (t, s) \in \mathcal{D}\},$$
$$\Delta_4 = \{(t, s) \in \Delta \mid t > s \text{ and } (t, s) \notin \mathcal{D}\}.$$

Let $\mathcal{F}_{\mathcal{D}}$ denote the class of $n \times n$ matrix valued functions f defined on Δ such that for $i = 1, 2, 3, 4$ the restriction of f to Δ_i can be extended continuously to the closure $\overline{\Delta}_i$. Now, let $k \in \mathcal{F}_{\mathcal{D}}$ have support in $\Delta_2 \cup \Delta_3$. For the present example the problem consists of finding all (or some) $k_{ext} \in \mathcal{F}_{\mathcal{D}}$ such that

---

[*] Partially supported by NASA contract NAS1-18347

$$k_{ext}(t, s) = k(t, s), \quad (t, s) \in \Delta_2 \cup \Delta_3,$$

and the operator $I+K_{ext}$, where K_{ext} is the Hilbert-Schmidt integral operator on $L_2^n([0, 1])$ with kernel function k_{ext}, is positive definite. The general theory developed in [GKW4], which is a variant of the band method from [GKW1, 2, 3], gave a natural necessary and sufficient condition for the existence of such an extension g and a formula for a special extension (namely, the one of maximum entropy). Also discrete versions of this example were treated in [GKW4].

In the present paper we analyze a general class of extension problems of the above mentioned type by reducing the problem to one which can be treated as a usual band problem. The class of problems considered here turns out to be smaller than the class of problems in [GKW4], nevertheless it covers the concrete examples referred to above. Furthermore, for the above examples the present paper gives the description of all solutions via a generalized linear fractional representation.

The paper consists of four sections. The second section contains the formulation of the problem in a general setting. The three main theorems are presented in Section 3. The results are applied in Section 4 to the class of Fredholm integral operator kernels introduced above.

§2. THE STATEMENT OF THE PROBLEM

Let $\mathcal{L}$ be an algebra with a unit $e_{\mathcal{L}}$ and involution $*$. A direct sum decomposition

$$\mathcal{L} = \mathcal{L}_-^0 \dotplus \mathcal{L}_d \dotplus \mathcal{L}_+^0 \tag{2.1}$$

will be called a *triangular structure* for $\mathcal{L}$ when $\mathcal{L}_-^0$, $\mathcal{L}_d$ and $\mathcal{L}_+^0$ are subalgebras with

$$(\mathcal{L}_-^0)^* = \mathcal{L}_+^0 \quad , \quad e_{\mathcal{L}} \in \mathcal{L}_d = (\mathcal{L}_d)^*,$$

and

$$\mathcal{L}_d \mathcal{L}_-^0 \subset \mathcal{L}_-^0 \quad , \quad \mathcal{L}_d \mathcal{L}_+^0 \subset \mathcal{L}_+^0 .$$

Note that the choice $\mathcal{L}_d = \mathcal{L}, \mathcal{L}_-^0 = (0) = \mathcal{L}_+^0$ always defines a triangular structure. We shall frequently use the notation

$$\mathcal{L}_- := \mathcal{L}_-^0 \dotplus \mathcal{L}_d, \mathcal{L}_+ = \mathcal{L}_+^0 \dotplus \mathcal{L}_d .$$

We shall assume throughout that we associated with $\mathcal{L}$ a triangular structure.

Let $\mathcal{M}$ be an algebra with a unit $e_{\mathcal{M}}$ and involution $*$. We shall associate with $\mathcal{M}$ a *band structure*, i.e., a direct sum decomposition

$$\mathcal{M} = \mathcal{M}_1 \dotplus \mathcal{M}_2^0 \dotplus \mathcal{M}_d \dotplus \mathcal{M}_3^0 \dotplus \mathcal{M}_4 \tag{2.2}$$

such that

$$e_{\mathcal{M}} \in \mathcal{M}_d = (\mathcal{M}_d)^*, \mathcal{M}_1^* = \mathcal{M}_4, (\mathcal{M}_2^0)^* = \mathcal{M}_3^0 ,$$

and the following multiplication rules are satisfied:

	$\mathcal{M}_1$	$\mathcal{M}_2^0$	$\mathcal{M}_d$	$\mathcal{M}_3^0$	$\mathcal{M}_4$
$\mathcal{M}_1$	$\mathcal{M}_1$	$\mathcal{M}_1$	$\mathcal{M}_1$	$\mathcal{M}_+^0$	$\mathcal{M}$
$\mathcal{M}_2^0$	$\mathcal{M}_1$	$\mathcal{M}_+^0$	$\mathcal{M}_2^0$	$\mathcal{M}_c$	$\mathcal{M}_-^0$
$\mathcal{M}_d$	$\mathcal{M}_1$	$\mathcal{M}_2^0$	$\mathcal{M}_d$	$\mathcal{M}_3^0$	$\mathcal{M}_4$
$\mathcal{M}_3^0$	$\mathcal{M}_+^0$	$\mathcal{M}_c$	$\mathcal{M}_3^0$	$\mathcal{M}_-^0$	$\mathcal{M}_4$
$\mathcal{M}_4$	$\mathcal{M}$	$\mathcal{M}_-^0$	$\mathcal{M}_4$	$\mathcal{M}_4$	$\mathcal{M}_4$

where we denote
$$\mathcal{M}_+^0 = \mathcal{M}_2^0 \dotplus \mathcal{M}_1 \;,\; \mathcal{M}_3 = \mathcal{M}_3^0 \dotplus \mathcal{M}_4 \;,\; \mathcal{M}_c = \mathcal{M}_2^0 \dotplus \mathcal{M}_d \dotplus \mathcal{M}_3^0 .$$

Note that
$$\mathcal{M} = \mathcal{M}_-^0 \dotplus \mathcal{M}_d \dotplus \mathcal{M}_+^0 \tag{2.3}$$

is a triangular structure on $\mathcal{M}$ which is implied by the band structure. Further, denote
$$\mathcal{M}_2 = \mathcal{M}_2^0 \dotplus \mathcal{M}_d \;,\; \mathcal{M}_3 = \mathcal{M}_3^0 \dotplus \mathcal{M}_d ,$$
$$\mathcal{M}_+ = \mathcal{M}_+^0 \dotplus \mathcal{M}_d \;,\; \mathcal{M}_- = \mathcal{M}_-^0 \dotplus \mathcal{M}_d .$$

Also, let $\mathcal{N}$ and $\mathcal{N}^*$ denote vector spaces, and
$$* : \mathcal{N} \to \mathcal{N}^*$$

an isomorphism between $\mathcal{N}$ and $\mathcal{N}^*$ with inverse also denoted by $*$, and assume that there exist multiplications acting
$$\mathcal{L} \times \mathcal{N} \to \mathcal{N} \;,\; \mathcal{N}^* \times \mathcal{L} \to \mathcal{N}^*,$$
$$\mathcal{N}^* \times \mathcal{N} \to \mathcal{M} \;,\; \mathcal{N} \times \mathcal{N}^* \to \mathcal{L},$$
$$\mathcal{M} \times \mathcal{N}^* \to \mathcal{N}^* \;,\; \mathcal{N} \times \mathcal{M} \to \mathcal{N},$$

satisfying all the natural associative and distributive laws (involving also the addition in $\mathcal{N}$, $\mathcal{M}$ and $\mathcal{L}$, and the multiplication in $\mathcal{M}$ and $\mathcal{L}$), and the rules

$$e_{\mathcal{L}} n = n \,,\; \hat{n} e_{\mathcal{L}} = \hat{n},\; e_{\mathcal{M}} \hat{n} = \hat{n},\; n e_{\mathcal{M}} = n,$$
$$(nm)^* = m^* n^* \;,\; (m\hat{n})^* = \hat{n}^* m^*,$$
$$(n\hat{n})^* = \hat{n}^* n^* \;,\; (n\hat{n})^* = \hat{n}^* n^*,$$
$$(\hat{n}\ell) = \ell^* \hat{n}^* \;,\; (\ell n)^* = n^* \ell^*,$$

where $n \in \mathcal{N}$, $\hat{n} \in \mathcal{N}^*$, $m \in \mathcal{M}$ and $\ell \in \mathcal{L}$.

We are now able to introduce an algebra of 2×2 block matrices for which we shall develop a scheme to handle certain types of positive extension problems. Let
$$\mathcal{F} = \left\{ \begin{bmatrix} \ell & n \\ \hat{n} & m \end{bmatrix} , \ell \in \mathcal{L}, n \in \mathcal{M}, \hat{n} \in \mathcal{M}^*, \hat{n} \in \mathcal{M} \right\},$$

and let the addition and multiplication in $\mathcal{F}$ be the usual matrix addition and multiplication. With these operations $\mathcal{F}$ is an algebra, and $\mathcal{F}$ has a unit

$$E := \begin{bmatrix} e & 0 \\ 0 & e \end{bmatrix},$$

and an involution $*$ defined by

$$\begin{bmatrix} \ell & n \\ \hat{n} & m \end{bmatrix}^{*} := \begin{bmatrix} \ell^{*} & \hat{n}^{*} \\ n^{*} & m^{*} \end{bmatrix}.$$

Note that

$$\mathcal{F} = \begin{bmatrix} \mathcal{L} & \mathcal{M} \\ \mathcal{M}^{*} & \mathcal{M} \end{bmatrix} = \begin{bmatrix} \mathcal{L}_{-}^{0} & 0 \\ \mathcal{M}^{*} & \mathcal{M}^{0} \end{bmatrix} \dotplus \begin{bmatrix} \mathcal{L}_{d} & 0 \\ 0 & \mathcal{M}_{d} \end{bmatrix} \dotplus \begin{bmatrix} \mathcal{L}_{+}^{0} & \mathcal{M} \\ 0 & \mathcal{M}_{+}^{0} \end{bmatrix} =$$
$$=: \mathcal{F}_{-}^{0} \dotplus \mathcal{F}_{d} + \mathcal{F}_{+}^{0} \tag{2.4}$$

defines a triangular structure on $\mathcal{F}$.

For an algebra $\mathcal{A}$ with an involution $*$ and unit e we say that a $\in \mathcal{A}$ is *positive semi-definite* (notation: $a \geq_{\mathcal{A}} 0$) when $a = c^{*}c$ for some c $\in \mathcal{A}$. When, in addition, we can choose c to be invertible, we say that a is *positive definite* (notation: $a > 0$). If, in addition, $\mathcal{A}$ has a triangular structure

$$\mathcal{A} = \mathcal{A}_{-}^{0} \dotplus \mathcal{A}_{d} \dotplus \mathcal{A}_{+}^{0}, \tag{2.4}$$

we say that the positive element a $\in \mathcal{A}$ allows a *right (left) spectral factorization* (with respect to (2.4)) if $a = a_{+}^{*}a_{+}$ ($a = a_{-}^{*}a_{-}$) with $a_{+}^{\pm 1} \in \mathcal{A}_{+} := \mathcal{A}_{+}^{0} + \mathcal{A}_{d}$ ($a_{-}^{\pm 1} \in \mathcal{A}_{-} = \mathcal{A}_{-}^{0} + \mathcal{A}_{d}$).

For an algebra $\mathcal{A}$ with a band or triangular structure we let $P_{\mathcal{A}_{c}}$, $P_{\mathcal{A}_{d}}$, $P_{\mathcal{A}_{-}}$, etc. denote the natural projection in $\mathcal{A}$ onto $\mathcal{A}_{c}$, $\mathcal{A}_{d}$, $\mathcal{A}_{-}$, etc. respectively.

We want to consider the following problem. Let $f_{c} = \begin{bmatrix} \ell & n \\ n^{*} & m_{c} \end{bmatrix}$ be a selfadjoint element in $\mathcal{F}_{c} = \begin{bmatrix} \mathcal{L} & \mathcal{N} \\ \mathcal{N}^{*} & \mathcal{M}_{c} \end{bmatrix}$. We say that $f = \begin{bmatrix} \ell & n \\ n^{*} & m \end{bmatrix}$ is a positive *extension* of f_{c} when $f \geq_{\mathcal{F}} 0$ and $m - f_{c} \in \mathcal{M}_{1} \dotplus \mathcal{M}_{4}$. We call f a *positive band extension* when, in addition, $f^{-1} \in \mathcal{F}_{c}$. The problem is to find a/all positive (band) extensions of a given $f \in \mathcal{F}_{c}$.

§3. MAIN THEOREMS

For the first result we need to introduce the following axiom.

Axiom (A0). If $f = \begin{bmatrix} \ell & n \\ n^{*} & m \end{bmatrix} >_{\mathcal{F}} 0$, then $\ell >_{\mathcal{L}} 0$ and $m >_{\mathcal{M}} 0$.

Note that when $\mathcal{F}$ satisfies Axiom (A0) then $\ell >_{\mathcal{L}} 0$ and $m >_{\mathcal{M}} 0$ imply that $m + n^*\ell^{-1}n >_{\mathcal{M}} 0$ and $\ell + nm^{-1}n^* >_{\mathcal{M}} 0$. Indeed, to prove the first inequality, note that when $\ell = \hat{\ell}^*\hat{\ell}$ and $m = \hat{m}^*\hat{m}$ we have that

$$\begin{bmatrix} \hat{\ell}^* & 0 \\ n^*\hat{\ell}^{*-1} & \hat{m}^* \end{bmatrix}\begin{bmatrix} \hat{\ell} & \hat{\ell}^{-1}n \\ 0 & \hat{m} \end{bmatrix} = \begin{bmatrix} * & * \\ * & m + n^*\ell^{-1}n \end{bmatrix}$$

is positive definite in $\mathcal{F}$. Now apply Axiom (A0).

THEOREM 3.1. *Let* $\mathcal{F} = \begin{bmatrix} \mathcal{L} & \mathcal{N} \\ \mathcal{N}^* & \mathcal{M} \end{bmatrix}$ *be as above and assume that* $\mathcal{F}$ *satisfies Axiom*

(A0). *Let* $\ell >_{\mathcal{L}} 0$, $n \in \mathcal{M}$ *and* $m_c = m_c^* \in \mathcal{M}_c$. *Then there exists a positive band extension of*

$$f_c = \begin{bmatrix} \ell & n \\ n^* & m_c \end{bmatrix}$$

which allows a right spectral factorization with respect to (2.4) *if and only if* ℓ *allows a right spectral factorization with respect to* (2.1), *and there exists an invertible solution* x *to the equation*

$$P_{\mathcal{M}_2}(P_{\mathcal{M}_c}(m_c - n^*\ell^{-1}n)x) = e_{\mathcal{M}}$$

with

$$x \in \mathcal{M}_2,\ x^{-1} \in \mathcal{M}_+,\ P_{\mathcal{M}_d}x >_{\mathcal{M}_d} 0.$$

In that case

$$f := \begin{bmatrix} \ell & n \\ n^* & m \end{bmatrix}^{-1}$$

with

$$m = n^*\ell^{-1}n + x^{*-1}(P_dx)\,x^{-1}$$

is a positive band extension of f_c, *and a right spectral factorization of* f *with respect to* (2.4) *is given via*

$$\begin{bmatrix} \ell & n \\ n^* & m \end{bmatrix}^{-1} = \begin{bmatrix} \ell_+^{-1} & -\ell^{-1}nxd^{-1} \\ 0 & xd^{-1} \end{bmatrix}\begin{bmatrix} \ell_+^{*-1} & 0 \\ -d^{*-1}x^*n^*\ell^{*-1} & d^{*-1}x^* \end{bmatrix}$$

where $\ell = \ell_+^*\ell_+$ *is a right spectral factorization for* ℓ *with respect to* (2.1), *and* d *is invertible in* $\mathcal{M}_d$ *with* $P_dx = d^*d$. *Moreover, all positive band extensions of* f_c *which allow a right spectral factorization with respect to* (2.4) *are obtained in this way.*

For an algebra $\mathcal{A}$ with unit e and involution $*$ which has the triangular structure

$$\mathcal{A} = \mathcal{A}_-^0 \dotplus \mathcal{A}_d \dotplus \mathcal{A}_+^0 \tag{3.0}$$

we introduce the following two axioms

   Axiom (A1) $P_d(c^*c) \geq_{\mathcal{A}_d} 0$ for all $c \in \mathcal{A}$.

   Axiom (A2) $P_d(c^*c) = 0$ implies $c = 0$.

   Further, we need to introduce the notion of multiplicative diagonal. When $a \in \mathcal{A}$ allows a right (left) spectral factorization $a = a_+^* a_+$ $(a = a_-^* a_-)$ with respect to (3.0) we define its *right (left) multiplicative diagonal* $\mu_r(a)(\mu_\ell(a))$ by

$$\mu_r(a) := (P_d a_+)^* P_d a_+ \; (: = (P_d a_-)^* P_d a_-).$$

Although a right (left) multiplicative factorization is not unique, the notion of right (left) multiplicative diagonal is well - defined. This is easy to check, but the argument can also be found in [GKW3].

   THEOREM 3.2. *Let* $\mathcal{F} = \begin{bmatrix} \mathcal{L} & \mathcal{M} \\ \mathcal{M}^* & \mathcal{M} \end{bmatrix}$ *be as above, and suppose that $\mathcal{F}$ satisfies* (A0), (A1) *and* (A2). *Assume that* $f_c = \begin{bmatrix} \ell & n \\ n^* & m \end{bmatrix} \in \mathcal{F}_c$ *has a positive band extension* f_{ext} *which allows a right spectral factorization with respect to* (2.4), *and let x be as in Theorem* 3.1. *Then the right multiplicative diagonal* $\mu_r(f_{ext})$ *of a positive extension f which allows a right spectral factorization with respect to* (2.4) *has the property that*

$$\mu_r(f_{ext}) \leq \begin{bmatrix} \mu_r(\ell) & 0 \\ 0 & P_d x \end{bmatrix}. \tag{3.1}$$

Furthermore, equality holds in (3.1) *if and only if* f_{ext} *is the unique band extension f of* f_c.

   To obtain a description for the set of all positive definite extensions of f_c we need to make some further assumptions on $\mathcal{M}$. We shall assume that $\mathcal{M}$ is a $*$-subalgebra of a B$*$-algebra $\mathcal{R}$ with norm $\|.\|_{\mathcal{R}}$, and $\mathcal{R}$ has a unit e which belongs to $\mathcal{M}$. In addition we assume that the following two axioms hold:

   Axiom (A3). If $m \in \mathcal{M}$ is invertible in $\mathcal{R}$, then $m^{-1} \in \mathcal{M}$

   Axiom (A4). If $m_n \in \mathcal{M}_+$, $m \in \mathcal{M}$ and $\lim_{n \to \infty} \|m_n - m\|_{\mathcal{R}} = 0$, then $m \in \mathcal{M}_+$.

   THEOREM 3.3. *Let $\mathcal{F}$ be as above, satisfying Axiom* (A0), *and suppose that $\mathcal{M}$ satisfies Axioms* (A3) *and* (A4). *Let* $f_c = \begin{bmatrix} \ell & n \\ n^* & m_c \end{bmatrix}$ *be a symmetric element of* $\mathcal{F}_c$, *and suppose that ℓ has a right factorization with respect to* (2.1), *and* $m_c - P_{\mathcal{M}_c}(n^* \ell^{-1} n)$ *has a band extension b which allows both right and left spectral factorizations with respect to* (2.3)

$$b = u^{*-1}u^{-1} = v^{*-1}v^{-1}, \; u^{\pm 1} \in \mathcal{M}_+, \; v^{\pm 1}\mathcal{M}_-.$$

Then each positive definite extension f_{ext} of f_c which allows a right spectral factorization with respect to (2.4) is of the form

$$\begin{bmatrix} \ell & n \\ n^* & T(g) + n^*\ell^{-1}n \end{bmatrix}$$

where

$$T(g) = (u + vg)^{*-1}(e - g^*g)(u + vg)^{-1},$$

and $g \in \mathcal{M}_1$ is so that $e - g^*g >_{\mathcal{M}} 0$ allows a right spectral factorization with respect to (2.3). Moreover, the correspondence is one-one.

The proofs of the three theorems above are based upon the results in [GKW1 – 3] and the following proposition

PROPOSITION 3.4. *Let $\mathcal{F}$ be as above and assume that Axiom (A0) is satisfied. Let*
$$f_c = \begin{bmatrix} \ell & n \\ n^* & m \end{bmatrix}_c$$
be a symmetric element of $\mathcal{F}_c$, and suppose that $\ell >_{\mathcal{L}} 0$ allows a right spectral factorization $\ell = \ell_+^* \ell_+$ with respect to (2.1). Then there exists a $1 - 1$ correspondence between positive extensions f of f_c which allow a right spectral factorization with respect to (2.4) and positive extensions q of $m_c - P_{\mathcal{M}_c}(n^*\ell^{-1}n)$ which allow a right spectral factorization with respect to (2.3). This one – one correspondence is described by

$$q \mapsto f := \begin{bmatrix} \ell & n \\ n^* & q + n^*\ell^{-1}n \end{bmatrix}.$$

Moreover, these right spectral factorizations of m and f are related as follows

$$q = u^{*-1}u^{-1}; \; f = \begin{bmatrix} \ell_+^* & 0 \\ -n^*\ell_+^{-1} & u^{*-1} \end{bmatrix}\begin{bmatrix} \ell_+ & -\ell_+^{-1}n \\ 0 & u^{-1} \end{bmatrix}. \tag{3.2}$$

Furthermore, q is in addition a band extension of $m_c - P_{\mathcal{M}_c}(n^*\ell^{-1}n)$ if and only if f is a band extension of f_c.

Proof. Let $f = \begin{bmatrix} \ell & n \\ n^* & m \end{bmatrix}$ be a positive extension of f_c. Then, using the invertibility of ℓ, we have that ∎

$$\begin{bmatrix} 1 & 0 \\ -n^*\ell^{-1} & 1 \end{bmatrix}\begin{bmatrix} \ell & n \\ n^* & m \end{bmatrix}\begin{bmatrix} 1 & -\ell^{-1}n \\ 0 & 1 \end{bmatrix} = \begin{bmatrix} \ell & 0 \\ 0 & m - n^*\ell^{-1}n \end{bmatrix}$$

is positive definite. By Axiom (A0) we obtain that $q := m - n^*\ell^{-1}n >_{\mathcal{M}} 0$. Further,

$$P_{\mathcal{M}_C}(q) = P_{\mathcal{M}_C}(m - n^*\ell^{-1}n) = m_C - P_{\mathcal{M}_C}(n^*\ell^{-1}n),$$

proving that q is an extenstion of $m_C - P_{\mathcal{M}_C}(n^*\ell^{-1}n)$. Moreover, if f has the right spectral factorization with respect to (2.4)

$$\begin{bmatrix} \ell & n \\ n^* & m \end{bmatrix} = f = \begin{bmatrix} \ell_+^* & 0 \\ -n^*\ell_+^{-1} & u^{*-1} \end{bmatrix}\begin{bmatrix} \ell_+^* & -\ell_+^{*-1}n \\ 0 & u^{-1} \end{bmatrix},$$

then $q = m - n^*\ell^{-1}n = \left(u^{*-1}u^{-1} + n^*\ell_+^{-1}\ell_+^{*-1}n\right) - n^*\ell^{-1}n = u^{*-1}u^{-1}$ has a right spectral factorization with respect to (2.3). Further, note that if $f^{-1} \in \mathcal{F}_C$ then $u \in \mathcal{M}_2$, and $q^{-1} = uu^* \in \mathcal{M}_C$. Thus if f is a band extension of f_C, the corresponding q is a band extension of $m_C - P_{\mathcal{M}_C}(n^*l^{-1}n)$.

Conversely, let q be a positive extension of $m_C - P_{\mathcal{M}_C}(n^*\ell^{-1}n)$ with right spectral factorization $q = u^{*-1}u^{-1}$ with respect to (2.3). Let f be defined via (3.2). Then clearly f has a right spectral factorization with respect to (2.4). Further

$$P_{\mathcal{F}_C}f = \begin{bmatrix} \ell & n \\ n^* & P_{\mathcal{M}_c}q + P_{\mathcal{M}_c}\left(n^*\ell^{-1}n\right) \end{bmatrix} = f_c$$

yielding that f is a positive extension of f_C. When q is a band extension of $m_C - P_{\mathcal{M}_C}(n^*\ell^{-1}n)$, then $u \in \mathcal{M}_2$. Using (3.2) it follows that $f^{-1} \in \mathcal{M}_C$, giving that f is a band extension of f_C.   ∎

The proofs of Theorems 3.1, 3.2 and 3.3 are direct applications of Proposition 3.4 in combination with Theorem I.1.1. in [GKW2], Theorem I.1.1 in [GKW3] and Theorem I.1.2 in [GKW2], respectively. Let us give the details.

Proof of Theorem 3.1. Let f_C be as in the theorem. By Proposition 3.4, f_C has a band extension which allows a right spectral factorization with respect to (2.4) if and only if $k := m_C - P_{\mathcal{M}_C}(n^*\ell^{-1}n) \in \mathcal{M}_C$ has a band extension which allows a right spectral factorization with respect to (2.3). By (half of) Theorem I.1.1. in [GKW2] the latter is equivalent to the existence of an invertible $x \in \mathcal{M}_2$ such that $P_{\mathcal{M}_2}(kx) = e_{\mathcal{M}}$, $x^{-1} \in \mathcal{M}_+$ and $P_{\mathcal{M}_d} x >_{\mathcal{M}_d} 0$. Furthermore, a band extension of k is given by

$$q = x^{*-1} P_d x \, x^{-1}.$$

Let $d \in M_d$ be so that $P_d x = d^* d$. Using again Proposition 3.4, we find now that

$$f = \begin{bmatrix} \ell_+^* & 0 \\ -n^* \ell_+^{-1} & x^{*-1} d^* \end{bmatrix} \begin{bmatrix} \ell_+ & -\ell_+^{*-1} n \\ 0 & dx^{-1} \end{bmatrix}$$

is a band extension of f_c which allows a spectral factorization with respect to (2.4). The one-one correspondence in Proposition 3.4 ensures that all band extensions of f_c with a right spectral factorization with respect to (2.4) are obtained in this way. $\blacksquare$

Proof of Theorem 3.2. Let f denote the positive band extension from Theorem 3.1 with factorization

$$f = \begin{bmatrix} \ell_+^* & 0 \\ n^* \ell_+^{-1} & x^{*-1} d^* \end{bmatrix} \begin{bmatrix} \ell_+ & \ell_+^{*-1} n \\ 0 & dx^{-1} \end{bmatrix},$$

and f_{ext} be a positive extension of f_c with factorization

$$f_{ext} = \begin{bmatrix} \ell_+^* & 0 \\ -n^* \ell_+^{-1} & u^{*-1} \end{bmatrix} \begin{bmatrix} \ell_+ & -\ell_+^{*-1} n \\ 0 & u^{-1} \end{bmatrix}.$$

Then, by Proposition 3.4, we have that $x^{*-1} d^* dx^{-1}$ and $u^{*-1} u^{-1}$ are a band extension and a positive extension of $m_c - P_{\mathcal{M}_c}(n^* \ell^{-1} n)$, respectively. From Theorem I.1.1 in [GKW3] we obtain that $P_d x = \mu_r(x^{*-1} d^* dx^{-1}) \geq \mu_r(u^{*-1} u^{-1})$, with equality if and only if $u = xd^{-1}$. But then it follows immediately that

$$\mu_r(f_{ext}) = \begin{bmatrix} \mu_r(\ell) & 0 \\ 0 & \mu_r(u^{*-1} u^{-1}) \end{bmatrix} \leq \begin{bmatrix} \mu_r(\ell) & 0 \\ 0 & P_d x \end{bmatrix}$$

with equality if and only if $f_{ext} = f$.

Proof of Theorem 3.3. Let $f_{ext} = \begin{bmatrix} \ell & n \\ n^* & m \end{bmatrix}$ be a positive definite extension of f_c which

allows a right spectral factorization with respect to (2.4). By Proposition 3.4 the element $m - n^*\ell^{-1}n$ is a positive extension of $m_c - P_{\mathcal{M}_c}(n^*\ell^{-1}n)$. By Theorem I.1.2 in [GKW2] we

have that

$$m - n^*\ell^{-1}n = T(g)$$

for some $g \in \mathcal{M}_1$ with $e_{\mathcal{M}} - g^*g >_{\mathcal{M}} 0$. But then

$$f_{ext} = \begin{bmatrix} \ell & n \\ n^* & T(g) + n^*\ell^{-1}n \end{bmatrix}.$$

Since both correspondences in Theorem I.1.2 in [GKW2] and in Proposition 3.4 are one-one, we also have that the above correspondence is one-one. This finishes the proof. ∎

§4. FREDHOLM INTEGRAL OPERATORS

In this section we apply the abstract result Theorem 3.3 to functions f which may be viewed as kernels of integral operators. We recall the definitions given in the introduction.

Let $\mathcal{D} \subset [0,1] \times [0,1]$ be the domain

$$\mathcal{D} = \{(t,s) \in [0,1]^2 \mid t - \tau < s < t + \tau \text{ or } s < \alpha \text{ or } t < \alpha\}.$$

Here α and τ are fixed numbers between 0 and 1.

Introduce $\mathcal{F} = \mathcal{F}_{\mathcal{D}}$ to be the class of $n \times n$ matrix valued functions $f(t,s)$ which are defined on the square $[0,1] \times [0,1]$, continuous on each of the open regions

$$\Delta_1 = \{(t,s) \in [0,1]^2 \mid t < s \text{ and } (t,s) \notin \mathcal{D}\},$$
$$\Delta_2 = \{(t,s) \in [0,1]^2 \mid t < s \text{ and } (t,s) \in \mathcal{D}\},$$
$$\Delta_3 = \{(t,s) \in [0,1]^2 \mid t > s \text{ and } (t,s) \in \mathcal{D}\},$$
$$\Delta_4 = \{(t,s) \in [0,1]^2 \mid t > s \text{ and } (t,s) \notin \mathcal{D}\},$$

and the restriction f_i of f to Δ_i extends continuously to the closure $\overline{\Delta}_i$. The set $\mathcal{F}$ is an algebra with the usual addition of functions, and the multiplication defined by

$$(f * g)(t,s) = \int_0^1 f(t,u)g(u,s)ds.$$

Also, $\mathcal{F}$ has a natural involution *, namely

$$f^*(t, s) := f(s,t)^*. \tag{4.1}$$

The * in the right hand side of (4.1) is the usual adjoint of a matrix. We shall say that $f \in \mathcal{F}$ is *regular* in $\mathcal{F}$ if there exists a $g \in \mathcal{F}$ such that

$$f + g + f^* g = 0, \ g + f + g^* f = 0.$$

In that case g is uniquely determined by f and denoted by $f^\dagger$.

Given $f \in \mathcal{F}$ we shall write F for the integral operator on $L_n^2[0, 1]$ with kernel f. Thus

$$(F\varphi)(t) = \int_0^1 f(t, s) \ \varphi(s)ds, \ 0 < t < 1.$$

If f is regular in $\mathcal{F}$, then $f^\dagger$ is precisely the kernel of the integral operator $(I + F)^{-1} - I$; in other words, $f^\dagger$ is the resolvent kernel. Furthermore, F^* is the integral operator with kernel f^*.

We shall deal with the following extension problem. Let

$$k \in \mathcal{F}_c := \{ \, f \in \mathcal{F} \mid f(t, s) = 0, \ (t, s) \in \Delta_1 \cup \Delta_4 \}.$$

A matrix valued function $k_{ext} \in \mathcal{F}$ is called a *positive extension* of k if $k(t, s) = k_{ext}(t, s)$, $(t, s) \in \Delta_2 \cup \Delta_3$, and $I + K_{ext}$ is a positive operator on $L_n^2[0, 1]$. We introduce the following notation. For $\xi \in [0, 1]$ let J_ξ denote the open set in [0, 1] given by

$$J_\xi = \{t : t < \xi \ , \ (t, \xi) \in \mathcal{D}\}.$$

For $k \in \mathcal{F}_c$ and $\xi \in [0, 1]$ let $A_{k,\xi}$ denote the integral operator on $L_n^2\left(\bar{J}_\xi\right)$ which is defined by

$$\left(A_{k,\xi}\varphi\right)(t) = \varphi(t) + \int_{J_\xi} k(t, s) \ \varphi(s)ds, \quad t \in \bar{J}_\xi. \tag{4.2}$$

Recall from [GKW4], Theorem II.2.1 that k has a positive extension of and only if for every ξ in the interval $0 < \xi < 1$ the operator $A_{k,\xi}$ in (4.2) is positive definite. Also a construction for the band extension was given there. In order to obtain the set of all positive extensions of k we apply Theorem 3.3 to obtain the following result.

THEOREM 4.1. *Let* $k \in \mathcal{F}_c$ *be given and suppose that for every* $0 < \xi < 1$ *the operator* $A_{k,\xi}$ *in (4.2) is positive definite. Introduce*

$$\psi = k + n^* * n + n^* * \ell^\dagger * n$$

where

$$\ell = k \, |_{[0,\alpha]\times[0,\alpha]}, \ n = k|_{[0,\alpha]\times[\alpha,1]}.$$

Let x *and* y *be given via*

$$x(t,s) + \int_{\max\{s-\tau,\alpha\}}^{s} \psi(t,u)\, x(u,s)\, ds = \psi(t,s),$$

for $s - \tau \le t \le s$, $(s,t) \in [\alpha,1]^2$, *and* $x(t,s) = 0$ *elsewhere,*

$$y(t,s) + \int_{s}^{\min\{s+\tau,1\}} \psi(t,u)\, y(u,s)\, du = \psi(t,s),$$

for $s \le t \le s + \tau$, $(s,t) \in [\alpha,1]^2$, *and* $y(t,s) = 0$ *elsewhere. Then* x *and* y *are regular.*

Furthermore, each positive extension k_{ext} *of* k *is of the form*

$$k_{ext}\Big|_{[0,\alpha]\times[0,1]\,\cup\,[0,1]\times[0,\alpha]} = k\Big|_{[0,a]\times[0,1]\,\cup\,[0,1]\times[0,\alpha]},$$

$$k_{ext}\Big|_{[\alpha,1]^2} = \left(g^* + g^* * y^* + x^*\right)^{\dagger} * \left(g + y*x + x\right)^{\dagger}$$

$$-\left(g^* + g^* * y^* + x^*\right)^{\dagger} * g^* * g * (g + y * g + x)^{\dagger} - (g + y * g + x)^{\dagger}$$

$$g^* * g * (g + y * g + x)^{\dagger} - (g^* + g^* * y^* + x^*)^{\dagger}$$

$$+ (g^* + g^* * y^* + x^*)^{\dagger} * g^* * g + g^* g - n^* * n - n^* * \ell^{\dagger} * n, \tag{4.3}$$

where g *is an element of* $\mathcal{F}$ *such that* $g(t,s) = 0$, $(t,s) \in \Delta_2 \cup \Delta_3 \cup \Delta_4$, *and the corresponding integral operator* G *has norm less than one. Moreover, the correspondence is one – one.*

Proof. Use Theorem 3.3 and Theorem I.5.2. in [GKW1]. ∎

It is useful to note that (4.3) is equivalent with

$$I + K_{ext}\Big|_{L_n^2[\alpha,1]} =$$

$$\left((I+Y)\,G + (I+X)\right)^{*-1}(I-G^*G)\left((I+Y)\,G + (I+X)\right)^{-1} - N^*(I + L)^{-1}N,$$

where G, X, Y, N and L denote the integral operator with kernel g, x, y, n and ℓ, respectively.

REFERENCES

[GKW1] I. Gohberg, M. A. Kaashoek and H. J. Woerdeman, The Band Method For Positive and Contractive Extension Problems. *J. Operator Theory* **22**: 109-155, 1989.

[GKW2] I. Gohberg, M. A. Kaashoek and H. J. Woerdeman, The Band Method For Positive and Contractive Extension Problems: an Alternative Version and New Applications. *Integral Equations Operator Theory* **12**: 343-382, 1989.

[GKW3] I. Gohberg, M. A. Kaashoek and H. J. Woerdeman, A Maximum Entropy Principle in the General Framework of the Band Method. *J. Funct. Anal.* **95**: 231-254, 1991.

[GKW4]   I. Gohberg, M. A. Kaashoek and H. J. Woerdeman, The Band Method for Positive
         Extension Problems of Non-Band Type, *J. Operator theory*, to appear.

I. Gohberg   H. J. Woerdeman
Department of Mathematics   Department of Mathematics
Tel Aviv University   College of William and Mary
Ramat Aviv 69978   Williamsburg, Virginia 23187-8795
ISRAEL   USA

M. A. Kaashoek
Department of Mathematics
Vrije Universiteit
De Boelelaan 1081a
1081 HV Amsterdam
THE NETHERLANDS

AMS (1991) subject classification: 47A20, 47B65, 46K99.

Operator Theory:
Advances and Applications, Vol. 62
© 1993 Birkhäuser Verlag Basel

L^p–DISTANCE BETWEEN UNITARY ORBITS IN TYPE III$_\lambda$ FACTORS

Fumio Hiai and Yoshihiro Nakamura

Dedicated to Professor Tsuyoshi Ando on his sixtieth birthday.

Let $\mathcal{M}$ be a factor of type III$_\lambda$, $0 < \lambda < 1$, and $L_0^p(\mathcal{M})$, $1 \leq p < \infty$, the discrete Haagerup L^p-spaces. We obtain the L^p-distance formula between the unitary orbits of selfadjoint elements in $L_0^p(\mathcal{M})$ in terms of their generalized s-numbers, and determine the diameter of the quotient space (or the closed unitary orbit space) of the positive unit sphere of $L_0^p(\mathcal{M})$.

INTRODUCTION

In [5] we discussed distance between unitary orbits of selfadjoint elements in a factor $\mathcal{M}$ and in the noncommutative L^p-spaces on $\mathcal{M}$. To obtain the L^p-distance formula when $\mathcal{M}$ is a semifinite factor, we used the notion of generalized s-numbers [2] of τ-measurable operators and the method of submajorization between generalized s-numbers. The Haagerup L^p-spaces $L^p(\mathcal{M})$ [3, 9] are defined for an arbitrary factor $\mathcal{M}$ as certain subspaces of the space of τ-measurable operators affiliated with the crossed product of $\mathcal{M}$ by the modular action. Using the submajorization method we obtained the L^p-distance formula also when $\mathcal{M}$ is a type III$_1$ factor.

In §2 of this paper we establish the L^p-distance formula between selfadjoint elements when $\mathcal{M}$ is a factor of type III$_\lambda$, $0 < \lambda < 1$. To do so we have to consider the "discrete" Haagerup L^p-spaces $L_0^p(\mathcal{M})$ associated with the discrete crossed product decomposition of $\mathcal{M}$. In fact, the continuous crossed product is too big for our purpose in the sense that the generalized s-numbers of $x \in L^p(\mathcal{M})$ (the "continuous" Haagerup L^p-spaces) depend only on the norm $\|x\|_p$ (see [2]). On the other hand, the generalized s-numbers $\mu_t(x)$, $\lambda \leq t \leq 1$, of $x \in L_0^p(\mathcal{M})$ supply us complete data. A brief survey on the discrete Haagerup L^p-spaces is given in §1.

We can introduce an equivalence relation on the positive cone $L^p(\mathcal{M})_+$ of $L^p(\mathcal{M})$ by dividing into the closed unitary orbits, i.e. the $\|\cdot\|_p$-closures of $\{uxu^* : u \in \mathcal{M} \text{ unitary}\}$, $x \in L^p(\mathcal{M})_+$. Then the quotient space $L^p(\mathcal{M})_+/\sim$ is a metric space with the L^p-distance between unitary orbits. Recently the space $\mathcal{M}_*^+/\sim = L^1(\mathcal{M})_+/\sim$ was completely characterized by Haagerup and Størmer [4] (also see the references therein).

An interesting problem is to determine the diameter of the quotient space $L^p(\mathcal{M})_{+,1}/\sim$ of $L^p(\mathcal{M})_{+,1} = \{x \in L^p(\mathcal{M})_+ : \|x\|_p = 1\}$. In particular when $p = 1$, $L^1(\mathcal{M})_{+,1}/\sim$ is the quotient space of the normal states of $\mathcal{M}$, whose diameter was computed by Connes, Haagerup and Størmer [1] for factors of all types. From the results in [5] we easily determine $\mathrm{diam}(L^p(\mathcal{M})_{+,1}/\sim)$ for factors of type I, II and III$_1$. For factors of type III$_\lambda$, $0 < \lambda < 1$, the result in §2 reduces the problem to some elementary extremal problem, which easily can be solved when $1 \leq p \leq 3$. In §3 we thus obtain

$$\mathrm{diam}(L^p(\mathcal{M})_{+,1}/\sim) = 2^{1/p}\frac{1 - \lambda^{1/2p}}{(1 + \lambda^{1/2})^{1/p}}, \qquad 1 \leq p \leq 3,$$

including the case $p = 1$ in [1]. But the diameter of $L^p(\mathcal{M})_{+,1}/\sim$, $p > 1$, for type III$_0$ factors is still open.

1. PRELIMINARIES ON DISCRETE HAAGERUP L^p-SPACES

First let $\mathcal{N}$ be a semifinite von Neumann algebra with a faithful normal semifinite trace τ. We denote by $\widetilde{\mathcal{N}}$ the set of all τ-measurable operators affiliated with $\mathcal{N}$ (see [9]). For each $x \in \widetilde{\mathcal{N}}$ the generalized s-numbers $\mu_t(x)$ [2] are defined by

$$\mu_t(x) = \inf\{s \geq 0 : \tau(e_{(s,\infty)}(|x|)) \leq t\}, \qquad t > 0,$$

where $e_E(|x|)$ denotes the spectral projection of $|x|$ corresponding to a Borel subset E of **R**.

Let $\mathcal{M}$ be a general von Neumann algebra with a faithful normal semifinite weight φ. Let $\mathcal{N} = \mathcal{M} \rtimes_{\sigma^\varphi} \mathbf{R}$, which has the canonical faithful normal semifinite trace τ and the dual action θ_s with $\tau \circ \theta_s = e^{-s}\tau$, $s \in \mathbf{R}$. Then the Haagerup L^p-spaces $L^p(\mathcal{M})$ [3] are defined by

$$L^p(\mathcal{M}) = \{x \in \widetilde{\mathcal{N}} : \theta_s(x) = e^{-s/p}x, \ s \in \mathbf{R}\}, \qquad 0 < p \leq \infty.$$

See [9] for details.

Now we assume that $\mathcal{M}$ is a factor of type III$_\lambda$, $0 < \lambda < 1$. As is well known in the structure theory of type III$_\lambda$ factors (see [8, §29]), $\mathcal{M}$ has a faithful normal semifinite weight φ_0 such that $\sigma_T^{\varphi_0} = \mathrm{id}$ where $T = -2\pi/\log\lambda$. Then $\mathcal{N}_0 = \mathcal{M} \rtimes_{\sigma^{\varphi_0}} (\mathbf{R}/T\mathbf{Z})$ becomes a factor of type II$_\infty$ and the dual action $\theta_0 \in \mathrm{Aut}(\mathcal{N}_0)$ of σ^{φ_0} satisfies $\tau_0 \circ \theta_0 = \lambda\tau_0$ (τ_0 is the canonical trace of $\mathcal{N}_0$) and $\mathcal{M} = \mathcal{N}_0^{\theta_0}$. In this setup, besides the "continuous" Haagerup L^p-spaces $L^p(\mathcal{M})$, the "discrete" Haagerup L^p-spaces $L_0^p(\mathcal{M})$ [3] are defined by

$$L_0^p(\mathcal{M}) = \{x \in \widetilde{\mathcal{N}}_0 : \theta_0(x) = \lambda^{1/p} x\}, \qquad 0 < p \le \infty,$$

which are appropriate for our purpose of computing the L^p-distance of unitary orbits. Note that $\mathcal{M} = L_0^\infty(\mathcal{M})$ as $\mathcal{M} = L^\infty(\mathcal{M})$ [9, II.10].

In the following let us mention some basics on the discrete Haagerup L^p-spaces $L_0^p(\mathcal{M})$, $0 < p < \infty$, for the convenience of the reader. See [6, 4] for proofs.

(1) For every normal semifinite weight ψ on $\mathcal{M}$, let $k_\psi = d\tilde{\psi}/d\tau_0$ where $\tilde{\psi}$ is the dual weight of ψ. Then $\psi \in \mathcal{M}_*^+$ if and only if $k_\psi \in \widetilde{\mathcal{N}}_0$ (hence $k_\psi \in L_0^1(\mathcal{M})$). In this case,

$$\psi(1) = \tau_0(k_\psi e_{(\lambda\alpha,\alpha]}(k_\psi))$$

independently of $\alpha > 0$.

(2) Let $x \in \widetilde{\mathcal{N}}_0$ and $0 < p < \infty$. If $x = v|x|$ is the polar decomposition of x, then $x \in L_0^p(\mathcal{M})$ if and only if $|x|^p \in L_0^1(\mathcal{M})$ and $v \in \mathcal{M}$. In particular, the support projection $s(|x|)$ of $|x|$ belongs to $\mathcal{M}$ if $x \in L_0^p(\mathcal{M})$.

(3) For each $x \in L_0^p(\mathcal{M})$, $0 < p < \infty$, define

$$\|x\|_p = \tau_0(|x|^p e_{(\lambda\alpha,\alpha]}(|x|^p))^{1/p}$$

independently of $\alpha > 0$. Then when $1 \le p < \infty$, $L_0^p(\mathcal{M})$ becomes a Banach space with the norm $\|x\|_p$. In particular, $\mathcal{M}_* \cong L_0^1(\mathcal{M})$ by the isomorphism extending $\psi \in \mathcal{M}_*^+ \mapsto k_\psi \in L_0^1(\mathcal{M})_+$.

(4) For every $x \in L_0^p(\mathcal{M})$, $0 < p < \infty$, we have

$$\tau_0(e_{(s,\infty)}(|x|)) = \lambda^{-1}\tau_0(e_{(\lambda^{-1/p}s,\infty)}(|x|)), \qquad s \ge 0,$$

$$\mu_{\lambda t}(x) = \lambda^{-1/p}\mu_t(x), \qquad t > 0,$$

$$\|x\|_p = \left\{ \int_{\lambda\beta}^{\beta} \mu_t(x)^p dt \right\}^{1/p}$$

independently of $\beta > 0$.

(5) For each $1 \le p < \infty$ there naturally exists a Banach space isomorphism $\Phi_p : L_0^p(\mathcal{M}) \to L^p(\mathcal{M})$ which preserves the positivity and the left and right multiplications by elements in $\mathcal{M}$, i.e. $\Phi_p(L_0^p(\mathcal{M})_+) = L^p(\mathcal{M})_+$ and $\Phi_p(axb) = a\Phi_p(x)b$ for all $x \in L_0^p(\mathcal{M})$ and $a, b \in \mathcal{M}$.

2. L^p-DISTANCE BETWEEN UNITARY ORBITS

Let $\mathcal{M}$ be a factor of type III$_\lambda$, $0 < \lambda < 1$, and $\mathcal{U}$ the set of unitaries in $\mathcal{M}$. The unitary orbit of $x \in L_0^p(\mathcal{M})$ is given by $\mathcal{U}(x) = \{uxu^* : u \in \mathcal{U}\}$, which is included in $L_0^p(\mathcal{M})$ (in fact, $\|uxu^*\|_p = \|x\|_p$ for all $u \in \mathcal{U}$).

In the next theorem we exactly estimate the L^p-distance and the anti-L^p-distance between the unitary orbits of selfadjoint elements in $L_0^p(\mathcal{M})$.

THEOREM 2.1. *Let $\mathcal{M}$ be a factor of type III$_\lambda$, $0 < \lambda < 1$. If $1 \le p < \infty$ and $x, y \in L_0^p(\mathcal{M})_{\mathrm{sa}}$, then*

$$\inf_{u \in \mathcal{U}} \|x - uyu^*\|_p = \left\{ \int_\lambda^1 |\mu_t(x_+) - \mu_t(y_+)|^p dt + \int_\lambda^1 |\mu_t(x_-) - \mu_t(y_-)|^p dt \right\}^{1/p}, \quad (2.1)$$

$$\sup_{u \in \mathcal{U}} \|x - uyu^*\|_p = \left\{ \int_\lambda^1 (\mu_t(x_+) + \mu_t(y_-))^p dt + \int_\lambda^1 (\mu_t(x_-) + \mu_t(y_+))^p dt \right\}^{1/p}, \quad (2.2)$$

where $x = x_+ - x_-$ is the Jordan decomposition of x.

Let us divide the proof of the theorem into several lemmas.

LEMMA 2.2. *If $x, y \in L_0^1(\mathcal{M})_{\mathrm{sa}}$, then*

$$\|x - y\|_1 \ge \int_\lambda^1 |\mu_t(x_+) - \mu_t(y_+)| dt + \int_\lambda^1 |\mu_t(x_-) - \mu_t(y_-)| dt, \quad (2.3)$$

$$\|x - y\|_1 \le \int_\lambda^1 (\mu_t(x_+) + \mu_t(y_-)) dt + \int_\lambda^1 (\mu_t(x_-) + \mu_t(y_+)) dt. \quad (2.4)$$

PROOF. (2.4) is immediate because the right-hand side of (2.4) is equal to $\|x\|_1 + \|y\|_1$. To show (2.3) let $a = x - y$ and $b = x + a_-$ $(= y + a_+)$. Then $|a| = a_+ + a_- = 2b - x - y$. Writing the isomorphism of $L_0^1(\mathcal{M}) \cong \mathcal{M}_*$ as $z \in L_0^1(\mathcal{M}) \mapsto \psi_z \in \mathcal{M}_*$, we have

$$\|x - y\|_1 = \psi_{|a|}(1) = \psi_{2b-x-y}(1)$$
$$= \psi_{2b_+}(1) - \psi_{x_+}(1) - \psi_{y_+}(1) - \psi_{2b_-}(1) + \psi_{x_-}(1) + \psi_{y_-}(1)$$
$$= \int_\lambda^1 (2\mu_t(b_+) - \mu_t(x_+) - \mu_t(y_+)) dt$$
$$\qquad + \int_\lambda^1 (\mu_t(x_-) + \mu_t(y_-) - 2\mu_t(b_-)) dt,$$

Since $b \ge x$ and $b \ge y$, it is easy to check that $\mu_t(b_+) \ge \mu_t(x_+)$, $\mu_t(b_+) \ge \mu_t(y_+)$, $\mu_t(b_-) \le \mu_t(x_-)$ and $\mu_t(b_-) \le \mu_t(y_-)$. Hence

$$2\mu_t(b_+) - \mu_t(x_+) - \mu_t(y_+) \ge |\mu_t(x_+) - \mu_t(y_+)|,$$
$$\mu_t(x_-) + \mu_t(y_-) - 2\mu_t(b_-) \ge |\mu_t(x_-) - \mu_t(y_-)|.$$

These imply (2.3). $\square$

LEMMA 2.3. *If* $1 < p < \infty$ *and* $x, y \in L_0^p(\mathcal{M})_{\text{sa}}$, *then*

$$\|x - y\|_p \geq \left\{ \int_\lambda^1 |\mu_t(x_+) - \mu_t(y_+)|^p dt + \int_\lambda^1 |\mu_t(x_-) - \mu_t(y_-)|^p dt \right\}^{1/p}, \qquad (2.5)$$

$$\|x - y\|_p \leq \left\{ \int_\lambda^1 (\mu_t(x_+) + \mu_t(y_-))^p dt + \int_\lambda^1 (\mu_t(x_-) + \mu_t(y_+))^p dt \right\}^{1/p}. \qquad (2.6)$$

PROOF. Define the functions $\lambda(x)$ and $\check{\lambda}(x)$ on $\mathbf{R}$ (with the Lebesgue measure) into $\mathbf{R}$ by

$$\lambda_t(x) = \begin{cases} \mu_t(x_+), & t > 0, \\ 0, & t = 0, \\ -\mu_{-t}(x_-), & t < 0, \end{cases}$$

$$\check{\lambda}_t(x) = -\lambda_t(-x) = \begin{cases} -\mu_t(x_-), & t > 0, \\ 0, & t = 0, \\ \mu_{-t}(x_+), & t < 0. \end{cases}$$

Then the following submajorizations hold by [5, Theorem 3.1 and Lemma 3.2]:

$$\int_0^s |\lambda(x) - \lambda(y)|^*(t)\, dt \leq \int_0^s \mu_t(x - y)\, dt \leq \int_0^s |\lambda(x) - \check{\lambda}(y)|^*(t)\, dt, \qquad s > 0,$$

where $|\lambda(x) - \lambda(y)|^*$ and $|\lambda(x) - \check{\lambda}(y)|^*$ denote the decreasing rearrangements of $|\lambda(x) - \lambda(y)|$ and $|\lambda(x) - \check{\lambda}(y)|$, respectively. Here we note that for $z \in L_0^p(\mathcal{M})$

$$\int_0^s \mu_t(z)\, dt < \infty, \qquad s > 0.$$

In fact, by (4) in §1

$$\int_0^s \mu_t(z)\, dt = \sum_{n=0}^\infty \int_{\lambda^{n+1}s}^{\lambda^n s} \mu_t(z)\, dt$$

$$= \sum_{n=0}^\infty \lambda^n \int_{\lambda s}^s \mu_{\lambda^n t}(z)\, dt$$

$$= \sum_{n=0}^\infty \lambda^{n(1-1/p)} \int_{\lambda s}^s \mu_t(z)\, dt$$

$$= \frac{1}{1 - \lambda^{1-1/p}} \int_{\lambda s}^s \mu_t(z)\, dt < \infty.$$

For every $1 \leq p' < p$, since $t \mapsto t^{p'}$ is an increasing convex function on $[0, \infty)$, we have by [7, Theorem 3.1]

$$\int_0^2 \mu_t(x - y)^{p'} dt \geq \int_0^2 \left(|\lambda(x) - \lambda(y)|^*(t) \right)^{p'} dt$$

$$\geq \int_0^1 |\mu_t(x_+) - \mu_t(y_+)|^{p'} dt + \int_0^1 |\mu_t(x_-) - \mu_t(y_-)|^{p'} dt, \qquad (2.7)$$

$$\int_0^1 \mu_t(x-y)^{p'}\,dt \leq \int_0^1 \left(|\lambda(x)-\check{\lambda}(y)|^*(t)\right)^{p'}\,dt$$
$$\leq \int_0^1 (\mu_t(x_+)+\mu_t(y_-))^{p'}\,dt + \int_0^1 (\mu_t(x_-)+\mu_t(y_+))^{p'}\,dt. \tag{2.8}$$

By the analogous computation as above, we have

$$\int_0^1 |\mu_t(x_+)-\mu_t(y_+)|^{p'}\,dt = \frac{1}{1-\lambda^{1-p'/p}}\int_\lambda^1 |\mu_t(x_+)-\mu_t(y_+)|^{p'}\,dt,$$

$$\int_0^1 |\mu_t(x_-)-\mu_t(y_-)|^{p'}\,dt = \frac{1}{1-\lambda^{1-p'/p}}\int_\lambda^1 |\mu_t(x_-)-\mu_t(y_-)|^{p'}\,dt,$$

$$\int_0^2 \mu_t(x-y)^{p'}\,dt = \frac{1}{1-\lambda^{1-p'/p}}\int_{2\lambda}^2 \mu_t(x-y)^{p'}\,dt.$$

These equalities and (2.7) altogether imply that

$$\int_{2\lambda}^2 \mu_t(x-y)^{p'}\,dt \geq \int_\lambda^1 |\mu_t(x_+)-\mu_t(y_+)|^{p'}\,dt + \int_\lambda^1 |\mu_t(x_-)-\mu_t(y_-)|^{p'}\,dt,$$

which yields (2.5) by letting $p' \to p$. Also similar estimates from (2.8) imply (2.6). $\square$

Since $\mu_t(uzu^*) = \mu_t(z)$ for every $z \in \tilde{\mathcal{N}}_0$ and $u \in \mathcal{U}$, Lemmas 2.2 and 2.3 imply that if $1 \leq p < \infty$ and $x,y \in L_0^p(\mathcal{M})_{\mathrm{sa}}$, then

$$\inf_{u\in\mathcal{U}} \|x-uyu^*\|_p \geq \left\{\int_\lambda^1 |\mu_t(x_+)-\mu_t(y_+)|^p dt + \int_\lambda^1 |\mu_t(x_-)-\mu_t(y_-)|^p dt\right\}^{1/p},$$

$$\sup_{u\in\mathcal{U}} \|x-uyu^*\|_p \leq \left\{\int_\lambda^1 (\mu_t(x_+)+\mu_t(y_-))^p dt + \int_\lambda^1 (\mu_t(x_-)+\mu_t(y_+))^p dt\right\}^{1/p}.$$

The rest of this section is devoted to the proof that the equalities indeed hold in the above estimates.

LEMMA 2.4. *For every nonzero $x \in L_0^p(\mathcal{M})_+$, $1 \leq p < \infty$, there exists an increasing family $\{e_t : t \geq 0\}$ of projections in $\mathcal{N}_0$ such that $\tau_0(e_t) = t$ for all $t \geq 0$, $s(x) = \bigvee_{t\geq 0} e_t$, $\theta_0(e_t) = e_{\lambda t}$ for all $t \geq 0$, and $x = \int_0^\infty \mu_t(x)de_t$.*

PROOF. It is enough to show the case $p = 1$ because we may consider x^p instead of x. Then [6, Proposition 3.3] says that x is represented as

$$x = \sum_{n=-\infty}^\infty \lambda^n \theta_0^{-n}(xe_{(\lambda,1]}(x)). \tag{2.9}$$

As in the proof of [5, Theorem 4.3], there exists an increasing family $\{q_t : \lambda \leq t \leq 1\}$ of projections in $\mathcal{N}_0$ such that $\tau_0(q_t) = t - \lambda$ for $\lambda \leq t \leq 1$, $e_{(\lambda,1]}(x) = q_1$, and $x e_{(\lambda,1]}(x) = \int_\lambda^1 \mu_t(x) dq_t$. Noting that $\theta_0^{-n}(e_{(\lambda,1]}(x)) = e_{(\lambda^{n+1},\lambda^n]}(x)$ for $n \in \mathbf{Z}$, we define $\{e_t : t \geq 0\}$ by

$$e_t = \sum_{k=-\infty}^{n-1} \theta_0^{-k}(q_1) + \theta_0^{-n}(q_{\lambda^n t})$$

if $\lambda^{-n+1} \leq t \leq \lambda^{-n}$, $n \in \mathbf{Z}$. Then by (2.9) and (4) in §1

$$
\begin{aligned}
x &= \sum_{n=-\infty}^{\infty} \lambda^n \int_\lambda^1 \mu_t(x) d\theta_0^{-n}(q_t) \\
&= \sum_{n=-\infty}^{\infty} \int_\lambda^1 \mu_{\lambda^{-n}t}(x) d\theta_0^{-n}(q_t) \\
&= \sum_{n=-\infty}^{\infty} \int_{\lambda^{-n+1}}^{\lambda^{-n}} \mu_t(x) d\theta_0^{-n}(q_{\lambda^n t}) \\
&= \int_0^\infty \mu_t(x) de_t.
\end{aligned}
$$

The other required properties are easy to check. $\quad\square$

To prove the equalities (2.1) and (2.2) we may assume by approximation that x_+, x_-, $e_{\{0\}}(x)$, y_+, y_-, and $e_{\{0\}}(y)$ are all nonzero. Then by Lemma 2.4 we can choose increasing families $\{e_t : t \geq 0\}$, $\{e'_t : t \geq 0\}$, $\{f_t : t \geq 0\}$, and $\{f'_t : t \geq 0\}$ of projections in $\mathcal{N}_0$ such that

$$\tau_0(e_t) = \tau_0(e'_t) = \tau_0(f_t) = \tau_0(f'_t) = t,$$

$$s(x_+) = \bigvee_{t \geq 0} e_t, \quad s(x_-) = \bigvee_{t \geq 0} e'_t, \quad s(y_+) = \bigvee_{t \geq 0} f_t, \quad s(y_-) = \bigvee_{t \geq 0} f'_t,$$

$$\theta_0(e_t) = e_{\lambda t}, \quad \theta_0(e'_t) = e'_{\lambda t}, \quad \theta_0(f_t) = f_{\lambda t}, \quad \theta_0(f'_t) = f'_{\lambda t},$$

$$x_+ = \int_0^\infty \mu_t(x_+) de_t, \qquad x_- = \int_0^\infty \mu_t(x_-) de'_t,$$

$$y_+ = \int_0^\infty \mu_t(y_+) df_t, \qquad y_- = \int_0^\infty \mu_t(y_-) df'_t.$$

Now define

$$y' = \int_0^\infty \mu_t(y_+) de_t - \int_0^\infty \mu_t(y_-) de'_t,$$

$$y'' = \int_0^\infty \mu_t(y_+) de'_t - \int_0^\infty \mu_t(y_-) de_t,$$

which belong to $L_0^p(\mathcal{M})_{\mathrm{sa}}$ by construction.

LEMMA 2.5. *Let y' and y'' be as above. Then y' and y'' belong to the $\|\cdot\|_p$-closure of $\mathcal{U}(y)$.*

PROOF. By approximation it is enough to assume that $\mu_t(y_+)$ and $\mu_t(y_-)$ on $[\lambda, 1]$ are step functions. In this case, since $\mathcal{N}_0$ is a factor of type II_∞, there exist partial isometries v and v' in $\mathcal{N}_0$ such that

$$v^*v = e_1 - e_\lambda, \quad vv^* = f_1 - f_\lambda, \quad \int_\lambda^1 \mu_t(y_+)df_t = v\left\{\int_\lambda^1 \mu_t(y_+)de_t\right\}v^*,$$

$$v'^*v' = e_1' - e_\lambda', \quad v'v'^* = f_1' - f_\lambda', \quad \int_\lambda^1 \mu_t(y_-)df_t' = v'\left\{\int_\lambda^1 \mu_t(y_-)de_t'\right\}v'^*.$$

Thanks to $\theta_0^n(e_1 - e_\lambda) = e_{\lambda^n} - e_{\lambda^{n+1}}$ and the same equalities for e_t', f_t, and f_t', we can define

$$w = \sum_{n=-\infty}^{\infty} \theta_0^{-n}(v), \qquad w' = \sum_{n=-\infty}^{\infty} \theta_0^{-n}(v'),$$

which are partial isometries in $\mathcal{M} = \mathcal{N}_0^{\theta_0}$. Then by the proof of Lemma 2.4

$$\begin{aligned}
y_+ &= \sum_{n=-\infty}^{\infty} \lambda^n \theta_0^{-n}\left(\int_\lambda^1 \mu_t(y_+)df_t\right) \\
&= w\left\{\sum_{n=-\infty}^{\infty} \lambda^n \theta_0^{-n}\left(\int_\lambda^1 \mu_t(y_+)de_t\right)\right\}w^* \\
&= wy_+'w^*
\end{aligned}$$

and $y_- = w'y_-'w'^*$. Moreover choose a partial isometry w'' in $\mathcal{M}$ such that $w''^*w'' = e_{\{0\}}(x)$ and $w''w''^* = e_{\{0\}}(y)$, and let $u = w + w' + w''$. Then $u \in \mathcal{U}$ and $y = uy'u^*$, so that we get the conclusion for y'. The proof for y'' is analogous. $\quad\Box$

LEMMA 2.6. *Let y' and y'' be as above. Then*

$$\|x - y'\|_p = \left\{\int_\lambda^1 |\mu_t(x_+) - \mu_t(y_+)|^p dt + \int_\lambda^1 |\mu_t(x_-) - \mu_t(y_-)|^p dt\right\}^{1/p},$$

$$\|x - y''\|_p = \left\{\int_\lambda^1 (\mu_t(x_+) + \mu_t(y_-))^p dt + \int_\lambda^1 (\mu_t(x_-) + \mu_t(y_+))^p dt\right\}^{1/p}.$$

PROOF. Since by definition

$$|x - y'|^p = \int_0^\infty |\mu_t(x_+) - \mu_t(y_+)|^p de_t + \int_0^\infty |\mu_t(x_-) - \mu_t(y_-)|^p de_t',$$

$$|x - y''|^p = \int_0^\infty (\mu_t(x_+) + \mu_t(y_-))^p de_t + \int_0^\infty (\mu_t(x_-) + \mu_t(y_+))^p de_t',$$

noting (4) in §1 we may show the following fact: if h is a non-negative function on $(0, \infty)$ satisfying $h(\lambda t) = \lambda^{-1} h(t)$ for all $t > 0$, then

$$\left\| \int_0^\infty h(t) \, de_t \right\|_1 = \int_\lambda^1 h(t) \, dt.$$

By (4) in §1 the left-hand side equals $\int_\lambda^1 h^*(t) dt$ because $\mu_t\left(\int_0^\infty h(t) de_t\right) = h^*(t)$. Further, by the basic result of the rearrangement and the assumption $h(\lambda t) = \lambda^{-1} h(t)$, we have

$$\begin{aligned}
\int_\lambda^1 h^*(t) \, dt &= \int_{\{s>0:\lambda<h(s)\leq 1\}} h(t) \, dt \\
&= \sum_{n=-\infty}^\infty \int_{\{s:\lambda^{n+1}<s\leq\lambda^n,\, \lambda<h(s)\leq 1\}} h(t) \, dt \\
&= \sum_{n=-\infty}^\infty \int_{\{s:\lambda^{n+1}<\lambda^n s\leq\lambda^n,\, \lambda<h(\lambda^n s)\leq 1\}} h(\lambda^n t)\lambda^n dt \\
&= \sum_{n=-\infty}^\infty \int_{\{s:\lambda<s\leq 1,\, \lambda^{n+1}<h(s)\leq\lambda^n\}} h(t) \, dt \\
&= \int_\lambda^1 h(t) \, dt.
\end{aligned}$$

This completes the proof. $\square$

Thus by Lemmas 2.5 and 2.6 we complete the proof of the theorem.

COROLLARY 2.7. *Let $\mathcal{M}$ be a factor of type III$_\lambda$, $0 < \lambda < 1$. If $1 \leq p < \infty$ and $x, y \in L_0^p(\mathcal{M})_{sa}$, then x is in the $\|\cdot\|_p$-closure of $\mathcal{U}(y)$ if and only if $\mu_t(x_+) = \mu_t(y_+)$ and $\mu_t(x_-) = \mu_t(y_-)$ for all $\lambda \leq t \leq 1$ (hence for all $t > 0$).*

3. DIAMETER OF THE QUOTIENT SPACE OF $L^p(\mathcal{M})_{+,1}$

For an arbitrary factor $\mathcal{M}$ and $1 \leq p < \infty$, let $L^p(\mathcal{M})_{+,1}/\sim$ denote the quotient space of $L^p(\mathcal{M})_{+,1} = \{x \in L^p(\mathcal{M})_+ : \|x\|_p = 1\}$ by the equivalence relation induced by the closed unitary orbits. In this section we are concerned with the diameter of $L^p(\mathcal{M})_{+,1}/\sim$:

$$\mathrm{diam}(L^p(\mathcal{M})_{+,1}/\sim) = \sup\left\{ \inf_{u\in\mathcal{U}} \|x - uyu^*\|_p : x, y \in L^p(\mathcal{M})_{+,1} \right\}.$$

The computation in case $p = 1$ was completely done in [1]. In particular,

(1) $\mathrm{diam}(L^1(\mathcal{M})_{+,1}/\sim) = 2(1 - \lambda^{1/2})/(1 + \lambda^{1/2})$ if $\mathcal{M}$ is a factor of type III$_\lambda$, $0 \leq \lambda \leq 1$.

On the other hand, the following are obtained from the L^p-distance formulas between unitary orbits [5]. In fact, (2) and (3) are easy from [5, Theorems 4.3 and 4.4]. (4) is a consequence of [5, Theorem 5.3].

(2) $\operatorname{diam}(L^p(\mathcal{M})_{+,1}/\sim) = \{1 - n^{-1} + (1 - n^{-1/p})^p\}^{1/p}$ if $\mathcal{M} = \mathbf{M}_n(\mathbf{C})$, the type I_n factor.

(3) $\operatorname{diam}(L^p(\mathcal{M})_{+,1}/\sim) = 2^{1/p}$ if $\mathcal{M}$ is a factor of type I_∞ or of type II.

(4) $\operatorname{diam}(L^p(\mathcal{M})_{+,1}/\sim) = 0$, i.e. $L^p(\mathcal{M})_{+,1}/\sim$ is a single point, if $\mathcal{M}$ is a factor of type III_1.

The aim of this section is to compute $\operatorname{diam}(L^p(\mathcal{M})_{+,1}/\sim)$ for factors of type III_λ, $0 < \lambda < 1$. By (5) in §1 we first note that

$$\operatorname{diam}(L^p(\mathcal{M})_{+,1}/\sim) = \operatorname{diam}(L_0^p(\mathcal{M})_{+,1}/\sim),$$

so that the formula in Theorem 2.1 is essentially useful.

We write $d(\lambda, p) = \operatorname{diam}(L^p(\mathcal{M})_{+,1}/\sim)$ when $\mathcal{M}$ is a factor of type III_λ, $0 < \lambda < 1$, and $1 \le p < \infty$. This indeed depends only on λ and p as is seen below. Let $\mathcal{F}_\lambda$ denote the set of all positive, non-increasing and right-continuous functions f on $[\lambda, 1]$ such that $f(1) = \lambda f(\lambda)$ and $\int_\lambda^1 f(t)dt = 1$.

LEMMA 3.1. For $0 < \lambda < 1$ and $1 \le p < \infty$,

$$d(\lambda, p)^p = \sup\left\{\int_\lambda^1 |f(t)^{1/p} - g(t)^{1/p}|^p dt : f, g \in \mathcal{F}_\lambda\right\}. \tag{3.1}$$

PROOF. If $x \in L_0^p(\mathcal{M})_{+,1}$, then $\mu_t(x)$ on $[\lambda, 1]$ belongs to $\{f^{1/p} : f \in \mathcal{F}_\lambda\}$ by (4) in §1. Conversely, for each $f \in \mathcal{F}_\lambda$ define $x = \int_0^\infty f(t)^{1/p} de_t$ where $\{e_t : t \ge 0\}$ is an increasing family of projections in $\mathcal{N}_0$ such that $\tau_0(e_t) = t$ and $\theta_0(e_t) = e_{\lambda t}$ for all $t \ge 0$ (see Lemma 2.4). Then $x \in L_0^p(\mathcal{M})_{+,1}$ and $\mu_t(x) = f(t)^{1/p}$. Hence $\{f^{1/p} : f \in \mathcal{F}_\lambda\}$ coincides with the set of $\mu_t(x)$ on $[\lambda, 1]$ for all $x \in L_0^p(\mathcal{M})_{+,1}$. Thus (3.1) is a consequence of Theorem 2.1. $\square$

LEMMA 3.2. For $0 < \lambda < 1$ and $1 \le p < \infty$,

$$d(\lambda, p)^p = \frac{1}{1 - \lambda} \max_{\lambda \le t \le 1}\left\{(1 - t^{1/p})^p\left(1 - \frac{\lambda}{t}\right) + (t^{1/p} - \lambda^{1/p})^p\left(\frac{1}{t} - 1\right)\right\}. \tag{3.2}$$

PROOF. Obviously $f, g \in \mathcal{F}_\lambda$ in the supremum of (3.1) can be restricted to step functions. The function $\int_\lambda^1 |f(t)^{1/p} - g(t)^{1/p}|^p dt$ is convex in $f, g \in \mathcal{F}_\lambda$ because so is $|\alpha^{1/p} - \beta^{1/p}|^p$ in $\alpha, \beta > 0$. But we know [1] that any step function in $\mathcal{F}_\lambda$ is a convex combination of the following extreme step functions:

$$f_r = \frac{1}{r(1-\lambda)}\chi_{[\lambda,r)} + \frac{\lambda}{r(1-\lambda)}\chi_{[r,1]}, \qquad \lambda \le r < 1.$$

Therefore

$$d(\lambda, p)^p = \sup\left\{\int_\lambda^1 |f_r(t)^{1/p} - f_s(t)^{1/p}|^p dt : \lambda \le r < s < 1\right\}.$$

Since for $\lambda \le r < s < 1$

$$\int_\lambda^1 |f_r(t)^{1/p} - f_s(t)^{1/p}|^p dt$$

$$= \frac{1}{1-\lambda}\left\{\left(\frac{1}{r^{1/p}} - \frac{1}{s^{1/p}}\right)^p (r - \lambda s) + \left(\frac{1}{s^{1/p}} - \frac{\lambda^{1/p}}{r^{1/p}}\right)^p (s - r)\right\}$$

$$= \frac{1}{1-\lambda}\left\{\left(1 - \left(\frac{r}{s}\right)^{1/p}\right)^p \left(1 - \frac{\lambda s}{r}\right) + \left(\left(\frac{r}{s}\right)^{1/p} - \lambda^{1/p}\right)^p \left(\frac{s}{r} - 1\right)\right\},$$

we obtain (3.2). $\square$

To calculate the maximum value of the right-hand side of (3.2), we need the following lemma.

LEMMA 3.3. *Let h be a real-valued C^3-function on an interval (a, b) and $a < \alpha < \beta < \beta + \delta < b$. If h is concave, i.e. $h'' \le 0$ on (a, b), then*

$$h(\beta + \delta) - h(\beta) \le h(\alpha + \delta) - h(\alpha).$$

If $h''' \le 0$ on (a, b), then

$$\big(h(\beta + \delta) - h(\beta)\big) - \big(h(\alpha + \delta) - h(\alpha)\big) \le \delta(\beta - \alpha)h''(\alpha).$$

PROOF. Since

$$h(\alpha + \delta) - h(\alpha) = \int_0^\delta h'(\alpha + t)\, dt,$$

we easily have

$$h(\beta + \delta) - h(\beta) \le h(\alpha + \delta) - h(\alpha),$$

if $h'' \le 0$ on (a, b). When $h''' \le 0$ on (a, b),

$$\big(h(\beta + \delta) - h(\beta)\big) - \big(h(\alpha + \delta) - h(\alpha)\big) = \int_0^\delta \big(h'(\beta + t) - h'(\alpha + t)\big)\, dt$$

$$= \int_0^\delta \int_\alpha^\beta h''(s + t)\, ds\, dt$$

$$\le \delta(\beta - \alpha)h''(\alpha).$$

This completes the proof. $\square$

For $0 < \gamma < 1$ and $1 \le p < \infty$ define

$$F_{\gamma,p}(t) = (1-t)^p\left(1 - \frac{\gamma^p}{t^p}\right) + (t-\gamma)^p\left(\frac{1}{t^p} - 1\right), \qquad \gamma \le t \le 1.$$

LEMMA 3.4. *For $0 < \gamma < 1$ and $1 \le p \le 3$, the function $F_{\gamma,p}$ attains its maximum at $\gamma^{1/2}$, i.e.*

$$\max_{\gamma \le t \le 1} F_{\gamma,p}(t) = F_{\gamma,p}(\gamma^{1/2}) = 2(1 - \gamma^{1/2})^p(1 - \gamma^{p/2}).$$

PROOF. By an easy calculation

$$F'_{\gamma,p}(t) = p\left\{(1-t)^{p-1}\left(\frac{\gamma^p}{t^{p+1}} - 1\right) - (t-\gamma)^{p-1}\left(1 - \frac{\gamma}{t^{p+1}}\right)\right\}$$

and $F'_{\gamma,p}(\gamma^{1/2}) = 0$. Since $F_{\gamma,p}(t) = F_{\gamma,p}(\gamma/t)$ for $\gamma \le t \le 1$, it suffices to show that $F'_{\gamma,p}(t) \ge 0$ for $\gamma \le t \le \gamma^{1/2}$. Putting

$$G(t) = (1-t)^{p-1}(\gamma^p - t^{p+1}) - (t-\gamma)^{p-1}(t^{p+1} - \gamma), \qquad \gamma \le t \le 1,$$

we show that $G(t) \ge 0$ for $\gamma \le t \le \gamma^{1/2}$. Here

$$\begin{aligned}
G(t) &= \gamma^p(1-t)^{p-1} - t^{p+1}(t-\gamma)^{p-1} - t^{p+1}(1-t)^{p-1} + \gamma(t-\gamma)^{p-1} \\
&= (\gamma - t^2)\gamma^{p-1}(1-t)^{p-1} + t^2\left\{\gamma^{p-1}(1-t)^{p-1} - t^{p-1}(t-\gamma)^{p-1}\right\} \\
&\quad + (\gamma - t^2)(t-\gamma)^{p-1} - t^2\left\{t^{p-1}(1-t)^{p-1} - (t-\gamma)^{p-1}\right\}.
\end{aligned}$$

First consider the case $1 \le p \le 2$. Then the function $t \mapsto t^{p-1}$ is concave. Since for $\gamma \le t \le \gamma^{1/2}$

$$\gamma(1-t) - t(t-\gamma) = t(1-t) - (t-\gamma) = \gamma - t^2 \ge 0$$

and $t(t-\gamma) \le t - \gamma$, by Lemma 3.3 we have

$$\gamma^{p-1}(1-t)^{p-1} - t^{p-1}(t-\gamma)^{p-1} \ge t^{p-1}(1-t)^{p-1} - (t-\gamma)^{p-1}.$$

Hence $G(t) \ge 0$ for $\gamma \le t \le \gamma^{1/2}$.

Next consider the case $2 < p \le 3$. In this case, by Lemma 3.3

$$\begin{aligned}
&\left\{t^{p-1}(1-t)^{p-1} - (t-\gamma)^{p-1}\right\} - \left\{\gamma^{p-1}(1-t)^{p-1} - t^{p-1}(t-\gamma)^{p-1}\right\} \\
&\qquad \le (p-1)(p-2)(\gamma - t^2)(1-t)(t-\gamma)^{p-2}t^{p-3}
\end{aligned}$$

for $\gamma \leq t \leq \gamma^{1/2}$. Further, since the function $(1-t)(t-\gamma)^{p-2} = (1-\gamma)(t-\gamma)^{p-2} - (t-\gamma)^{p-1}$ is concave and

$$\frac{d}{dt}\left\{(1-t)(t-\gamma)^{p-2}\right\} = (t-\gamma)^{p-3}\left\{(p-2+\gamma) - (p-1)t\right\},$$

we have

$$(1-t)(t-\gamma)^{p-2} \leq \left(1 - \frac{p-2+\gamma}{p-1}\right)\left(\frac{p-2+\gamma}{p-1} - \gamma\right)^{p-2}$$

$$= \frac{1}{p-2}\left(\frac{(p-2)(1-\gamma)}{p-1}\right)^{p-1}$$

$$\leq \frac{1}{p-2}\left(\frac{1-\gamma}{2}\right)^{p-1}, \qquad \gamma \leq t \leq 1.$$

On the other hand, since the function t^{p-1} is convex,

$$\frac{\gamma^{p-1}(1-t)^{p-1} + (t-\gamma)^{p-1}}{2} \geq \left\{\frac{\gamma(1-t)+(t-\gamma)}{2}\right\}^{p-1} = \left(\frac{1-\gamma}{2}\right)^{p-1} t^{p-1}.$$

Hence

$$G(t) \geq 2(\gamma - t^2)\left(\frac{1-\gamma}{2}\right)^{p-1} t^{p-1} - (p-1)(\gamma - t^2)\left(\frac{1-\gamma}{2}\right)^{p-1} t^{p-1}$$

$$= (3-p)(\gamma - t^2)\left(\frac{1-\gamma}{2}\right)^{p-1} t^{p-1} \geq 0$$

for $\gamma \leq t \leq \gamma^{1/2}$. $\quad\square$

Now the next theorem follows from Lemmas 3.2 and 3.4.

THEOREM 3.5. *For $0 < \lambda < 1$ and $1 \leq p \leq 3$,*

$$d(\lambda, p) = 2^{1/p}\frac{1 - \lambda^{1/2p}}{(1 + \lambda^{1/2})^{1/p}}. \tag{3.3}$$

Remark that, when $p > 3$, $F_{\gamma,p}$ does not necessarily attain its maximum at $\gamma^{1/2}$. In fact, by a direct calculation

$$F_{\gamma,p}''(\gamma^{1/2}) = -2p\gamma^{-1/2}(1 - \gamma^{1/2})^{p-2}(1 - \gamma^{p/2} - p\gamma^{1/2} + p\gamma^{(p-1)/2}).$$

Since for $p > 2$

$$1 - \gamma^{p/2} < \frac{p}{p-2}(1 - \gamma^{p/2-1}),$$

$F_{\gamma,p}''(\gamma^{1/2}) > 0$ if $(p-2)\gamma^{1/2} \geq 1$. Hence $F_{\gamma,p}(\gamma^{1/2})$ is not the maximum, when $\gamma \geq (p-2)^{-2}$. This means that (3.3) is not valid for $p > 3$ and $\lambda \geq (p-2)^{-2p}$.

Although the exact values of $d(\lambda, p)$ for $p > 3$ are not known, we obtain

THEOREM 3.6. *For $0 < \lambda < 1$ and $3 < p < \infty$,*

$$2^{1/p}\frac{1 - \lambda^{1/2p}}{(1 + \lambda^{1/2})^{1/p}} \leq d(\lambda, p) < \max\left\{\frac{1 - \lambda^{1/p}}{(1 - \lambda)^{1/p}}, 2^{1/p}\frac{1 - \lambda^{1/p}}{1 + \lambda^{1/p}}\right\} \; (< 2^{1/p}).$$

PROOF. The lower estimate follows from Lemma 3.2 as Theorem 3.5. To prove the upper estimate, by Lemma 3.2 it suffices to show that for $0 < \gamma < 1$

$$F_{\gamma,p}(t) < \max\left\{(1 - \gamma)^p, 2\left(\frac{1 - \gamma}{1 + \gamma}\right)^p(1 - \gamma^p)\right\}, \qquad \gamma \leq t \leq \gamma^{1/2}.$$

Let $\eta = 2\gamma/(1 + \gamma)$. When $\gamma \leq t < \eta$, we have

$$\begin{aligned}
F_{\gamma,p}(t) &= (1 - t)^p - (t - \gamma)^p + \left(1 - \frac{\gamma}{t}\right)^p - \left(\frac{\gamma}{t} - \gamma\right)^p \\
&< (1 - t)^p - (t - \gamma)^p \\
&\leq (1 - \gamma)^p.
\end{aligned}$$

When $\eta \leq t \leq \gamma^{1/2}$, we have

$$\begin{aligned}
F_{\gamma,p}(t) &\leq (1 - \eta)^p - (\eta - \gamma)^p + \left(1 - \frac{\gamma}{\gamma^{1/2}}\right)^p - \left(\frac{\gamma}{\gamma^{1/2}} - \gamma\right)^p \\
&= \left(\frac{1 - \gamma}{1 + \gamma}\right)^p(1 - \gamma^p) + (1 - \gamma^{1/2})^p(1 - \gamma^{p/2}) \\
&< 2\left(\frac{1 - \gamma}{1 + \gamma}\right)^p(1 - \gamma^p),
\end{aligned}$$

which completes the proof. $\square$

We see by Theorems 3.5 and 3.6 that for each $1 \leq p < \infty$

$$\lim_{\lambda \to 0} d(\lambda, p) = 2^{1/p}, \qquad \lim_{\lambda \to 1} d(\lambda, p) = 0,$$

and for each $0 < \lambda < 1$

$$\lim_{p \to \infty} d(\lambda, p) = 0.$$

The computation of $\mathrm{diam}(L^p(\mathcal{M})_{+,1}/\sim)$ as well as the L^p-distance formula between unitary orbits for type III_0 factors is left open.

REFERENCES

1. A. Connes, U. Haagerup and E. Størmer, *Diameters of state spaces of type III factors*, Operator Algebras and their Connections with Topology and Ergodic Theory (H. Araki et al., eds.), Lecture Notes in Math., 1132, Springer-Verlag, 1985, pp. 91–116.

2. T. Fack and H. Kosaki, *Generalized s-numbers of τ-measurable operators*, Pacific J. Math. **123** (1986), 269–300.

3. U. Haagerup, *L^p-spaces associated with an arbitrary von Neumann algebra*, Colloq. Internat. CNRS, no. 274, CNRS, Paris, 1979, pp. 175–184.

4. U. Haagerup and E. Størmer, *Equivalence of normal states on von Neumann algebras and the flow of weights*, Adv. in Math. **83** (1990), 180–262.

5. F. Hiai and Y. Nakamura, *Distance between unitary orbits in von Neumann algebras*, Pacific J. Math. **138** (1989), 259-294.

6. V. Kaftal and R. Mercer, *Spectral projections of L^1 operators in type III$_\lambda$ von Neumann algebras*, Integral Equations Operator Theory **9** (1986), 679–693.

7. Y. Sakai, *Weak spectral order of Hardy, Littlewood and Pólya*, J. Math. Anal. Appl. **108** (1985), 31–46.

8. Ş. Strătilă, *Modular Theory in Operator Algebras*, Editura Academiei and Abacus Press, Tunbridge Wells and Bucureşti, 1981.

9. M. Terp, *L^p spaces associated with von Neumann algebras*, Notes, Copenhagen Univ., 1981.

Fumio Hiai
Department of Mathematics
Ibaraki University
Mito, Ibaraki 310, Japan

Yoshihiro Nakamura
Research Institute for Electronic Science
Hokkaido University
Sapporo 060, Japan

MSC 1991: 46L10, 46L50

Operator Theory:
Advances and Applications, Vol. 62
© 1993 Birkhäuser Verlag Basel

Finite Dimensional Solution Sets of Extremal Problems in H^1

Jyunji Inoue* and Takahiko Nakazi *

Dedicated to Professor Tuyoshi Ando on his 60th birthday

For a non-zero function f in H^1, the classical Hardy space on the unit circle, put

$$S_{|f|/f} = \{g \in H^1 : \|g\|_1 = 1, \arg f(e^{it}) = \arg g(e^{it})\ a.e.t\ \},$$

then $S_{|f|/f}$ is the set of extremal functions of a well known linear extremal problem in H^1. It is known and easy to see that if f^{-1} belongs to H^1 then the dimension of $< S_{|f|/f} >$, the linear span of $S_{|f|/f}$, is one. A simple example shows that even if f^{-1} belongs to H^p for some $p\,(0 < p < 1)$, the dimension of $< S_{|f|/f} >$ may be infinite. On the other hand, a sophisticated example (will be shown in this paper) shows that even if f^{-1} locally belongs to H^1 on the unit circle except a finite set, the dimension of $< S_{|f|/f} >$ may be infinite. In this paper it is shown that if $f \in H^1$ has the properties such that f^{-1} locally belongs to H^1 on the unit circle except a finite set and that $f^{-1} \in H^p$ for some $p > 0$, then the dimension of $< S_{|f|/f} >$ is finite.

1   Introduction

Let U be the open unit disc with the boundary $T = \partial U$. An analytic function f on U is said to belong to N if $\sup_{0 \le r < 1} \int_{-\pi}^{\pi} \log^+ | f(re^{it}) |\, dt < \infty$.

Each function f in N has a boundary value $f(e^{it}) = \lim_{r \to 1-} f(re^{it})$ a.e.t. The set of all f in N satisfying

$$\lim_{r \to 1-} \int_{-\pi}^{\pi} \log^+ | f(re^{it}) |\, dt = \int_{-\pi}^{\pi} \log^+ | f(e^{it}) |\, dt$$

---

*This research was partially supported by Grant-in-Aid for Scientific Research, Ministry of Education.

will be denoted by N^+. The Hardy space $H^p(U)$ $(0 < p \leq \infty)$ is the subspace of all f in N^+ whose boundary function $f(e^{it})$ belongs to $L^p(T)$, the usual Lebesgue space on T. Let $1 \leq p, q \leq \infty$, $1/p + 1/q = 1$ and $\varphi \in L^q$. The bounded linear functional T_φ on H^p with kernel function φ is defined by

$$T_\varphi(f) = \int_{-\pi}^{\pi} f(e^{it})\, \varphi(e^{it})\, dt/2\pi \qquad (f \in H^p),$$

and the norm of T_φ is given by $\|T_\varphi\| = \sup\{|\, T_\varphi(f)\,| : f \in H^p, \|f\|_p \leq 1\}$. When T_φ is nonzero, we put $S_\varphi = \{f \in H^p : T_\varphi(f) = \|T_\varphi\|, \|f\|_p \leq 1\}$. The set S_φ is the intersection of the hyperplane $\{f : T_\varphi(f) = \|T_\varphi\|\}$ with the unit ball of H^p. When $1 < p \leq \infty$, the structure of the set S_φ is simple since the set S_φ consists of exactly one point. But when $p=1$ the situation is quite different from the former case.

In 1958, deLeeuw and Rudin [1] studied the structure of S_φ, and among many fundamental results concerning extremal problems, they completely described the set S_φ when the function φ can be extended analytically to the set $\{|\, z\, | \geq r\}$ for some $r < 1$.

Since then, the study of S_φ has been continued by several authors. Nakazi, the one of the authors of this paper, showed in [7] that if S_φ is a $weak^*$- compact subset of H^1, $< S_\varphi >$ has finite dimension, and that the set S_φ can be described completely in the same manner as deLeeuw and Rudin carried out in [1]. For continuous kernel φ, the set S_φ is $weak^*$–compact, and hence Nakazi's result above can be applied. It is easy to see that the above deLeeuw-Rudin's case is contained in this case.

Let $\varphi = |\, f\, |\, /f$ for some nonzero function $f \in H^1$. Even if f is a polynomial, φ may not be continuous but $< S_\varphi >$ is finite dimensional. We are interested in the problem to decide the functions $f \in H^1$ such that $< S_\varphi >$ for $\varphi = |\, f\, |\, /f$ has finite dimension. If f^{-1} belongs to H^1 then $S_{|f|/f} = \{f\}$, that is the dimension of $< S_{|f|/f} >$ is one ([7]). We want to relax the hypothesis on f^{-1}. It is known that f^{-1} belongs to N_+ if and only if f is outer. Hence without loss of generality we may assume that f^{-1} belongs to N_+. Regarding $N_+ \cap L^p = H^p$, we want to define a local version of the Hardy space H^p.

For each $c \in T$, I_c denotes an arc in T which contains c at the center of I_c and put $L^p(I_c) = \{f : f \text{ is measurable on } T \text{ and } \int_{I_c} |\, f(e^{it})\, |^p\, dt < \infty\}$.

Definition. For the function $g \in N_+$ we say that g locally belongs to H^p at $c \in T$ when $g \in N_+ \cap L^p(I_c)$ for some I_c.

If g locally belongs to H^p at every point on T, then it is easy to see that g belongs to H^p . We can conjecture that if $f \in H^1$ and $f^{-1} \in N_+$ locally belongs to H^1 on the unit circle except a finite set, then the linear span of $S_{|f|/f}$ is finite dimensional. However, we have a counter example which will be given in §3. In the above conjecture, if $f^{-1} \in H^p$ for some $0 < p < 1$ instead of $f^{-1} \in N_+$ then the conjecture is valid. This is our main result which will be proved in §2 .

In the proof of the main result (Theorem 1) the following Lemma 1 , which may be some independent interest, plays an essential role. If $A = \emptyset$ in (b) of Lemma 1, it follows easily that F must be constant. This is the Neuwirth-Newman Theorem [8].

Lemma 1. *Let F be a function in N^+ such that*

(a) *F belongs to H^p for some $p > 0$,*

(b) *F locally belongs to $H^{1/2}$ except a finite set A of T,*

(c) *F is outer and $F(e^{it}) \geq 0$ a.e. on T.*

Then F can be extended to a rational function.

2 The main theorem and its corollaries

In this section the main theorem in this paper and its corollaries will be proved. The proof of Lemma 1 which plays an essential role in the proof will be given in §4.

Notation : In the rest of this paper , the notation $[a]$ for a real number a is used to express the largest integer which is equal or less than a.

Theorem 1. *Suppose f is a nonzero function in H^1 . If f^{-1} locally belongs to H^1 on the unit circle except a finite set A of T and $f^{-1} \in H^p$ for some $p > 0$, then the linear span of $S_{|f|/f}$ has finite dimension.*

*Moreover, if $A = \{\alpha_j : j = 1, 2, ..., n\}$ and if $p_j = \sup\{q : f^{-1} \text{ locally belongs to}$

H^q at α_j } *for* $j = 1, 2, ..., n$, *then we have*

$$\dim < S_{|f|/f} > \leq 1 + \sum_{j=1}^{n} 2 \left[\frac{p_j + 1}{2p_j}\right]$$

Proof.   If $S_{|f|/f} = \{f/\|f\|\}$, the theorem is trivial, and we suppose that $S_{|f|/f} \neq \{f/\|f\|\}$. Let g be an arbitrary outer function in $S_{|f|/f}$ and put $F(z) = g(z)/f(z)$ $(z \in U)$. We claim that $F(z)$ has the following properties:

(a) F locally belongs to $H^{1/2}$ except a finite set A of T,

(b) At $z = \alpha_j \in A$, F locally belongs to H^q for each q $(0 < q < \frac{p_j}{1+p_j})$,

(c) F is outer and $F(e^{it}) \geq 0$ a.e. on T.

(a) and (b) are deduced from the conditions on f and g by using the Hölder's inequality, and (c) is obvious from the conditions on f and g.

By (a), (b) and (c) using Lemma 1 of §4, F can be extended to a rational function with poles only at points in A. We also express this extension by the same symbol $F(z)$. Note that if α_j is a pole of order n_j of F, n_j must be an even integer by (c), and hence we have $n_j = 2m_j$ with $m_j \leq [\frac{p_j+1}{2p_j}]$. Since $F(z)$ has no zeros on U, its zeros are all on T. Therefore, with some $\gamma > 0$, we can represent $F(z) = g(z)/f(z)$ in the form

$$F(z) = \frac{\gamma \prod_{k=1}^{N}(z - z_k)(\bar{z}_k z - 1)}{\prod_{j=1}^{n}(z - \alpha_j)^{m_j}(1 - \bar{\alpha}_j z)^{m_j}}, \tag{1}$$

where $N = m_1 + ... + m_n$ and $z_k \in T$ $k = 1, ..., N$. From (1), it follows that each outer function g in $S_{|f|/f}$ has the expression of the form

$$g(z) = \gamma f(z) \frac{\prod_{k=1}^{N}(z - z_k)(\bar{z}_k z - 1)}{\prod_{j=1}^{n}(z - \alpha_j)^{m_j}(1 - \bar{\alpha}_j z)^{m_j}} \qquad (z \in U),$$

with the relations $m_j \leq [\frac{p_j+1}{2p_j}]$ $j = 1, 2, ..., n$.

By using the Nakazi's results in [7], we can conclude that

$$\dim < S_{|f|/f} > \leq 1 + \sum_{j}^{n} 2 \left[\frac{p_j + 1}{2p_j}\right]$$

and this completes the proof.

In the following three corollaries of Theorem 1, we need not assume $f^{-1} \in H^p$ for some $p > 0$.

Corollary 1. *Suppose f is a schlicht function in H^1. If f^{-1} locally belongs to H^1 on T except a finite set A, then $\dim < S_{|f|/f} > \leq \max\{3, 2 \mid A \mid + 1\}$, where $\mid A \mid$ denotes the number of elements of A.*

Proof. We divide the situation into two cases: (i) $f(z) \neq 0$ for each $z \in U$, (ii) There exists $\alpha \in U$ such that $f(\alpha) = 0$. In the case (i), f^{-1} is a schlicht function and hence belongs to H^p for all $p\,(0 < p < 1/2)$ at each point of A. Thus we have $\dim < S_{|f|/f} > \leq 2 \mid A \mid + 1$ by Theorem 1. In the case (ii), we can represent f in the form $f(z) = (z - \alpha)\,g(z)$ with g^{-1} is a bounded analytic function on U, and hence $g/\|g\|_1$ is an exposed point of the unit ball of H^1. Hence we get $\dim < S_{|f|/f} > = 3$ (cf. [7]). Thus the proof is complete.

Corollary 2. *Suppose f is a nonzero function in H^1. If there exists a nonzero function $h \in H^1$ such that hf is an analytic polynomial of degree n, then $\dim < S_{|f|/f} > \leq 2n + 1$.*

Proof. Let $f = qg$ be a factorization with an inner function q and an outer function g, and put $s = hf$. Since s is an analytic polynomial of degree n, we can represent s in the form $s(z) = \beta \prod_{j=1}^{L}(z - \alpha_j)^{m_j}$, where $\alpha_1, ..., \alpha_L, \beta \in C$, and $m_1 + ... + m_L = n$. It is easy to see that q is a finite Blaschke product whose zeros are contained in $A \cap U$, where $A = \{\alpha_1, ..., \alpha_L\}$, and α_j may be a zero of q with order at most m_j. On the other hand, $g^{-1}(= qh/s)$ locally belongs to H^1 at each point of $T \backslash A$, and locally belongs to H^q with $0 < q < m_j^{-1}/(m_j^{-1} + 1)$ at each point of $\alpha_j \in T \cap A$. Hence we can apply Theorem 1 to g, and get $\dim < S_{|g|/g} > \leq 1 + \sum_{|\alpha_j|=1} 2[\frac{m_j+2}{2}]$. Thus we have

$$\dim < S_{|f|/f} > \leq \sum_{\alpha_j \in U} 2m_j + 1 + \sum_{|\alpha_j|=1} 2\,[\frac{m_j + 2}{2}] \leq 1 + 2n.$$

This completes the proof.

Corollary 3. *Suppose $f \in H^1$ is continuous and $\Re[f(z)] \geq 0$ on $\bar{U}$. If f has at most a finite number of zeros on T, the dimension of $< S_{|f^n|/f^n} >$ is finite for any positive integer n.*

Proof. Since $\Re[f(z)] \geq 0$ on U, we have $\Re[f^{-1}(z)] \geq 0$ on U, and hence f^{-1} belongs to H^p for each $p < 1$. Therefore, f^{-n} belongs to H^p for some $p > 0$, and f^{-n} locally belongs to H^1 except finite points on T. Thus the result follows from Theorem 1.

If q is an inner function and $f = 1 - q$ then $\Re f$ is nonnegative a.e. on T and hence $f/\|f\|_1$ is an exposed point of the unit ball of H^1, that is dim $< S_{|f|/f} > = 1$. If q is a finite Blaschke product then by Corollary 3 dim $< S_{|f^2|/f^2} >< \infty$. However, it is easy to see that if q is not a finite Blaschke product then dim $S_{|f^2|/f^2} = \infty$.

If $g \in H^1$ is continuous and $| g(z) |\le 1$ on $\bar{U}$, and $f = 1 - g$, then $\Re f$ is nonnegative. If the peak set of g is a finite set on the unit circle, then by Corollary 3 dim $S_{|f^n|/f^n} < \infty$ for any positive integer $n > 0$. A well known Rudin-Carleson theorem [3] tells us how to construct such a g.

Corollary 4. *If $q(z)$ is an inner function such that* i) $q(z)$ *is not a finite Blaschke product* ii) *The singular support of $q(z)$ is a finite set A. Then for each number $\alpha \in T$, $B = \{z \in T\backslash A : q(z) = \alpha\}$ is an infinite set.*

Proof. If B were a finite set, $f = (\alpha - q)^2$ would satisfy the conditions of Theorem 1 and hence dim $< S_{|f|/f} >$ must be finite. But it is well known and easy to prove that if $b(z)$ is an inner function which is not a finite Blaschke product, we have dim $< S_{|(\alpha-b)^2|/(\alpha-b)^2} >= \infty$. Thus we get a contradiction, which completes the proof.

3   Examples

In this section we give counter examples which we mentioned in the abstract and §1. That is, to get the conclusion we cannot omit in Theorem 1 either of (a) and (b) below:

(a) $f(z)$ locally belongs to H^1 except finite points of T,

(b) $f^{-1} \in H^p$ for some $p > 0$.

Example 1. Let $0 < p < 1$ and let b be an infinite Blaschke product on the unit disc such that its zeros converge to 1. Choose p' such that $0 < p < p' < 1$ and put $f = (1 - b)^{1/p'}$. We claim that f satisfies (b) of the conditions above, but dim $< S_{|(1-b)^2|/(1-b)^2} >= \infty$.

Indeed, since $(f^{-1})^p = (1 - b)^{-p/p'}$ with $p/p' < 1$, we have $f^{-1} \in H^p$, and since $f = (1 - b)^2(1 - b)^{-2+(1/p')}$ with $(1 - b)^{-2+(1/p')} \in H^1$ we have dim $< S_{|f|/f} >= \infty$. Note that f doesn't satisfy (a) of the conditions above by Corollary 4.

Example 2. (cf. [5]) Let $f(z)$ be a function defined by

$$F(z) = \prod_{k=1}^{\infty} \frac{(z - \alpha_k)(1 - \bar{\alpha}_k z)}{(z - 1)(1 - z)} \qquad (z \in U) \qquad (2)$$

where $\alpha_k = e^{-i/k^2}$, $k = 1, 2, 3, \dots$. It is easy to see that the right hand side infinite product in (2) converges uniformly on each compact set of $C \backslash \{1\}$, and $F(e^{it}) \geq 0$ on $T \backslash \{1\}$. Moreover $F(z)$ is an outer function. To see this, we consider $\log F(z)$ on $\bar{U} \backslash \{1\}$ such that $\Im[\log F(-1)] = 0$. Since $\Im[\log F(e^{it})]$ is a monotone decreasing step function on $(0, 2\pi)$, with jumps 2π at $t = 1/k^2$ $(k = 1, 2, \dots)$, and hence $\Im[\log F]$ is a real harmonic function belonging to the Zygmund's class. Therefore the harmonic conjugates of $\Im[\log F]$ are contained in h^1 (cf. [6]), which implies that $F(z)$ is outer.

Choose $\varepsilon_k > 0$ so that

$$\tfrac{1}{k^2} > \tfrac{1}{k^2} - \varepsilon_k > \tfrac{1}{(k+1)^2} + \varepsilon_{k+1},$$

$$k = 1, 2, \dots$$

$$|F(e^{it})| \leq 1 \qquad t \in \left(\tfrac{1}{k^2} - \varepsilon_k, \tfrac{1}{k^2} + \varepsilon_k\right)$$

and put $\Omega = \bigcup_{k=1}^{\infty} \left(\tfrac{1}{k^2} - \varepsilon_k, \tfrac{1}{k^2} + \varepsilon_k\right)$. If we define a function g on $T \backslash \{1\}$ by

$$g(e^{it}) = \begin{cases} \min\left\{ \frac{1}{|F(e^{it})|}, 1 \right\} : & t \in [0, 2\pi) \backslash \Omega \\ \frac{1}{\varepsilon_k k^4} : & t \in \left(\tfrac{1}{k^2} - \varepsilon_k, \tfrac{1}{k^2} + \varepsilon_k\right), \qquad k = 1, 2, \dots \end{cases}$$

then $\log g(e^{it})$ belongs to $L^1(T)$. Using this $g(e^{it})$ we define an outer function $f(z) \in H^1$ by

$$f(z) = \exp \int_0^{2\pi} \frac{e^{it} + z}{e^{it} - z} \log g(e^{it})\, dt / 2\pi. \qquad (z \in U)$$

Then it is easy to see that $f(z)$ locally belongs to H^1 except one point $z = 1$. But since $f(z)F(z) \in H^1$ and $\arg f(e^{it}) = \arg f(e^{it})F(e^{it})$ a.e. on $[0, 2\pi)$ we can see that

$$\dim < S_{|f|/f} >= \dim < S_{|fF|/fF} >= \infty.$$

4 Proof of Lemma 1

In this section, we prove Lemma 1 used in the proof of the main theorem in §2. In the following, C'^+ and C_+ denote the half planes given by $C'^+ = \{z : \Im z > 0\}$ and

$C_+ = \{z : \Re z > 0\}$. $H^p(C^+)$ (resp. $H^p(C_+)$) is the Hardy space on C^+ (resp. C_+) in the usual sense (cf. [3, p.51]). $\psi(z)$ denotes the linear fractional transformation defined by $\psi(z) = (z - i)/(z + i)$, which maps C^+ conformally onto U.

Lemma 1. *Let F be a function in N^+ such that*

(a) *F belongs to H^p for some $p > 0$,*

(b) *F locally belongs to $H^{1/2}$ except a finite subset A of T,*

(c) *F is outer and $F(e^{it}) \geq 0$ a.e. on T.*

Then F can be extended to a rational function.

Proof. By conditions on F, $F(z)$ can be extended beyond every boundary point in $T \backslash A$ to a holomorphic function on $C \cup \{\infty\} \backslash A$, which we also express by $F(z)$. Therefore if we can prove that each point $\alpha \in A$ is at most a pole of $F(z)$, it follows that F is a rational function.

In the rest of the proof of this lemma, $M_i(i = 1, 2, ..., 5)$ stands for an appropriate positive constant. Let α be a point of A. To prove that α is a pole of F, we can assume without loss of generality that $\alpha = 1$. If we put $G(z) = F(z)^p(1 - z)^2$, then trivially $G(z)/(1 - z)^2 = F(z)^p \in H^1(U)$, and we have $G(\psi(z)) = F(\psi(z))^p(1 - \psi(z))^2 \in H^1(C^+)$ (cf. [4] p.130). From this we get

$$| F(\psi(z)) |^p | 1 - \psi(z) |^2 \leq M_1 \qquad (\Im z \geq 1)$$

and hence

$$| F(\psi(z)) | \leq \frac{M_1^{1/p}}{| 1 - \psi(z) |^{2/p}} \leq M_2 | z |^{2/p} \qquad (\Im z \geq 1) \qquad (3)$$

By the conditions (b) and (c), it follows that $\bar{F}(\psi(z)) = F(\psi(\bar{z}))$ on $C \backslash A$, and hence we get

$$| F(\psi(z)) | \leq M_2 | z |^{2/p} \qquad (\Im z \leq -1) \qquad (4)$$

Next, since $F(\psi(z + a))^p(1 - \psi(z + a))^2 \in H^1(C^+)$ for each $a > 0$, we have by Fejer-Riesz inequality (cf. [2], p.46) modified to hold for a function in $H^p(C^+)$

$$\int_0^\infty | F(\psi(it + a)) |^p | 1 - \psi(it + a) |^2 \frac{1}{(1 + t)^2} dt$$

$$\leq \frac{1}{2} \int_{-\infty}^\infty | F(\psi(x + a)) |^p | 1 - \psi(x + a) |^2 \frac{1}{1 + x^2} dx$$

$$\leq \sup_{a>0} \int_{-\infty}^{\infty} \mid F(\psi(x+a)) \mid^p \mid 1 - \psi(x+a) \mid^2 dx \leq \| G \circ \psi \|_1 < \infty \tag{5}$$

Since

$$\mid 1 - \psi(it+a) \mid = \mid 1 - (it+a-i)/(it+a+i) \mid$$

$$= \mid 2i/(it+a+i) \mid \geq \mid 1/(it+a+1) \mid \quad (t>0, a>0),$$

we get

$$M_3 \geq \int_0^{\infty} \mid F(\psi(it+a)) \mid^p \mid 1 - \psi(it+a)) \mid^2 \frac{1}{(t+1)^2} dt$$

$$\geq \frac{1}{2} \int_0^{\infty} \mid F(\psi(it+a)) \mid^p \mid \frac{1}{it+a+1} \mid^4 dt \qquad (a>0) \tag{6}$$

Therefore if we choose $a > 0$ large enough to assure $\Re\alpha < a$ $(\alpha \in A)$, we get that

$F(\psi(z+a))/(z+a+1)^{4/p} \mid_{C_+}$ belongs to $H^p(C_+)$. From this we get

$$\mid F(\psi(z)) \mid \leq M_4 \mid z \mid^{4/p} \qquad (\Re z \geq a) \tag{7}$$

In the same way, we have

$$\mid F(\psi(z)) \mid \leq M_5 \mid z \mid^{4/p} \qquad (\Re z \leq -b) \tag{8}$$

for a large positive number b. From (3), (4), (7) and (8), we can conclude that ∞ is at most a pole of $F(\psi)$, that is $\alpha = 1$ is at most a pole of F. This completes the proof of Lemma 1.

REFERENCES

1. deLeeuw, K. and Rudin, W., Extreme points and extremum problems in H^1, Pacific J. Math. 8 (1958), 467-485.

2. Duren, P. L., Theory of H^p spaces, Academic Press Inc., New York and London 1970.

3. Garnett, J. B., Bounded analytic functions, Academic press. Inc. 1981.

4. Hoffman, K., Banach spaces of analytic functions, Prentice-Hall Inc., Englewood Cliffs, New Jersey 1962.

5. Inoue, J., An example of a non-exposed extreme function on the unit ball of H^1, preprint.

6. Koosis, P., introduction to H^p spaces, Cambridge University Press, 1980

7. Nakazi, T., Exposed points and extremal problems in H^1, J. Funct. Anal. 53 (1983), 224-230.

8. Neuwirth, J. and Newman, D. J., Positive $H^{1/2}$ functions are constants, Proc. Amer. Math. Soc. 18 (1967) 958

Jyunji Inoue   Takahiko Nakazi

Department of Mathematics   Department of Mathematics

Faculty of Science   Faculty of Science

Hokkaido University   Hokkaido University

Sapporo 060   Sapporo 060

Japan   Japan

AMS Subject Classification: Primary 30D55, 46J15, Secondary 47B35

Operator Theory:
Advances and Applications, Vol. 62
© 1993 Birkhäuser Verlag Basel

FACTORIZATION OF OPERATORS WITH ANGULARLY CONSTRAINED SPECTRA

Charles R. Johnson[1] and Ilya M. Spitkovsky[2]

We are pleased to dedicate this paper to T. Ando, who is not only a mathematical inspiration but also a friend to all the community.

For arbitrary angular sectors W_1 and W_2 of the complex plane, we characterize $\{AB : F(A) \subseteq W_1 \text{ and } F(B) \subseteq W_2\}$. This encompasses a number of classical results, such as Lyapunov theorems.

0. INTRODUCTION AND NOTATION.

Let H be a Hilbert space, and let $\mathscr{B}(H)$ be the algebra of all bounded linear operators on H. For $A \in \mathscr{B}(H)$ let

$$F(A) = \{(Ax, x) : x \in H, \|x\| = 1\}$$

denote the field of values of A and let $\sigma(A)$ denote the spectrum of A. Denote also by Re A and Im A the real (Hermitian) and imaginary parts of A:

$$\text{Re } A = \frac{1}{2}(A + A^*), \quad \text{Im } A = \frac{1}{2i}(A - A^*).$$

Since notation Im A is already occupied, the range A(H) of A will be denoted by $\mathscr{R}(A)$. Finally, $\text{Ker} A = \{x : Ax = 0\}$.

It is well known that F(A) is convex and bounded. See [I, Chapter 6] for these and other properties of the numerical range. In the finite dimensional case (dim $H < \infty$) this set is also closed, which is not necessarily the case for operators A on infinite dimensional spaces. (See [HJ2, Chapter 1] for a discussion of the field of values/numerical range in the finite dimensional ease.)

---

[1] Partially supported by NSF Grant DMS - 92-00899 and by ONR contract N00014-90-J-1739.

[2] Partially supported by NSF Grant DMS-91-01143.

The theorem of Lyapunov [HJ2] says that for an n-by-n matrix A, $\sigma(A)$ is located in the open right half plane

$$\Pi_+ = \{z : \text{Re } z > 0\}$$

if and only if there is a positive definite (Hermitian) matrix G such that $H = \text{Re } (GA)$ is positive definite. Based on simple observations [J1], Lyapunov's theorem may be rephrased in the following way. Let dim $H = n < \infty$. Then for $A \in \mathscr{B}(H)$, we have $\sigma(A)$ in the open right half plane if and only if A may be written as $A = KB$, in which F(K) lies in the positive real axis (K is positive definite) and F(B) lies in the open right half plane. (Identify K as G^{-1} and B as GA in our more traditional statement of Lyapunov's theorem.) This point of view was generalized in [J1] by showing that the open right half plane could be replaced (for both $\sigma(A)$ and F(B)) by any open angular sector of the complex plane of opening no more than π.

One of the results in [BJ] may be phrased in a similar manner. The spectrum of A avoids the nonpositive real axis if and only if A may be written as $A = BC$, in which each of F(B) and F(C) lie in the open right half plane. This fact was rediscovered in its equivalent "Lyapunov form" as the result of [N] : A has no nonpositive real eigenvalues if and only if there is a B such that both Re(B) and Re(BA) are positive definite.

The classical theorem of Lyapunov may also be generalized to consider spectra in the *closed* right half plane; however, in this event an additional Jordan structure condition arises. In our language, the result [CS] may be stated as follows. We have $\sigma(A)$ in the closed right half plane *and* any eigenvalues on the imaginary axis are geometrically simple (all Jordan blocks are one-by-one) if and only if A may be written as $A = KB$, with F(K) in the positive real axis and F(B) in the closed right half plane.

Finally, let us mention that Lyapunov's theorem is valid in the infinite dimensional case if one substitutes F(K) and F(B) in its wording by their closures cl F(K) and cl F(B) [DK, Theorem 5.1]. We use notations cl, bd, int for topological operations of taking the closure, boundary and interior respectively; a bar — is reserved for complex conjugation.

All of these results suggest a close connection between angular spectral location and angular location of the fields of values of factors. It is our purpose here to give general results that naturally encompass the finite dimensional special cases we have cited and also indicate a further infinite dimensional generalization. As is indicated by the results cited, Jordan structure conditions play a role on the boundary of angular sectors. As we wish to consider all possible meaningful sectors, we need some elaborate notation. For an interval I of one of the following types:

$$(\theta_1, \theta_2) = \{\theta : \theta_1 < \theta < \theta_2\};$$
$$(\theta_1, \theta_2] = \{\theta : \theta_1 < \theta \leq \theta_2\};$$
$$[\theta_1, \theta_2) = \{\theta : \theta_1 \leq \theta < \theta_2\}; \text{ and}$$
$$[\theta_1, \theta_2] = \{\theta : \theta_1 \leq \theta \leq \theta_2\},$$

let

$$W_I = \{z \in \mathbb{C} : \arg z \in I\}$$

(The complex number 0 may or may not be included in W_I).

Finally, for a given sector $W = W_I$ let

$$\Lambda_W = \{A : \sigma(A) \subset W\}, \ \Pi_W = \{A : \operatorname{cl} F(A) \subset W\}.$$

Since $\operatorname{cl} F(A)$ is convex and compact, we need not consider sets Π_W for sectors $W = W_I$ having opening $|I|$ more than π, or such that I is a closed interval of length π and $0 \notin W_I$. We do consider such regions (up to $|I| = 2\pi$) for sets Λ_W.

Let us mention also that there are two cases in which it's not necessary to take a closure of $F(A)$ when defining Π_W, i.e., when

$$\Pi_W = \{A : F(A) \subset W\}.$$

The first such case occurs when W is closed, and the second when $\dim H < \infty$, and therefore $F(A)$ is closed for all A.

1. PRELIMINARY RESULTS.

It is well known [I, Theorem 6.2.1] that $\operatorname{cl} F(A) \supset \sigma(A)$. Therefore directly from the definitions it follows that

$$\Pi_W \subset \Lambda_W. \tag{1.1}$$

Of course, this inclusion is strict. Nevertheless, it turns into equality if one restricts oneself to the class $N(H)$ of normal operators on H:

$$\Pi_W \cap \mathcal{N}(H) = \Lambda_W \cap \mathcal{N}(H), \tag{1.2}$$

since for normal operators A $\operatorname{cl} F(A)$ coincides with the convex hull of $\sigma(A)$. Let H_1 be a Hilbert space isomorphic to H (in particular, $H_1 = H$). For an arbitrary invertible operator $S : H_1 \to H$

$$(S^*ASx, x) = (Ay, y) \, \|Sx\|^2, \ \text{where} \ y = \frac{Sx}{\|Sx\|},$$

and therefore

$$S^*\Pi_W S = \Pi_W \tag{1.3}$$

Finally,

$$(\Pi_W)^* = \Pi_{\overline{W}}, \tag{1.4}$$

since

$$F(A^*) = \overline{F(A)}.$$

This paper is devoted to the description of sets $\Pi_{W_1} \Pi_{W_2}$ (that is, all the products $A_1 A_2$ with $A_j \in \Pi_{W_j}$) where $W_j \ (= W_{I_j})$ are given sectors of the form (0.1).

LEMMA 1.1. *For any* W_1, W_2 : i) *the set* $\Pi_{W_1} \Pi_{W_2}$ *is closed under similarity,* ii) *the finite direct sum of operators from* $\Pi_{W_1} \Pi_{W_2}$ *is again in* $\Pi_{W_1} \Pi_{W_2}$.

PROOF. i) Let $A \in \Pi_{W_1} \Pi_{W_2}$:

$$A = A_1 A_2, \text{ with } A_j \in \Pi_{W_j} (j = 1, 2).$$

If $S : H_1 \to H$ is invertible, then

$$S^{-1}AS = (S^{-1}A_1 S^{-1*}) \cdot (S^* A_2 S).$$

According to (1.3), $S^{-1}A_1 S^{-1*} \in \Pi_{W_1}$, and $S^* A_2 S \in \Pi_{W_2}$, hence the latter equality means that $S^{-1}AS \in \Pi_{W_1} \Pi_{W_2}$.

ii) It sufficies to consider the case of two direct summands. Let $A \in \mathscr{B}(H)$, H be a direct sum of subspaces $\mathscr{L}$ and $\mathfrak{M}$ invariant with respect to A, $A|\mathscr{L} = B$, $A|\mathfrak{M} = C$, and B, $C \in \Pi_{W_1} \Pi_{W_2}$. Let $B = B_1 B_2$ and $C = C_1 C_2$ be the corresponding factorizations of B and C (B_j, $C_j \in \Pi_{W_j}$, $j = 1, 2$). Let us introduce the new Hilbert space $H' = \mathscr{L} \oplus \mathfrak{M}$ (the orthogonal sum of $\mathscr{L}$ and $\mathfrak{M}$), and let us denote by S the identity operator considered as an operator from H to H'. Then

$$A = S^{-1}(B \oplus C)S = S^{-1}(B_1 \oplus C_1)(B_2 \oplus C_2)S. \tag{1.5}$$

It's clear that *orthogonal* sums $B_1 \oplus C_1$ and $B_2 \oplus C_2$ belong to Π_{W_1} and Π_{W_2} correspondingly, and therefore $A' = (B_1 \oplus C_1)(B_2 \oplus C_2) \in \Pi_{W_1} \Pi_{W_2}$. According to (1.5), A is similar to A'. Using already proved statement i) we find out from here that $A \in \Pi_{W_1} \Pi_{W_2}$. ∎

REMARK. Classes Π_{W_1}, Π_{W_2} themselves are *not* closed under similarity (when this similarity is not unitary equivalence) or taking direct sums (when these sums are not orthogonal). The significance of Lemma 1.1 is in the fact that statements i), ii) are valid for products $\Pi_{W_1} \Pi_{W_2}$ without being valid for factors.

We will use the natural notation $W_1 W_2$ for the product

$$W = \{\zeta = \zeta_1 \zeta_2 : \zeta_j \in W_j, j = 1, 2\},$$

and $I_1 + I_2$ for the sum

$$I = \{x = x_1 + x_2 : x_j \in I_j, j = 1, 2\}.$$

It is clear that $W_{I_1} W_{I_2} = W_{I_1 + I_2}$ with $0 \in W_{I_1 + I_2}$ if and only if $0 \in W_{I_1} \cup W_{I_2}$, and that $I = I_1 + I_2$ is an interval with the left (right) endpoint equal to the sum of the left (right) endpoints of I_1, I_2; this endpoint belongs to I if and only if it belongs to *both* intervals I_1, I_2.

LEMMA 1.2. $\Pi_{W_1} \Pi_{W_2} \subset \Lambda_{W_1 W_2}$. $\tag{1.6}$

To prove (1.6) we need one more lemma. Recall that the *approximate point* spectrum $\sigma_{ap}(A)$ of an operator $A \in \mathscr{B}(H)$ is the set of all $\lambda \in \mathbb{C}$ such that there exist $x_n \in H$, $\|x_n\| = 1$, and $\|Ax_n - \lambda x_n\| \to 0$. Of course, $\sigma_{ap}(A)$ contains the set $\sigma_p(A)$ of all eigenvalues of A.

LEMMA 1.3. *Let* $A \in \mathscr{B}(H)$. *Then all angular boundary points of* F(A) *belong to the approximate point spectrum of* A.

PROOF. Translating A by λ, without loss of generality one can suppose that a point under consideration is 0. (By the way, it's the only one case in which we are interested.) From 0 being an angular boundary point of F(A) it follows that $A \in \Pi_I$ with $|I| < \pi$. Multiplying A by a nonzero scalar, one can suppose further that $I \subset \left[-\dfrac{\pi}{2} + \varepsilon, \dfrac{\pi}{2} - \varepsilon \right]$ for a sufficiently small $\varepsilon > 0$. Then

$$\left| ((\operatorname{Im} A)x, x) \right| \leq \tan\left(\frac{\pi}{2} - \varepsilon \right) \cdot ((\operatorname{Re} A)x, x), \tag{1.7}$$

in particular, $\operatorname{Re} A \geq 0$. Let us denote by X the non-negative square root of Re A. According to [K, Theorem 6.3.2], it follows from (1.7) that

$$\operatorname{Im} A = XYX,$$

where $Y \in \mathscr{B}(H)$, $Y = Y^*$.

As $0 \in \operatorname{cl} F(A)$, there exists a sequence $\{x_n\} \subset H$ such that $\|x_n\| = 1$ and $(Ax_n, x_n) \to 0$. But $(Ax_n, x_n) = ((X^2 + iXYX)x_n, x_n) = \|Xx_n\|^2 + i(YXx_n, Xx_n)$. So, $\|Xx_n\| \to 0$ simultaneously with (Ax_n, x_n). It follows from here that

$$Ax_n = (X^2 + iXYX)x_n = (X + iXY)Xx_n \to 0.$$

in other words, that $0 \in \sigma_{ap}(A)$. ∎

REMARK. Actually we proved that if λ is an angular boundary point of F(A) then $\|Ax_n - \lambda x_n\| \to 0$ for *all* sequences x_n such that $\|x_n\| = 1$ and $(Ax_n, x_n) \to \lambda$. In particular, if $\lambda \in F(A)$, and therefore there exists $x \in H$ such that $(Ax, x) = \lambda$, $\|x\| = 1$, then $Ax = \lambda x$, that is, λ is an eigenvalue of A (and x is a corresponding eigenvector). The latter result belongs to [Do].

PROOF OF LEMMA 1.2. Denote $W = W_1 W_2$. Let us consider the case of closed W_1, W_2 at first. If $|I_1| = |I_2| = \pi$ then $\Lambda_W = \mathscr{B}(H)$, and there is nothing to prove. Suppose that $|I_2| < \pi$, $A_j \in \Pi_{W_j}$ ($j = 1, 2$), but $A_1 A_2 \notin \Lambda_W$, in other words,

$$\sigma(A_1 A_2) \backslash W \neq \emptyset.$$

Then there exists $\zeta \in (\operatorname{bd} \sigma(A_1 A_2)) \backslash W$, in particular, $\zeta \neq 0$. Since the boundary of a spectrum belongs to the approximate spectrum, there exists a sequence $\{x_n\} \subset H$, $\|x_n\| = 1$, such that

$$A_1 A_2 x_n - \zeta x_n \to 0. \tag{1.8}$$

Taking a scalar product with $A_2 x_n$ yields :

$$\left(A_1 A_2 x_n, A_2 x_n \right) - \zeta \overline{\left(A_2 x_n, x_n \right)} \to 0. \tag{1.9}$$

If $(A_2 x_n, x_n) \to 0$, then $A_2 x_n \to 0$ according to Remark to Lemma 1.3, and therefore $A_1 A_2 x_n \to 0$. It contradicts (1.8), because $\zeta \neq 0$ and $\|x_n\| = 1$. Hence, 0 is not a limit of the sequence $(A_2 x_n, x_n)$. In other words, there exists a subsequence x_{n_k} such that

$$(A_2 x_{n_k}, x_{n_k}) \to \lambda_2 \neq 0. \tag{1.10}$$

It follows from (1.9) and (1.10) that

$$(A_1 A_2 x_{n_k}, A_2 x_{n_k}) \to \zeta \bar{\lambda}_2 \stackrel{\text{def}}{=} \lambda_1.$$

So, $\zeta = \lambda_1 \bar{\lambda}_2^{-1}$, where $\lambda_j \in W_j$, and therefore $\zeta \in W$. The contradiction obtained means that actually $\sigma(A_1 A_2) \subset W$.

If $|I_1| < \pi$ then, using (1.4) :

$$\Pi_{W_1} \Pi_{W_2} = \left(\left(\Pi_{W_2} \right)^* \left(\Pi_{W_1} \right)^* \right)^* = \left(\Pi_{\overline{W}_2} \Pi_{\overline{W}_1} \right)^*,$$

and, according to already proved case of (1.6),

$$= \left(\Lambda_{\overline{W}} \right)^* = \Lambda_W.$$

This completes the proof of Lemma 1.2 in the case of closed W_1, W_2.

Let now W_1, W_2 be not necessarily closed, but both contain 0. Then let us represent I_j as a union $\bigcup_{m=1}^{\infty} I_{j,m}$ of the increasing sequence of closed intervals, and put $W_{j,m} = W_{I_{j,m}} \cup \{0\}$, $W_m = W_{1,m} W_{2,m}$. According to the already proved part of Lemma 1.2,

$$\Pi_{W_{1,m}} \Pi_{W_{2,m}} \subset \Lambda_{W_m}, \tag{1.11}$$

since the sectors $W_{j,m}$ are closed. But $W_j = \bigcup_{m=1}^{\infty} W_{j,m}$, and so $\Pi_{W_j} = \bigcup_{m=1}^{\infty} \Pi_{W_{j,m}}$. Using (1.11), we derive from here that $\Pi_{W_1} \Pi_{W_2} = \bigcup_{m=1}^{\infty} \Pi_{W_{1,m}} \Pi_{W_{2,m}} \subset \bigcup_{m=1}^{\infty} \Lambda_{W_m} = \Lambda_W.$

If 0 is not supposed to be contained in W_j, then the result already proved, when applied to sectors $W_j \cup \{0\}$ instead of W_j, yields

$$\Pi_{W_1} \Pi_{W_2} \subset \Lambda_{W \cup \{0\}}. \tag{1.12}$$

When $0 \in W_1 \cup W_2$, then $W \cup \{0\} = W$, and (1.12) is exactly what we want.

Finally, if $0 \notin W_1 \cup W_2$, then both Π_{W_1} and Π_{W_2} consist of invertible operators only. Hence, all the operators in $\Pi_{W_1} \Pi_{W_2}$ are also invertible, and (1.12) actually means (1.6)

2. THE OPEN SECTOR CASE.

The natural conjecture based on Lemmas 1.1, 1.2 is that
$$\Pi_{W_1} \Pi_{W_2} = \Lambda_W, \text{ where } W = W_1 W_2 \tag{2.1}$$
Unfortunately, it is not true in general : not all operators $A \in \Lambda_W$ admit a representation
$$A = A_1 A_2, \text{ where } A_j \in \Pi_{W_j} (j = 1, 2) \tag{2.2}$$
The following lemma gives the complete description of the situation for normal $A \in \Lambda_W$.

LEMMA 2.1. *Let* $W_j = W_{I_j}$ *be the sectors of the form* (0.1), $W(= W_I) = W_1 W_2$.

Then:

i) *if* $\dim H = \infty$ *and* i_1) $0 \in W_1 \backslash W_2$, I_2 *is not closed, or* i_2) $0 \in W_2 \backslash W_1$, I_1 *is not closed, or* i_3) $W = \mathbb{C} \backslash \{0\}$, *then there exist normal operators A not having decomposition* (2.2),

ii) *in all other cases* (2.2) *holds for all* $A \in \Lambda_W \cap \mathcal{N}(H)$, *moreover,*
$$\Lambda_W \cap \mathcal{N}(H) = (\Pi_{W_1} \cap \mathcal{N}(H))(\Pi_{W_2} \cap \mathcal{N}(H)), \tag{2.3}$$
that is, A_j *in* (2.2) *can be chosen to be normal* $(j = 1, 2)$.

PROOF. i) Let us consider a normal operator A with the spectrum $\sigma(A) \subset W$ such that for every $\delta > 0$ the set $\Omega_\delta = \{re^{i\theta} : r > 0, \text{dist}(\theta, \text{bd } I) < \delta\}$ has a nonempty intersection with $\sigma(A)$. Suppose that A admits a representation (2.2), in particular, cl $F(A_2) \subset W_2$. In the case i_1) let $\ell (\ni 0)$ be that side of W_2 which is disjoint with W_2. Then cl $F(A_2) \cap \ell = \emptyset$, and from compactness of cl $F(A_2)$ it follows that there exists an interval $I_\varepsilon \subset I_2$ such that $|I_\varepsilon| < |I_2|$ and $F(A_2) \subset \text{cl } W_{I_\varepsilon} = W_\varepsilon$. Then, according to Lemma 1.2, $A \in \Pi_{W_1} \Pi_{W_3} \subset \Lambda_{W_3}$, where $W_3 = W_{I_3}, I_3 = I_1 + I_\varepsilon$, and therefore $|I_3| < |I|$. Hence, $\sigma(A) \cap \Omega_\delta = \emptyset$ when $\delta < |I| - |I_3|$, which is a contradiction. In the case i_2) the sector W_1 can be shrunk to $W_\varepsilon \supset F(A_1)$, and then the same reasoning applies. Finally, in the case i_3) $0 \notin W_1 \cup W_2$, $|I_1| = |I_2| = \pi$, and therefore both I_1 and I_2 are not closed. Now the reasoning of i_1) again can be applied.

ii) Let N be a normal operator with a spectrum $\sigma(N) \subset W$, and let
$$N = \int_{\sigma(N)} z \, d E(z)$$
be its spectral representation. Let us introduce bounded measurable functions φ_1, φ_2 defined on $\sigma(N)$ in such a way that

$$\varphi_1(z)\varphi_2(z) = z, \ \varphi_j(z) \in W_j. \tag{2.4}$$

Then the operators

$$N_j = \int_{\sigma(N)} \varphi_j(z) \ \mathrm{cl} \ E(z), \ j = 1, 2,$$

are normal, $\sigma(N_j) = c\ell \, \varphi_j(\sigma(N))$, and

$$N = N_1 N_2 \ (= N_2 N_1). \tag{2.5}$$

According to (1.2), to complete the proof of (2.3) it remains to choose φ_j in such a way that

$$\mathrm{cl} \ \varphi_j(\sigma(N)) \subset W_j (j = 1, 2). \tag{2.6}$$

The last condition is satisfied, of course, when $\varphi_j(\sigma(N))$ are closed. It is obviously true for

$\dim H < \infty$. Passing to the case of $\dim H = \infty$, let us mention that all the possibilities different
from $i_1) - i_3)$ are the following:

$\mathrm{ii}_1) \ 0 \in W_1 \cap W_2$

$\mathrm{ii}_2) \ 0 \notin W_1 \cup W_2, W \neq \mathbb{C}\backslash\{0\}$

$\mathrm{ii}_3) \ 0 \in W_1 \backslash W_2, W_2 \cup \{0\}$ is closed

$\mathrm{ii}_4) \ 0 \in W_2 \backslash W_1, W_1 \cup \{0\}$ is closed.

In all these cases it is possible to choose continuous φ_j satisfying (2.4). For example, for

$z = r \, e^{i\theta}, \theta \in I$ one can put $|\varphi_1(z)| = |\varphi_2(z)| = r^{1/2}$ in cases $\mathrm{ii}_1) - \mathrm{ii}_2)$, $|\varphi_1(z)| = r$, $|\varphi_2(z)| = 1$ in

case $\mathrm{ii}_3)$, $|\varphi_1(z)| = 1$, $|\varphi_2(z)| = r$ in case $\mathrm{ii}_4)$, and $\arg \varphi_j(z) = \dfrac{|I_j|}{|I_1|+|I_2|} \theta$ in all these cases.

Since φ_j are continuous, the sets $\varphi_j(\sigma(N))$ are closed, and therefore (2.6) holds.

      REMARK. If the sets W_j are closed, then condition (2.6) follows directly from

(2.4), and functions φ_j need not be chosen continuous. For example, when W_j are closed half

planes, one can put

$$\varphi_1(z) = \begin{cases} \zeta \\ -\zeta \end{cases}, \quad \varphi_2(z) = \begin{cases} \zeta^{-1}z \\ -\zeta^{-1}z \end{cases} \text{ when } \begin{cases} z \in W_2 \\ z \notin W_2 \end{cases},$$

where ζ is an arbitrary positive (negative) number if $\mathbb{R}_+ \, (-\mathbb{R}_+) \subset W_1$. Hence, in this case one of

multiples in the representation (2.5) can be chosen Hermitian and invertible.

      The main purpose of this section is to formulate the additional conditions on

W_1, W_2, under which the conjecture (2.1) is valid. These conditions are delivered by the

following.

      THEOREM 2.1. *Let* $W_j = W_{I_j}$ $(j = 1, 2)$ *be such that* $0 \notin W_1 \cup W_2$ *and* $I_1 + I_2$ *is*

open. Then (2.1) holds.

PROOF. Due to Lemma 1.2 we are left with a proof of the inclusion
$$\Pi_{W_1}\Pi_{W_2} \supset \Lambda_{W},$$
where $W = W_1 W_2$.

As $I_1 + I_2$ is open, at least one of the intervals I_j (let it be I_2) has positive length, and
$$I_1 + I_2 = I_1 + \text{int } I_2.$$
For this reason, without loss of generality, one can suppose that I_2 itself is open.

To prove (2.7) in this case, let us consider an arbitrary $A \in \Lambda_W$. According to [AFHV, Proposition 10.1], there exists a normal operator N and invertible operators T_n such that $\sigma(N) = \sigma(A)$, $N = \lim\limits_{n \to \infty} T_n^{-1}AT_n$. According to the statement ii) of Lemma 2.1, there exists a representation (2.5) for N with $N_j \in \mathcal{N}(H) \cap \Pi_{W_j}$ $(j = 1, 2)$. The set Π_{W_2} is open due to openness of W_2, and therefore

$$N_2 + N_1^{-1}(T_n^{-1}AT_n - N) \in \Pi_{W_2} \tag{2.8}$$

for n sufficiently large.

According to (1.3), it follows from (2.8) that
$$A_2 = T_n^{*-1}(N_2 + N_1^{-1}(T_n^{-1}AT_n - N))T_n^{-1} \in \Pi_{W_2}.$$

Again using (1.3), we find that

$$A_1 = T_n N_1 T_n^* \in \Pi_{W_1}. \tag{2.9}$$

But
$$A_1 A_2 = T_n N_1 T_n^* \cdot T_n^{*-1}(N_2 + N_1^{-1}(T_n^{-1}AT_n - N))T_n^{-1} = T_n(N_1 N_2 + T_n^{-1}AT_n - N)T_n^{-1} = A,$$
and thus $A \in \Pi_{W_1}\Pi_{W_2}$.

The case when int $I_1 \neq \emptyset$ can be considered analogously, or by taking adjoints to both parts of (2.2) in the already considered case int $I_2 \neq \emptyset$.

3. THE GENERAL CASE.

For a sector $W = W_I$ let us denote
$$W' = \{r\, e^{i\, \text{bd}\, I}, r \geq 0\}.$$
Of course, W' coincides with the bd W except the case when $|I| = 2\pi$ and I isn't open. In the latter case W' is a ray, while bd $W = \{0\}$ if $0 \notin W$ and bd $W = \emptyset$ otherwise.

For a given operator $A \in \mathcal{B}(H)$ a point λ_0 is called its *normally splitting eigenvalue* [GK], if the whole space H can be represented as a direct sum
$$H = \mathcal{L}_{\lambda_0} + \mathfrak{M}_{\lambda_0}$$

of invariant with respect to A subspaces $\mathcal{L}_{\lambda_0}$ and $\mathfrak{M}_{\lambda_0}$ such that $(A - \lambda_0 I)|\mathfrak{M}_{\lambda_0}$ is invertible, $\dim \mathcal{L}_{\lambda_0} < \infty$, and $\sigma(A|\mathcal{L}_{\lambda_0}) = \{\lambda_0\}$.

Normally splitting eigenvalues are isolated points of $\sigma(A)$. We'll say that a normally splitting eigenvalue $\lambda_0 \in \sigma(A)$ has *height* k if $k \times k$ is the maximal size of Jordan blocks of A corresponding to λ_0. Let's call normally splitting eigenvalue λ_0 *simple* if it has height 1, in other words, if $A|\mathcal{L}_{\lambda_0} = \lambda_0 I$.

THEOREM 3.1. *Let* $W_1 = W_{I_1}$, $W_2 = W_{I_2}$, *and* $W = W_1 W_2$. *Let* $A \in \mathcal{B}(H)$ *be such that* $\sigma(A) \cap W'$ *consists of normally splitting eigenvalues only.*

1) *If* W_1, W_2 *both are closed half planes, then* $A \in \Pi_{W_1} \Pi_{W_2}$,

2) *If one of the sectors* W_1, W_2 *is a closed half plane, another one has an opening strictly less than* π *and contains* 0*, then* $A \in \Pi_{W_1} \Pi_{W_2}$ *if and only if* $A \in \Lambda_W$, *all the eigenvalues* $\lambda \in \sigma(A) \cap W' \backslash \{0\}$ *are simple, and* 0 *has height not greater than* 2,

3) *For all other admissible choices of* W_1, W_2 $A \in \Pi_{W_1} \Pi_{W_2}$ *if and only if* $A \in \Lambda_W$, *and all the eigenvalues* $\lambda \in \sigma(A) \cap W'$ *are simple.*

PROOF. All the statements of Theorem 3.1 are invariant under rotations of W_j. Hence without loss of generality we can suppose that left endpoints of intervals I_1, I_2 (and, therefore, of their sum I) are equal 0.

SUFFICIENCY. Due to requirements on A, there is only a finite number of points in $\sigma(A) \cap W'$. If $\lambda_1, \ldots, \lambda_n$ are all of them, then, splitting them out one by one, we obtain the decomposition

$$H = \mathcal{L}_{\lambda_1} \dotplus \mathcal{L}_{\lambda_2} \dotplus \cdots \dotplus \mathcal{L}_{\lambda_n} \dotplus \mathfrak{M},$$

where all the subspaces $\mathcal{L}_{\lambda_1}, \ldots, \mathcal{L}_{\lambda_n}$, $\mathfrak{M}$ are invariant with respect to A, $(A - \lambda_k I)^{m_k}\mathcal{L}_{\lambda_k} = \{0\}$ (m_k is a geometrical multiplicity of an eigenvalue λ_k), and $\sigma(A|\mathfrak{M}) = \sigma(A)\backslash W'$.

According to the statement ii) of Lemma 1.1, it suffices to show that all the direct summands $A|\mathfrak{M}$, $A|\mathcal{L}_{\lambda_j}$ ($j = 1, \ldots, n$) belong to $\Pi_{W_1} \Pi_{W_2}$.

The summand $A|\mathfrak{M}$ has its spectrum in $W\backslash W'$. Being compact, $\sigma(A|\mathfrak{M})$ is then actually located in an open sector $W^0 \subset W$ such that $W^0 \cap W' = \emptyset$. Let us choose subintervals $I_j^0 \subseteq I_j$ ($j = 1, 2$) such that $W_{I_1^0} W_{I_2^0} = W^0$. Then, according to Theorem 2.1,

$$A|\mathfrak{M} \in \Pi_{W_{I_1^0}} \Pi_{W_{I_2^0}} \subset \Pi_{W_1} \Pi_{W_2}.$$

The summands $A|\mathcal{L}_j$ corresponding to simple eigenvalues are normal operators $\lambda_j I$, and therefore they are taken care of by Lemma 2.1. This completes the proof of sufficiency in case 3). Dealing with multiple eigenvalues in cases 1) and 2), according to Lemma 1.1 we can suppose

that $A|\mathcal{L}$ is just a Jordan $k \times k$ block $J_{\lambda,k} = \begin{bmatrix} \lambda & 1 & & \\ & \ddots & \ddots & \\ & & \ddots & 1 \\ 0 & & & \lambda \end{bmatrix}$, $\lambda \in W'$, and the

corresponding basis is orthonormal.

In case 1) W_1 and W_2 are closed upper half planes, and therefore W' is the non-negative ray $\mathbb{R}_+$. The desired representation of $J_{\lambda,k}$ is then delivered by

$$J_{\lambda,k} = \begin{bmatrix} & & 1 \\ & \cdot^{\cdot^{\cdot}} & \\ 1 & & \end{bmatrix} \begin{bmatrix} & & & \lambda \\ & & \lambda & 1 \\ & \cdot^{\cdot^{\cdot}} & \cdot^{\cdot^{\cdot}} & \\ \lambda & 1 & & \end{bmatrix}, \tag{3.1}$$

where both factors are Hermitian.

In case 2) we are left with 0 eigenvalues of height 2. If W_1 is the upper half plane, and ζ is a point in W_2, $|\zeta| = 1$, then one of suitable factorizations is delivered by

$$\begin{bmatrix} 0 & 1 \\ 0 & 0 \end{bmatrix} = \begin{bmatrix} 0 & \bar{\zeta} \\ \zeta & 0 \end{bmatrix}\begin{bmatrix} 0 & 0 \\ 0 & \zeta \end{bmatrix}.$$

If W_2 is the upper half plane, $\zeta \in W_1$, then, of course,

$$\begin{bmatrix} 0 & 1 \\ 0 & 0 \end{bmatrix} = \begin{bmatrix} \zeta & 0 \\ 0 & 0 \end{bmatrix}\begin{bmatrix} 0 & \bar{\zeta} \\ \zeta & 0 \end{bmatrix}$$

works.

We need two more auxiliary results before the proof of necessity in Theorem 3.1 can be carried out.

LEMMA 3.1. *Let for* $A \in \Pi_W$ *there exists* $x \in H$ *such that* $\|x\| = 1$ *and* $(Ax, x) \in \mathrm{bd}\, W$. *Then for all* $y \in H$, $y \perp x$: $|(Ax, y)| = |(Ay, x)|$, *and* $\arg(Ax, y) + \arg(Ay, x) = 2\arg(Ax, x)$ $(= 0$ *if* $(Ax, x) = 0)$.

PROOF. The statement of Lemma 3.1 is invariant under multiplication of A by nonzero scalars. Therefore without loss of generality one may suppose that

$$\zeta = (Ax, x) \geq 0$$

and $F(A)$ is located in the upper or lower half plane, in other words, Im A is semi-definite. We can also suppose that $\|y\| = 1$. Let us consider the 2-dimensional subspace $H_0 = \text{Span}\ \{x, y\}$. Let P be the orthoprojection of H onto H_0, and

$$A_0 = PA|H_0.$$

Then A_0 has a matrix representation $\begin{pmatrix} \xi & \zeta_1 \\ \zeta_2 & \eta \end{pmatrix}$, where $\zeta_1 = (Ax, y), \eta = (Ay, y), \zeta_2 = (x, Ay)$,

with respect to a basis $\{x, y\}$. Therefore, a matrix representation for Im A_0 is $\begin{pmatrix} 0 & z \\ \bar{z} & \text{Im } \eta \end{pmatrix}$ with

$z = \dfrac{\zeta_1 - \zeta_2}{2i}$. But Im A_0 is semi-definite together with Im A. It means that $z = 0$, i.e.,

$(Ax, y) = \overline{(Ay, x)}$.

Recall (see, for example, [K, Chapter 4] or [CPU]) that an operator $A \in \mathscr{B}(H)$ is called semi-Fredholm if $\mathscr{R}(A)$ is closed and at least one number $n(A) = \dim \text{Ker } A$ or $d(A) = \dim \text{Ker } A^*$ is finite; the difference ind $A = n(A) - d(A)$ is the so called index of A. A semi-Fredholm operator A is called Fredholm if and only if ind A (or, in other words, both $n(A)$ and $d(A)$) are finite.

LEMMA 3.2. *If A is a semi-Fredholm operator, and $A \in \Pi_W$ then H is an orthogonal sum of* Ker A *and* $\mathscr{R}(A)$:

$$H = \text{Ker } A \oplus \mathscr{R}(A),$$

and $A|\mathscr{R}(A)$ is invertible.

PROOF. Without loss of generality we can suppose that W lies in the (closed) upper half plane, in other words, that Im $A \geq 0$. In any neighborhood of A there are invertible operators. It follows from here and stability of the index ([K, Theorem 4.5.17] or [CPU, Theorem 4.4.1]) that ind $A = 0$.

Suppose that Ker $A \neq$ Ker A^2. Then there exist $x, y \in H$ such that $Ay = x, Ax = 0$, $y \perp x$ and $\|x\| = 1$. According to Lemma 3.1, $(Ax, y) = \overline{(Ay, x)}$. But $(Ax, y) = (0, y) = 0$, and $(Ay, x) = \|x\|^2 = 1$. The contradiction obtained means that Ker $A^2 =$ Ker A. According to the index theorem ([CPU, Theorem 3.2.7]), the operator A^2 is Fredholm, and ind $A^2 = 2$ ind $A = 0$. Hence, $d(A^2) = n(A^2) = n(A) = d(A)$, that is, $\mathscr{R}(A^2) = \mathscr{R}(A)$ and Ker $A \cap \mathscr{R}(A) = \{0\}$. It follows from here that $H = H_1 + H_2$, where $H_1 =$ Ker A is finite dimensional, $H_2 = \mathscr{R}(A)$, and $A|H_2$ is invertible.

Suppose that H_1 is not orthogonal to H_2, in other words, that there exist $x \in H_1$, $y \in H_2$ such that $(Ay, x) \neq 0$. Then, by choosing $\lambda \in \mathbb{C}$,

$$(A(\lambda x + y), \lambda x + y) = (Ay, y) + \bar{\lambda}(Ay, x)$$

can be made an arbitrary complex number, which contradicts the condition Im $A \geq 0$.

PROOF OF THEOREM 3.1. NECESSITY. There is nothing to prove in case 1). Hence, we can suppose that in representation (2.2) at least one of sectors W_1, W_2 is not a closed half plane. We can switch W_1 and W_2 taking adjoints, and therefore without loss of generality we can let it be W_2. Finally, $F(A_2)$ is then located in a sector with an opening smaller than π and, again without loss of generality, $W_2 = W_{I_2}$ with $|I_2| < \pi$.

Necessity of condition $A \in \Lambda_W$ follows from Lemma 1.2. Therefore, we are left with a proof of the following.

LEMMA 3.3. *Let* $W_1 = W_{I_1}$, $W_2 = W_{I_2}$ *be sectors such that* $|I_1| \leq \pi$, $|I_2| < \pi$, $W = W_1 W_2$, *and* $\sigma(A) \cap$ bd W *consists of normally splitting eigenvalues only. Then for representation (2.2) to exist it is necessary that:*

    i) *all eigenvalues* $\lambda \in \sigma(A) \cap$ bd $W \backslash \{0\}$ *are simple*

    ii) *0 is an eigenvalue of* A *with height not greater than 2*

    iii) *If* $|I_1| < \pi$ *or* A_2 *is invertible, then the height of 0 is also not greater than 1.*

PROOF. First of all, A is Fredholm operator because 0 is its normally splitting eigenvalue. Hence [CPU, Corollary 1.3.6], in (2.2) A_1 and A_2 are semi-Fredholm. According to Lemma 3.2, from $A_j \in \Pi_{W_j}$ it follows then that A_1 and A_2 are both Fredholm (with index 0), $H = \text{Ker } A_1 \oplus \mathscr{R}(A_1)$, and $A_1 | \mathscr{R}(A_1)$ is invertible.

Let us denote by $A_1^{(-1)}$ the generalized inverse to A_1, that is, $(x_1 \in \text{Ker } A_1$, $x_2 \in \mathscr{R}(A_1))$, where $y_2 \in \mathscr{R}(A_1)$, $A_1 y_2 = x_2$. Then $P = A_1^{(-1)} A_1$ is an orthoprojection onto $\mathscr{R}(A_1)$.

    i) Suppose $\lambda \neq 0$ is an eigenvalue of A, $\lambda \in$ bd W. Then there exists $x \in H$, $x \neq 0$, such that

$$A_1 A_2 x = \lambda x. \tag{3.2}$$

Since $\lambda \neq 0$, $x = \lambda^{-1} A_1 A_2 x \in \mathscr{R}(A_1)$.

Applying $A_1^{(-1)}$ to (3.2) yields:

$$PA_2 x = \lambda A_1^{(-1)} x,$$

or

$$A_2 x = \lambda A_1^{(-1)} x + z_1, \, z_1 \in \operatorname{Ker} A_1. \tag{3.3}$$

Taking a scalar product with x yields

$$(A_2 x, x) = \lambda (A_1^{(-1)} x, x) \tag{3.4}$$

since $(z_1, x) = 0$ due to Lemma 3.2. According to (3.2), $A_2 x \neq 0$. Then (see Remark to Lemma 1.3) $(A_2 x, x) \neq 0$, and therefore in (3.4) $(A_1^{(-1)} x, x) \neq 0$. Then it follows from (3.4) that

$$\lambda = (A_1^{(-1)} x, x)^{-1} (A_2 x, x). \tag{3.5}$$

It's clear from the definition of $A_1^{(-1)}$ that $A_1^{(-1)} \in \Pi_{\overline{W_1}}$. Hence, $(A_1^{(-1)} x, x)^{-1} \in W_1$, $(A_2 x, x) \in W_2$, and, since $\lambda \in \operatorname{bd} W$, the representation (3.5) is possible only if

$$(A_1^{(-1)} x, x)^{-1} \in \operatorname{bd} W_1, \, (A_2 x, x) \in \operatorname{bd} W_2.$$

Multiplying A_1 and A_2 by nonzero scalars if necessary, we can, without loss of generality, suppose that

$$(A_1^{(-1)} x, x) > 0, \, (A_2 x, x) > 0$$

(in this case, of course, automatically $\lambda > 0$).

If λ is not a simple eigenvalue of $A_1 A_2$, then there exists $y \perp x$ such that

$$A_1 A_2 y - \lambda y = x. \tag{3.6}$$

It's clear from (3.6) that $y \in \mathcal{R}(A_1)$, and applying $A_1^{(-1)}$ to (3.6) yields:

$$A_2 y = A_1^{(-1)} x + \lambda A_1^{(-1)} y + z_2, \, z_2 \in \operatorname{Ker} A_1,$$

or, taking (3.3) into consideration:

$$A_2 y = \lambda^{-1} A_2 x + \lambda A_1^{(-1)} y + z_3, \, z_3 \in \operatorname{Ker} A_1. \tag{3.7}$$

Taking a scalar product of both parts of (2.7) by y, and both parts of (3.7) by x yields correspondingly:

$$(A_2 x, y) = \lambda (A_1^{(-1)} x, y), (A_2 y, x) = \lambda^{-1} (A_2 x, x) + \lambda (A_1^{(-1)} y, x).$$

But, accordingly to Lemma 2.1, $(A_2 x, y) = \overline{(A_2 y, x)}$, $(A_1^{(-1)} x, y) = \overline{(A_1^{(-1)} y, x)}$, and therefore $(A_2 x, x) = 0$, which is a contradiction.

Hence, the eigenvalue λ is simple, and the proof of i) is complete.

iii) Suppose now that $\lambda = 0$ is a nonsimple eigenvalue of A. Then there exist $x, y \in H$ such that $x \neq 0$, $y \perp x$, $A_1 A_2 y = x$, and $A_1 A_2 x = 0$.

Therefore, $A_2 x \in \operatorname{Ker} A_1$ and $x = A_1 A_2 y \in \operatorname{Im} A_1$. According to Lemma 3.2 $\operatorname{Ker} A_1 \perp \mathcal{R}(A_1)$, and so $(A_2 x, x) = 0$.

Applying now Remark to Lemma 1.3 again, we find that $A_2 x = 0$, that is, $x \in \operatorname{Ker} A_2$. We want to conclude from here, using additional conditions of statement iii), that $x = 0$. It is clear when A_2 is invertible. If this is not the case, but $|I_1| < \pi$, let us apply Lemma 3.2 to operator A_2, concluding that $\operatorname{Ker} A_2$ is orthogonal to $\mathscr{R}(A_2)$. In particular, $x = A_1 A_2 y \perp A_2 y$, or $(A_1 A_2 y, A_2 y) = 0$. When $|I_1| < \pi$ it follows from here and Remark to Lemma 1.3 that $A_2 y = 0$. Hence, $x = A_1 A_2 y = 0$. This contradiction completes the proof of iii).

ii) According to Lemma 3.2, $H = \operatorname{Ker} A_2 \oplus \mathscr{R}(A_2)$, and an operator $Y = A_2 | \mathscr{R}(A_2)$ is invertible. Since 0 is an angular point of W_2, and $0 \notin (\sigma(Y)$, 0 is not a point of $F(Y)$ either (Lemma 1.3). Let

$$A_1 = \begin{bmatrix} X_{11} & X_{12} \\ X_{21} & X_{22} \end{bmatrix}, \quad A_2 = \begin{bmatrix} 0 & 0 \\ 0 & Y \end{bmatrix}$$

be the matrix representations corresponding to the above mentioned decomposition of H.

Further, $F(X_{22}) \subset F(A_1) \subset W_1$, and, being a compression of A_1 onto a subspace of finite codimension, the operator X_{22} is Fredholm. Applying Lemma 3.2 to X_{22}, we obtain the decomposition

$$\mathscr{R}(A_2) = \operatorname{Ker} X_{22} \oplus \mathscr{R}(X_{22})$$

with respect to which

$$X_{22} = \begin{pmatrix} 0 & 0 \\ 0 & Z \end{pmatrix}$$

with Z invertible. In the corresponding representation

$$Y = \begin{pmatrix} Y_{11} & Y_{12} \\ Y_{21} & Y_{22} \end{pmatrix}$$

of Y $F(Y_{22}) \subset F(Y)$, and therefore Y_{22} is invertible. Direct computations show that

$$(A_1 A_2)^2 = \begin{bmatrix} I & 0 \\ 0 & Y^{-1} \end{bmatrix} \begin{bmatrix} 0 & X_{12} Y X_{22} \\ 0 & (Y X_{22})^2 \end{bmatrix} \begin{bmatrix} I & 0 \\ 0 & Y \end{bmatrix},$$

and, with respect to the decomposition

$$H = \operatorname{Ker} A_2 \oplus \operatorname{Ker} X_{22} \oplus \mathscr{R}(X_{22}),$$

$$\begin{bmatrix} 0 & X_{12} Y X_{22} \\ 0 & (Y X_{22})^2 \end{bmatrix} = \begin{bmatrix} 0 & 0 & * \\ 0 & 0 & * \\ 0 & 0 & (Y_{22} Z)^2 \end{bmatrix},$$

where * stands for blocks exact values of which are not important for our purposes. Hence, the operator $(A_1 A_2)^2$ is similar to $\begin{bmatrix} 0 & S \\ 0 & T \end{bmatrix}$, where T is invertible.

Since operators $\begin{bmatrix} 0 & S \\ 0 & T \end{bmatrix}$ and $\begin{bmatrix} 0 & S \\ 0 & T \end{bmatrix}^2$ have the same kernel, $\lambda = 0$ is a simple

eigenvalue for $(A_1 A_2)^2$. In other words, the height of 0 as an eigenvalue of $A_1 A_2 = A$ does not exceed 2.

It completes the proof of Lemma 3.3, and, therefore, of Theorem 3.1.

The first statement of Theorem 3.1 can be strengthened the following way.

THEOREM 3.2. *Let for an operator* $A \in \mathcal{B}(H)$ *there exists a ray* $\ell = \{re^{i\theta}, r \geq 0\}$ *such that all the points of* $\sigma(A) \cap \ell$, *if any, are normally splitting eigenvalues. Then there exists a representation* $A = A_1 A_2$ *such that* A_1 *is Hermitian and invertible, and* $F(A_2)$ *is located in one of closed half planes into which* $\mathbb{C}$ *is divided by the line* $\ell \cup (-\ell)$.

PROOF. Multiplying A and A_2 by $e^{-i\theta}$, one can reduce the general case to the situation when $\theta = 0$. Then the statement i) of Theorem 3.1 can be applied when $W_1 = W_2 = \{z : \text{Im } z \geq 0\}$. Following the proof of this statement, taking into consideration Remark to Lemma 2.1, and using formulas (3.1) and (2.9) for the blocks of A having spectrum on and outside of ℓ correspondingly, one can easily check that in the factorization $A = A_1 A_2$ delivered by Theorem 3.1, A_1 can be chosen to be Hermitian and invertible.

4. CONCLUDING REMARKS.

1. According to Theorem 2.1, the equality (2.1) holds when $W = W_1 W_2$ is open, except the cases $W = \mathbb{C}$ or $W = \mathbb{C}\setminus\{0\}$. The following example shows that these two cases really have to be excluded.

Let A be an arbitrary invertible operator such that the set

$$\{ \lambda : A - \lambda I \text{ is Fredholm, } \text{ind}(A - \lambda I) \neq 0 \}$$

contains a closed curve γ passing around 0 (one can choose $\gamma = \{z : |z| = 1\}$ for simplicity). Then, according to the Stability Theorem for Fredholm operators (See [K, Theorem 4.5.17]) for all $\lambda \in \gamma$ there exists $\varepsilon_\lambda > 0$ such that for all $B \in \mathcal{B}(H)$, $\|B - A\| < \varepsilon_\lambda$ implies that $B - \lambda I$ is a Fredholm operator with nonzero index. Since $\|B + (\lambda - \lambda')I - A\| \leq \|B - A\| + |\lambda - \lambda'|$, for all λ' from the $\frac{\varepsilon_\lambda}{2}$-neighborhood of λ $\varepsilon_{\lambda'}$ can be chosen to be not less than $\frac{\varepsilon_\lambda}{2}$. Selecting a finite subcovering from the covering $\gamma \subset \bigcup_{\lambda \in \gamma} \{\zeta : |\zeta - \lambda| < \varepsilon_\lambda \}$, one can find $\varepsilon > 0$ such that for all

$\lambda \in \gamma$ and $\|B - A\| < \varepsilon$ $B - \lambda I$ is a Fredholm operator with nonzero index. In particular, $\sigma(B) \supset \gamma$.

From the other hand, if A admits a representation (2.2) for a certain pair of half planes W_1, W_2, then by an arbitrary small perturbations of A_1, A_2 we can obtain operators A_1', A_2' having their fields of values in sectors with openings less than π. According to Lemma 1.2, it means that $\sigma(A)$ is located in a sector having opening strictly less than 2π. This contradicts the property $\sigma(B) \supset \gamma$ mentioned above.

2. The condition on $\sigma(A) \cap W'$ imposed in Section 3 seems to be unnecessary restrictive. Nevertheless, the following examples show that it cannot be omitted (neither in sufficient nor in necessary part of the statement).

For a given $\alpha \in (0, \frac{\pi}{2})$ let $K = \dfrac{\pi \tan \alpha}{2\ell} \, \mathscr{T}$, where $\mathscr{T}$ is a classical Volterra operator

$$(\mathscr{T}f)(x) = 2i \int_x^\ell f(t)dt$$

in the space $H = L_2(0, \ell)$. Then (see, for example, [D]) $\|K\| = \tan \alpha$, Im K is a one dimensional non-negative operator, and therefore for $A = I + K$ $F(A) \subset W_{[0,\alpha]}$. Putting $W_1 = W_{[0,0]}$, $W_2 = W_{[0,\alpha]}$, we see that the representation $A = I(I + K)$ has all the properties of (2.2) in spite of the fact that in this case $W' \supset \mathbb{R}_+$, and 1 is a point of $\sigma(A) \cap W'$ which is not a normally splitting eigenvalue.

From another point of view, analyzing the proof of Lemma 3.3 one can see that statement i) of this lemma actually can be sharpened the following way:

If A is a Fredholm operator having a 2-dimensional invariant subspace $\mathscr{L}$ such that $A|\mathscr{L} = \begin{pmatrix} \lambda & 1 \\ 0 & \lambda \end{pmatrix}$, $\lambda \in W'$, then representation (2.2) does not exist.

For $W_1 = W_2 = W_{[0,\alpha)}$ a possible example of such an A is a direct sum of an infinite dimensional identity operator with a Jordan block $J_{1,2} = \begin{pmatrix} 1 & 1 \\ 0 & 1 \end{pmatrix}$. For such an A $\sigma(A) = \{1\} \subset W'$, and 1 is not a normally splitting eigenvalue.

Hence, the complete description of the sets Π_{W_1} Π_{W_2} is not clear when conditions of Theorem 3.1 are not satisfied.

One very simple partial result, supplementing Theorem 3.1, is provided by the following.

THEOREM 4.1. *Let* $I_j = \{\theta_j\}$, $j = 1, 2$, *and* $0 \notin W_1 \cap W_2$. *Then* $\Pi_{W_1} \Pi_{W_2} = e^{i(\theta_1 + \theta_2)}P$, *where* P *is a class of all operators* $A \in \mathscr{B}(H)$ *similar to Hermitian nonnegative.*

PROOF. Without loss of generality one can suppose that $\theta_1 = \theta_2 = 0$. The latter case was considered, for instance, in [W2].

3. Finally, let us consider the case dim $H < \infty$. In this case $\sigma(A)$ consists of normally splitting eigenvalues only, and therefore Theorem 3.1 gives the complete description of $\Pi_{W_1} \Pi_{W_2}$.

In particular, if $W_1 = \mathbb{R}_+ \backslash \{0\}$, $W_2 = \mathrm{cl}\, \Pi_+ = \{z : \mathrm{Re}\, z \geq 0\}$, then statement 3) of Theorem 3.1 yields that an $n \times n$ matrix A admits a representation
$$A = A_1 A_2 \tag{4.1}$$
with A_1 positive and $F(A_2) \subset \mathrm{cl}\, \Pi_+$ if and only if $\sigma(A) \subset \mathrm{cl}\, \Pi_+$ and $\sigma(A) \cap i\mathbb{R}$ consists of simple eigenvalues only. This is a result of D. Carlson and H. Schneider [CS, Corollary III.1].

If $W_1 = W_2 = \mathbb{R}_+$, then the same statement yields that a representation (4.1) with Hermitian nonnegative A_1 and A_2 exists if and only if all the eigenvalues of A are nonnegative and simple, in other words, when A is similar to a nonnegative matrix. This is a result of P. Y. Wu [W1, Theorem 2.2].

According to statement 2) of Theorem 3.1, a representation (4.1) with nonnegative Hermitian A_1 and $F(A_2) \subset \mathrm{cl}\, \Pi_+$ exists if and only if $\sigma(A) \subset \mathrm{cl}\, \Pi_+$, all the eigenvalues of A lying on $i\mathbb{R} \backslash \{0\}$ are simple, and 0 as an eigenvalue of A has a multiplicity not exceeding 2. The latter property of products (4.1) with A_1 nonnegative Hermitian and A_2 Hermitian is established in [HH].

Choosing $W_1 = \mathbb{R}_+ \backslash \{0\}$, $W_2 = \Pi_+ \cup \{0\}$, we obtain from statement 3) of Theorem 3.1 that all products (4.1) have their spectrum in Π_+, if A_1 is positive Hermitian, A_2 is nonsingular, and $F(A_2) \subset \Pi_+ \cup \{0\}$. This is a "sufficient" part of the result from [J2] saying that a matrix X is H-stable (that is, a spectrum of its product by an arbitrary positive matrix remains in Π_+) if and only if X is nonsingular, and $F(X) \subset \Pi_+ \cup \{0\}$.

Theorem 3.2 means that an arbitrary $n \times n$ matrix A admits a representation (4.1) with A_1 Hermitian and invertible, and $F(A_2)$ lying in an a priori prescribed closed half plane. In particular, one can choose A_2 such that $\mathrm{Re}(A_2) \geq 0$. This is Corollary II.1 in [CS].

Finally, Theorem 4.1 for dim $H < \infty$ means that a matrix A can be represented as a product of two positive Hermitian matrices if and only if all its eigenvalues are positive and simple. This is a classical fact, see [HJ], Theorem 7.6.3. See also [C, Wi] for related results.

REFERENCES

[AFHV] C. Apostol, L. Fialkow, D. Herrero, D. Voiculescu. Approximation of Hilbert space operators, Pitman, 2(1984).

[BJ] C. Ballantine and C. R. Johnson. Accretive Matrix Products, Linear and Multilinear Algebra 3(1975), 169-185.

[C] D. Carlson, A New Criterion for H-Stability of Complex Matrices, Linear Algebra and its Applications, 1(1968), 59-64.

[CPU] S. R. Caradus, W. E. Pfaffenberger, B. Yood. Calkin Algebras and Algebras of Operators on Banach Spaces. Lecture Notes in Pure and Applied Mathematics, Marcel Dekker, 9(1974).

[CS] D. Carlson, H. Schneider. Inertia Theorems for Matrices: The semidefinite case, J. Math. Anal. and Appl. 6(1963), 430-446.

[D] K. R. Davidson. Nest Algebras. Pitman (1988).

[Do] W. Donoghue, On the Numerical Range of a Bounded Operator, Michigan Mathematics Journal 4(1957), 261-263.

[DK] Yu L. Daletski, M. G. Krein. Stability of Solutions of Differential Equations in Banach Space, Transl. Math. Monographs, AMS, Providence, RI, 43(1974).

[GK] I. C. Gohberg, M. G. Krein. Introduction to the Theory of Linear Nonselfadjoint Operators. Transl. Math. Monographs, AMS, Providence, RI, 18(1969).

[HH] Y. Hong and R. Horn. The Jordan Canonical Form of a Product of a Hermitian and a Positive Semidefinite Matrix, Linear Algebra and Its Applications 147(1991), 373-386.

[HJ] R. Horn and C. R. Johnson. Matrix Analysis. Cambridge University Press (1985).

[HJ2] R. Horn and C. R. Johnson. Topics in Matrix Analysis. Cambridge University Press (1991).

[I] V. Istrăţescu. Introduction to Linear Operator Theory. Marcel Dekker (1981).

[J1] C. R. Johnson. A Lyapunov Theorem for Angular Cones, J. Res. NBS 78B(1974), 7-10.

[J2] C. R. Johnson. The Field of Values and Spectra of Positive Definite Multiples, J. Res. NBS 78B(1974), 197-8.

[K] T. Kato. Perturbation Theory for Linear Operators. Springer Verlag (1966).

[N] M. Neumann. Weak Stability for Matrices, Linear and Multilinear Algebra 7(1979), 257-262.

[OS] A. Ostrowski and H. Schneider. Some Theorems on the Inertia of General Matrices, J. Math. Anal. Appl. 4(1962), 72-84.

[W1] P. Y. Wu. Products of Positive Semidefinite Matrices, Linear Algebra and Its Applications 111(1988), 53-61.

[W2] P. Y. Wu. The Operator Factorization Problems, Linear Algebra and Its Applications 117(1989), 35-63.

[Wi] H. Wielandt, On the Eigenvalues of A + B and AB, J. Research NBS 77B(1973), 61-63.

Department of Mathematics
The College of William and Mary
Post Office Box 8795
Williamsburg, VA 23187-8795
U.S.A.

AMS Subject Classification: Primary 47A12, Secondary 15A23, 15A60, 47A68

Operator Theory:
Advances and Applications, Vol. 62
© 1993 Birkhäuser Verlag Basel

ON THE COEFFICIENTS OF RIEMANN MAPPINGS
OF THE UNIT DISK INTO ITSELF

Kin Y. Li and James Rovnyak[1]

To Tsuyoshi Ando, with best wishes on the occasion of his 60-th birthday.

Two methods to construct inequalities are compared and investigated for their potential to characterize initial segments of coefficients. One is based on special function theory as in the proof of the Bieberbach conjecture and its power generalizations. While this method produces the best coefficient estimates to date, it is shown by an example that the method cannot, in its present form, characterize initial segments of coefficients.

1 Introduction

By a *Riemann mapping* we mean a function which is analytic and one-to-one on the open unit disk $|z| < 1$. Such a function is said to be *normalized* if it has value zero and positive derivative at the origin. We are concerned with normalized Riemann mappings of the unit disk into itself. The conformal contraction of the unit disk by such a mapping, say $B(z)$, implies analytical properties of the function. Of particular interest are inequalities involving the coefficients in the expansion $B(z) = B_1 z + B_2 z^2 + \cdots$. Such inequalities impose constraints on the numbers $B_1, \ldots, B_r$. It is an open problem to find necessary and sufficient conditions that given numbers arise as an initial segment of coefficients of a normalized Riemann mapping of the unit disk into itself.

We compare two existing methods which at first glance appear to be different but on closer inspection turn out to have a similar form. The first is of a geometric nature and draws on an analogy with the Carathéodory-Fejér interpolation problem [3,4]. It is based on a study of operators of the form $T : f(z) \mapsto f(B(z))$ acting on generalized power series $f(z) = \sum_{n=1}^{\infty} a_n z^{\nu+n}$. If $B(z)$ is a normalized Riemann mapping of the unit disk into itself, then substitution by $B(z)$ is contractive relative to certain quadratic forms. A proof using Loewner's differential equation is given in [15]. In this paper we give a new and more elementary argument in Theorem 2.2. Theorem 2.2 implies necessary conditions that given numbers $B_1, \ldots, B_r$ $(B_1 > 0)$ occur as the initial segment of coefficients of a normalized Riemann mapping of the unit disk into itself. When $r = 2$, the conditions have a symmetry, and this is useful in proving that the necessary conditions are sufficient

---

[1]The authors were supported by a grant from the National Science Foundation.

[10,11]. We extend the symmetry to an arbitrary positive integer r, but it is unknown if the necessary conditions are sufficient in general.

The second method is analytical and based on de Branges' proof of Milin's conjecture [5] which depends on Loewner's differential equation and special function theory. We use, in fact, stronger inequalities for powers or Riemann mappings [6,15]. Nevertheless the resulting conditions on coefficients have a similar character to those derived from the first method [2]. We answer, in the negative, the question if necessary conditions derived in this way are capable of characterizing initial segments of coefficients. This is done by exhibiting numbers B_1, B_2, B_3 $(B_1 > 0)$ which satisfy all conditions obtainable by the method in its present form, yet which are not the first three coefficients of a normalized Riemann mapping of the unit disk into itself because they do not satisfy the conditions obtained by the first method.

The main part of this work was done while the first author was a Whyburn Postdoctoral Instructor at the University of Virginia. The second author thanks the Mathematics Department of University College London for its hospitality during a sabbatical stay when the work was completed.

2   Contractive substitution property

For any real ν and sequence $\sigma = \{\sigma_1, \sigma_2, \ldots\}$ of real numbers, the *Grunsky space* $\mathcal{G}_\sigma^\nu$ is the space of all generalized power series $f(z) = \sum_{n=1}^\infty a_n z^{\nu+n}$ with complex coefficients such that $\sum_{n=1}^\infty |(\nu+n)\sigma_n||a_n|^2 < \infty$. A scalar product is defined in the space by

$$\langle f(z), g(z) \rangle_{\mathcal{G}_\sigma^\nu} = \sum_{n=1}^\infty (\nu+n)\sigma_n a_n \overline{c_n}$$

if $f(z) = \sum_{n=1}^\infty a_n z^{\nu+n}$ and $g(z) = \sum_{n=1}^\infty c_n z^{\nu+n}$. We identify two elements if their coefficients coincide for all indices n such that $(\nu+n)\sigma_n \neq 0$. Then $\mathcal{G}_\sigma^\nu$ has the structure of a Kreĭn space [1]. Let $\mathcal{D}_\nu$ be $\mathcal{G}_\sigma^\nu$ with $\sigma_n = 1$ for all n, so $\mathcal{D}_0$ is the *Dirichlet space*. For any real number ν and positive integer r, let $\mathcal{D}_\nu^r$ be $\mathcal{G}_\sigma^\nu$ with $\sigma_n = 1$ or 0 according as $1 \leq n \leq r$ or $n > r$.

THEOREM 2.1 *Let $\mathcal{H}$ and $\mathcal{K}$ be Kreĭn spaces having finite and equal negative indices, and let T be a continuous operator on $\mathcal{H}$ to $\mathcal{K}$. If T is a contraction, so is T^*.*

Our notation for Kreĭn spaces follows [8]. In Theorem 2.1, the *adjoint* T^* of T is defined so that $\langle Tf, g \rangle_\mathcal{K} = \langle f, T^*g \rangle_\mathcal{H}$ for all f in $\mathcal{H}$ and g in $\mathcal{K}$. We call T a *contraction* if $\langle Tf, Tf \rangle_\mathcal{K} \leq \langle f, f \rangle_\mathcal{H}$ for all f in $\mathcal{H}$. Theorem 2.1 is a well-known result proved in [8, Theorem 1.3.7] (for the essential case $\mathcal{H} = \mathcal{K}$).

We apply Theorem 2.1 to operators defined by substituting a formal power series $B(z) = B_1 z + B_2 z^2 + \cdots$ $(B_1 > 0)$ in a generalized power series $f(z) = \sum_{n=1}^\infty a_n z^{\nu+n}$. The result is a series $f(B(z)) = \sum_{n=1}^\infty a_n B(z)^{\nu+n} = \sum_{n=1}^\infty b_n z^{\nu+n}$ of the same form. It is computed from the expansion

$$(1) \qquad\qquad B(z)^\nu \;=\; B_0(\nu)z^\nu + B_1(\nu)z^{\nu+1} + \cdots,$$

where [15]

$$(2) \qquad \begin{cases} B_0(\nu) &= B_1^\nu, \\ B_1(\nu) &= \nu B_1^{\nu-1} B_2, \\ B_2(\nu) &= \nu B_1^{\nu-1}[B_3 + \tfrac{1}{2}(\nu-1)B_1^{-1}B_2^2], \\ \quad \cdots \end{cases}$$

In our applications, we shall not distinguish between a normalized Riemann mapping $B(z)$ and the formal power series determined from its sequence of Taylor coefficients.

THEOREM 2.2 *Let $B(z)$ be a normalized Riemann mapping of the unit disk into itself, and let $\mathcal{G}_\sigma^\nu$ be any Grunsky space with ν real and $\sigma_1 \geq \sigma_2 \geq \cdots \geq 0$. Then the operator*

$$(3) \qquad T : f(z) \;\mapsto\; f(B(z))$$

is an everywhere defined and continuous contraction of $\mathcal{G}_\sigma^\nu$ into itself.

Previous versions appear in [2,15]. The novel feature in our proof is the connection with Theorem 2.1. Interesting functional analysis aspects of such results and more complex methods have recently been discussed by V. I. Vasyunin and N. K. Nikol'skiǐ[13].

PROOF. Continuity is automatic by the closed graph theorem once it is known that the operator is everywhere defined. We first prove the result for $\mathcal{D}_\nu$ (following [2]). Let C_ρ be the positively oriented circle $|z| = \rho$, $0 < \rho < 1$. By Schwarz's lemma, $B(C_\rho)$ lies inside C_ρ. Choose a radial slit S from a point of maximum modulus on $B(C_\rho)$ to C_ρ and a branch of z^ν on the region G bounded by $\Gamma = S + C_\rho - S - B(C_\rho)$. Then any element $f(z) = z^\nu \sum_{n=1}^\infty a_n z^n$ of $\mathcal{D}_\nu$ represents a function which is analytic on G and has a continuous extension to $\overline{G}$, and the derivative of the function has the same properties. Writing $f(B(z)) = z^\nu \sum_{n=1}^\infty b_n z^n$, we obtain by Green's Theorem [12, p. 241],

$$\begin{aligned} 0 \;\leq\; & \frac{1}{\pi} \iint_G |f'(z)|^2 \, dx \, dy \\ = \;& \frac{1}{2\pi i} \int_\Gamma f'(z)\overline{f(z)} \, dz \\ = \;& \frac{1}{2\pi i} \int_{C_\rho} f'(z)\overline{f(z)} \, dz - \frac{1}{2\pi i} \int_{B(C_\rho)} f'(z)\overline{f(z)} \, dz \\ = \;& \frac{1}{2\pi i} \int_{C_\rho} f'(z)\overline{f(z)} \, dz - \frac{1}{2\pi i} \int_{C_\rho} f'(B(w))\overline{f(B(w))} B'(w) \, dw \\ = \;& \sum_{n=1}^\infty (\nu + n)\rho^{2\nu+2n}(|a_n|^2 - |b_n|^2). \end{aligned}$$

Passage to the limit with $\rho \uparrow 1$ yields

$$(4) \qquad \langle f(B(z)), f(B(z))\rangle_{\mathcal{G}_\sigma^\nu} \;\leq\; \langle f(z), f(z)\rangle_{\mathcal{G}_\sigma^\nu}$$

when $\mathcal{G}_\sigma^\nu$ is $\mathcal{D}_\nu$.

Next let $\mathcal{G}_\sigma^\nu$ be $\mathcal{D}_\nu^r$ where r is a positive integer. Let T be the operator (3) viewed as acting on $\mathcal{D}_\nu$ and T_r the same operator but viewed as acting on $\mathcal{D}_\nu^r$. In a natural way, we may think of $\mathcal{D}_\nu^r$ as a subspace of $\mathcal{D}_\nu$. Let P_r be the projection of $\mathcal{D}_\nu$ on $\mathcal{D}_\nu^r$. Then $T_r = P_r T | \mathcal{D}_\nu^r$. The orthogonal complement of $\mathcal{D}_\nu^r$ in $\mathcal{D}_\nu$ is invariant under T, and so $\mathcal{D}_\nu^r$ is invariant under T^*. Therefore $T_r^* = T^* | \mathcal{D}_\nu^r$. Since T is a contraction by the first part of the proof, T^* is a contraction by Theorem 2.1. Hence its restriction T_r^* is a contraction, and so T_r is a contraction by Theorem 2.1. Thus the result holds for $\mathcal{D}_\nu^r$.

In the general case, it is enough to prove (4) when $\sigma_n = 0$ for large n. This follows from the identity

$$(5) \qquad \sum_{n=1}^{\infty} (\nu + n)\sigma_n |c_n|^2 = \sum_{n=1}^{\infty} (\sigma_n - \sigma_{n+1}) \sum_{j=1}^{n} (\nu + j)|c_j|^2$$

and the special case of (4) for the spaces $\mathcal{D}_\nu^r$ proved above.

$\square$

It is known that Theorem 2.2 applied only to the Dirichlet space is a weak result in the sense that it is not characteristic of the class of normalized Riemann mappings of the unit disk into itself. An example is $B(z) = \epsilon z(1 + \rho z)$, where $\epsilon > 0, \rho > 1$, and $\epsilon^2(1+\rho)^2 \le \frac{1}{3}$. This function is bounded by 1 in the unit disk but not a Riemann mapping. The operator (3) is a contraction on $\mathcal{D}_0$. For if $f(z) = \sum_{n=1}^{\infty} a_n z^{\nu+n}$ is in $\mathcal{D}_0$, then

$$f(B(z)) = \sum_{n=1}^{\infty} \sum_{j=1}^{n} \epsilon^j \rho^{n-j} \binom{j}{n-j} a_j z^n.$$

Writing $c_n = \sqrt{n} a_n$, we bring the inequality to be proved to the form

$$\sum_{n=1}^{\infty} \left| \sum_{j=1}^{n} \left(\frac{n}{j}\right)^{\frac{1}{2}} \epsilon^j \rho^{n-j} \binom{j}{n-j} c_j \right|^2 \le \sum_{n=1}^{\infty} |c_n|^2.$$

In fact, the Hilbert-Schmidt norm of the coefficient matrix is at most 1:

$$\sum_{n=1}^{\infty} \sum_{j=1}^{n} \left| \left(\frac{n}{j}\right)^{\frac{1}{2}} \epsilon^j \rho^{n-j} \binom{j}{n-j} \right|^2 = \sum_{j=1}^{\infty} \sum_{n=j}^{2j} \frac{n}{j} \epsilon^{2j} \rho^{2n-2j} \binom{j}{n-j}^2$$

$$\le 2 \sum_{j=1}^{\infty} \sum_{n=j}^{2j} \epsilon^{2j} \rho^{2n-2j} \binom{j}{n-j}^2 = 2 \sum_{j=1}^{\infty} \epsilon^{2j} \sum_{n=0}^{j} \rho^{2n} \binom{j}{n}^2 \le 2 \sum_{j=1}^{\infty} \epsilon^{2j}(1+\rho)^{2j} \le 1.$$

Therefore (3) is a contraction on $\mathcal{D}_0$. But $f(z) = z^{-1}$ is in $\mathcal{D}_{-2}$ and $f(B(z))$ is not in $\mathcal{D}_{-2}$, so the conclusion of Theorem 2.2 breaks down for $\mathcal{D}_{-2}$. In contrast, de Branges [2] has shown that the conclusion of Theorem 2.2 holds for every space $\mathcal{D}_\nu$ with ν a nonpositive integer only if $B(z)$ represents a normalized Riemann mapping of the unit disk into itself.

The contractive substitution property has an equivalent form as an estimate of $[B(z)^\nu - B'(0)^\nu z^\nu]/\nu$ for any real ν. If $\nu = 0$ this expression reduces to $\log B(z)/[zB'(0)]$, which appears in a related estimate equivalent to Milin's conjecture [5].

THEOREM 2.3 *Let* $B(z) = B_1 z + B_2 z^2 + \cdots$ $(B_1 > 0)$ *be a formal power series, and consider sequences* $\sigma = \{\sigma_1, \sigma_2, \ldots\}$ *and* $\tau = \{\sigma_2, \sigma_3, \ldots\}$ *where* $\sigma_1, \sigma_2, \ldots$ *are real numbers. The following conditions are equivalent.*

(i) For any real ν, $f(z)$ *in* $\mathcal{G}_\sigma^\nu$ *implies* $f(B(z))$ *in* $\mathcal{G}_\sigma^\nu$ *and*

$$(6) \qquad \langle f(B(z)), f(B(z)) \rangle_{\mathcal{G}_\sigma^\nu} \leq \langle f(z), f(z) \rangle_{\mathcal{G}_\sigma^\nu}.$$

(ii) For any real ν, $g(z)$ *in* $\mathcal{G}_\tau^\nu$ *implies* $\nu^{-1}[B(z)^\nu - B'(0)^\nu z^\nu] + g(B(z))$ *in* $\mathcal{G}_\tau^\nu$ *and*

$$(7) \quad \left\langle \frac{B(z)^\nu - B'(0)^\nu z^\nu}{\nu} + g(B(z)), \frac{B(z)^\nu - B'(0)^\nu z^\nu}{\nu} + g(B(z)) \right\rangle_{\mathcal{G}_\tau^\nu} - \langle g(z), g(z) \rangle_{\mathcal{G}_\tau^\nu}$$
$$\leq \frac{1 - B'(0)^{2\nu}}{\nu} \sigma_1.$$

PROOF. Set $\mu = \nu + 1$. If $g(z)$ is in $\mathcal{G}_\tau^\mu$ and c_1 is a constant, then

$$(8) \qquad f(z) \;=\; c_1 z^{\nu+1} + g(z)$$

belongs to $\mathcal{G}_\sigma^\nu$. Every element of $\mathcal{G}_\sigma^\nu$ has this form, and $\langle f(z), f(z) \rangle_{\mathcal{G}_\sigma^\nu}$ may be computed as $\mu \sigma_1 |c_1|^2 + \langle g(z), g(z) \rangle_{\mathcal{G}_\tau^\mu}$. Assume (i). If $g(z)$ is in $\mathcal{G}_\tau^\mu$ and $\nu \neq -1$, we may choose $c_1 = 1/(\nu + 1)$ in (8). Then by (i),

$$f(B(z)) = \frac{1}{\nu+1} B(z)^{\nu+1} + g(B(z)) = \frac{1}{\mu} B'(0)^\mu z^\mu + \frac{B(z)^\mu - B'(0)^\mu z^\mu}{\mu} + g(B(z))$$

belongs to $\mathcal{G}_\sigma^\nu$ and (6) holds. Therefore $\mu^{-1}[B(z)^\mu - B'(0)^\mu z^\mu] + g(B(z))$ belongs to $\mathcal{G}_\tau^\mu$ and

$$\frac{1}{\mu} B'(0)^{2\mu} \sigma_1 + \left\langle \frac{B(z)^\mu - B'(0)^\mu z^\mu}{\mu} + g(B(z)), \frac{B(z)^\mu - B'(0)^\mu z^\mu}{\mu} + g(B(z)) \right\rangle_{\mathcal{G}_\tau^\mu}$$
$$\leq \frac{1}{\mu} \sigma_1 + \langle g(z), g(z) \rangle_{\mathcal{G}_\tau^\mu} \quad .$$

This is equivalent to (7) with ν replaced by μ. When $\nu = -1$, that is $\mu = 0$, the same inequality holds as a limiting case. Therefore (ii) holds.

Assume (ii). Apply (ii) with ν replaced by μ and $g(z)$ by $(\nu + 1)^{-1} c_1^{-1} g(z)$. Multiplying the inequality derived from (7) in this way by $(\nu + 1)^2 |c_1|^2$, we get

$$\langle [B(z)^{\nu+1} - B'(0)^{\nu+1} z^{\nu+1}] c_1 + g(B(z)), [B(z)^{\nu+1} - B'(0)^{\nu+1} z^{\nu+1}] c_1 + g(B(z)) \rangle_{\mathcal{G}_\tau^\mu}$$
$$- \langle g(z), g(z) \rangle_{\mathcal{G}_\tau^\mu} \leq (\nu + 1) \sigma_1 |c_1|^2 - (\nu + 1) \sigma_1 B'(0)^{2\nu+2} |c_1|^2.$$

Thus if $f(z)$ is the element (8) of $\mathcal{G}_\sigma^\nu$, then

$$f(B(z)) \;=\; c_1 B(z)^{\nu+1} + g(B(z))$$
$$=\; c_1 B'(0)^{\nu+1} z^{\nu+1} + [B(z)^{\nu+1} - B'(0)^{\nu+1} z^{\nu+1}] c_1 + g(B(z))$$

belongs to $\mathcal{G}_\sigma^\nu$ and (6) holds. The case $(\nu + 1)c_1 = 0$ is handled by an approximation argument. Thus (ii) implies (i).

□

The preceding results may be used to simplify some arguments in [15]. Let $B(z)$ be a normalized Riemann mapping of the unit disk into itself. Let ν be any real number, and let $\mathcal{G}_\sigma^\nu$ be a Grunsky space such that $\sigma_1 \geq \sigma_2 \geq \cdots \geq 0$. Define $\mathcal{G}(\nu, \sigma, B)$ to be the space of all elements $f(z)$ of $\mathcal{G}_\sigma^\nu$ such that

$$\sup \left\{ \langle f(z) + g(B(z)), f(z) + g(B(z)) \rangle_{\mathcal{G}_\sigma^\nu} - \langle g(z), g(z) \rangle_{\mathcal{G}_\sigma^\nu} \right\} < \infty,$$

where the supremum is over all $g(z)$ in $\mathcal{G}_\sigma^\nu$. Then $\mathcal{G}(\nu, \sigma, B)$ is a Hilbert space in a norm such that the supremum is $\|f(z)\|_{\mathcal{G}(\nu,\sigma,B)}^2$. We can simplify the proof that $\mathcal{G}(\nu, \sigma, B)$ is a Hilbert space given in [15, **Cor. 3.8**]. By the discussion preceding [15, **Theorem 3.4**] or general properties of complementation [7,9], all that is needed is to show that $1 - TT^*$ is a nonnegative operator, where $T : f(z) \to f(B(z))$ in $\mathcal{G}_\sigma^\nu$. But T is contractive by Theorem 2.2, hence T^* is contractive by Theorem 2.1, and this gives the result.

Another simplification is to [15, **Theorem 3.4**]:

COROLLARY 2.4 *Let $B(z)$ be a normalized Riemann mapping of the unit disk into itself. For any real number ν, let $\mathcal{G}_\sigma^\nu$ be a Grunsky space such that $\sigma_1 \geq \sigma_2 \geq \cdots \geq 0$. Then $[B(z)^\nu - B'(0)^\nu z^\nu]/\nu$ belongs to $\mathcal{G}(\nu, \sigma, B)$ and*

$$(9) \qquad \left\| \frac{B(z)^\nu - B'(0)^\nu z^\nu}{\nu} \right\|_{\mathcal{G}(\nu,\sigma,B)}^2 \leq \frac{1 - B'(0)^{2\nu}}{\nu} \sigma_1.$$

PROOF. The hypotheses imply that conditions (i) and (ii) of Theorem 2.3 hold relative to the sequences $\{\sigma_1, \sigma_1, \sigma_2, \ldots\}$ and $\{\sigma_1, \sigma_2, \ldots\}$. The norm inequality (9) then follows on taking the supremum of the left hand side of (7) over all $g(z)$ in the space. $\qquad\square$

Every formal power series $B(z) = B_1 z + B_2 z^2 + \cdots$ $(B_1 > 0)$ has a *compositional inverse*, which is defined as the unique formal power series $A(z)$ of the same form such that $A(B(z)) = B(A(z)) = z$. The coefficients in the expansions

$$(10) \qquad \begin{cases} A(z)^\nu & = \sum_{n=0}^\infty A_n(\nu) z^{\nu+n}, \\ B(z)^\nu & = \sum_{n=0}^\infty B_n(\nu) z^{\nu+n} \end{cases}$$

are connected by the Lagrange-Bürmann formula [15, **Theorem 3.5**]:

$$(11) \qquad A_n(\nu) = \frac{\nu}{\nu + n} B_n(-\nu - n),$$

$n = 0, 1, 2, \ldots$. Write $B^*(z) = \overline{B}_1 z + \overline{B}_2 z^2 + \cdots$.

THEOREM 2.5 *Let $A(z)$ and $B(z)$ be as above. Let $\mathcal{G}_\sigma^\nu$ be a Grunsky space such that $\sigma_n = 0$ for $n > r$ and $\sigma_1 \cdots \sigma_r \neq 0$, and consider $\mathcal{G}_\tau^\mu$ where $\mu = -\nu - r - 1$ and $\tau = \{-\sigma_r^{-1}, -\sigma_{r-1}^{-1}, \ldots, -\sigma_1^{-1}, 0, 0, \ldots\}$. Then the mapping*

$$W : \sum_{n=1}^r a_n z^{\nu+n} \mapsto \sum_{n=1}^r \sigma_n a_n z^{-\nu-n}$$

is a Kreĭn space isomorphism from $\mathcal{G}_\sigma^\nu$ onto $\mathcal{G}_\tau^\mu$. If operators S and T are defined by $T : f(z) \mapsto f(B(z))$ on $\mathcal{G}_\sigma^\nu$ and $S : g(z) \mapsto g(A^*(z))$ on $\mathcal{G}_\tau^\mu$, then $T^* = W^{-1} SW$.

PROOF. The mapping W is an isomorphism from $\mathcal{G}_\sigma^\nu$ onto $\mathcal{G}_\tau^\mu$ since for any numbers $a_1, \ldots, a_r$,

$$
\left\langle \sum_{n=1}^{r} \sigma_n a_n z^{-\nu-n}, \sum_{n=1}^{r} \sigma_n a_n z^{-\nu-n} \right\rangle_{\mathcal{G}_\tau^\mu}
$$
$$
= \left\langle \sum_{n=1}^{r} \sigma_{r+1-n} a_{r+1-n} z^{-\nu-r-1+n}, \sum_{n=1}^{r} \sigma_{r+1-n} a_{r+1-n} z^{-\nu-r-1+n} \right\rangle_{\mathcal{G}_\tau^\mu}
$$
$$
= \sum_{n=1}^{r} (-\nu - r - 1 + n)(-\sigma_{r+1-n}^{-1})|\sigma_{r+1-n} a_{r+1-n}|^2
$$
$$
= \sum_{n=1}^{r} (\nu + n)\sigma_n |a_n|^2
$$
$$
= \left\langle \sum_{n=1}^{r} a_n z^{\nu+n}, \sum_{n=1}^{r} a_n z^{\nu+n} \right\rangle_{\mathcal{G}_\sigma^\nu}.
$$

We show that WT^* and SW have the same action on any $f(z) = \sum_{n=1}^{r} a_n z^{\nu+n}$ in $\mathcal{G}_\sigma^\nu$. By [15, Theorem 3.6], $T^* : f(z) \mapsto \sum_{n=1}^{r} b_n z^{\nu+n}$, where for $n = 1, \ldots, r$,

$$
(\nu + n)\sigma_n b_n = \sum_{n=1}^{r} (\nu + j)\sigma_j \overline{B}_{j-n}(\nu + n)a_j.
$$

Since $WT^* : f(z) \mapsto \sum_{n=1}^{r} \sigma_n b_n z^{-\nu-n}$ and $SW : f(z) \mapsto \sum_{n=1}^{r} \sigma_n a_n A^*(z)^{-\nu-n}$, it is required to show that, as elements of $\mathcal{G}_\tau^\mu$,

$$
(12) \qquad \sum_{n=1}^{r} \sigma_n b_n z^{-\nu-n} = \sum_{n=1}^{r} \sigma_n a_n A^*(z)^{-\nu-n},
$$

that is, the coefficients of $z^{-\nu-p}$ coincide for $p = 1, \ldots, r$. By (10) and (11),

$$
\sum_{n=1}^{r} \sigma_n a_n A^*(z)^{-\nu-n} = \sum_{n=1}^{r} \sigma_n a_n \sum_{j=0}^{\infty} \overline{A}_j(-\nu - n) z^{-\nu-n+j}
$$
$$
= \sum_{n=1}^{r} \sigma_n a_n \sum_{j=0}^{\infty} \frac{-\nu - n}{-\nu - n + j} \overline{B}_j(\nu + n - j) z^{-\nu-n+j}.
$$

For any $p = 1, \ldots, r$, the coefficient of $z^{-\nu-p}$ in the last term on the right side is

$$
\sum_{n-j=p} \frac{-\nu - n}{-\nu - n + j} \overline{B}_j(\nu + n - j)\sigma_n a_n = \sum_{n=p}^{r} \frac{\nu + n}{\nu + p} \sigma_n \overline{B}_{n-p}(\nu + p)a_n = \sigma_p b_p.
$$

This proves (12). The possibility of division by zero when ν is a negative integer must be considered. The coefficient of $z^{-\nu-p}$ on the right side of (12) is a polynomial in ν, so this case follows by continuity.

$\square$

The proof of Theorem 2.2 uses the full strength of Theorem 2.1. By taking advantage of Theorem 2.5, we can rewrite the argument so as to use only the finite-dimensional case of Theorem 2.1 given in [**14**].

VARIANT ON THE PROOF OF THEOREM 2.2. Prove the result for $\mathcal{D}_\nu$ as before. The case of any space $\mathcal{D}_\nu^r$ such that $\nu > -r - 1$ then follows by truncation. For if $f(z) = \sum_{n=1}^r a_n z^{\nu+n}$ is in $\mathcal{D}_\nu^r$ and $f(B(z)) = \sum_{n=1}^\infty b_n z^{\nu+n}$, then by the result for $\mathcal{D}_\nu$,

$$\sum_{n=1}^\infty (\nu + n)|b_n|^2 \le \sum_{n=1}^r (\nu + n)|a_n|^2.$$

The terms for $n > r$ on the left are positive because $\nu + r + 1 > 0$, and so these terms can be dropped, yielding the result for $\mathcal{D}_\nu^r$ when $\nu > -r - 1$.

In particular, the result holds for $\mathcal{D}_\nu^r$ when $\nu \ge -\frac{1}{2}(r+1)$. Now apply Theorem 2.5 with $\mathcal{G}_\sigma^\nu$ equal to $\mathcal{D}_\nu^r$. The operator T in Theorem 2.5 is a contraction, so T^* is a contraction by the finite-dimensional case of Theorem 2.1 . Therefore S is a contraction on $\mathcal{G}_\tau^\mu$, and S^{-1} is a contraction on the anti-space of $\mathcal{G}_\tau^\mu$, which is $\mathcal{D}_\mu^r$, $\mu = -\nu - r - 1 \le -\frac{1}{2}(r + 1)$. But S^{-1} is substitution by $B^*(z)$. Equivalently, substitution by $B(z)$ is contractive in $\mathcal{D}_\nu^r$ when $\nu \le -\frac{1}{2}(r + 1)$ and hence for all real ν. Once this is known, we may complete the argument as before using the identity (5).

□

3   Necessary conditions on segments of coefficients

Any property of the class of normalized Riemann mappings $B(z) = B_1 z + B_2 z^2 + \cdots$ of the unit disk into itself which depends only on an initial segment $B_1, \ldots, B_r$ of coefficients may be taken as a necessary condition that given numbers arise in this way. Theorem 2.2 may be used to generate one set of conditions, and similar but more complicated conditions occur in the theory of the Bieberbach conjecture. In this section we formulate these conditions and observe that either approach may be used to characterize the first two coefficients. It is unknown if the first approach can be used to characterize initial segments in general. For the second, this question is answered negatively in §3. In view of Theorem 2.3, the conditions derived from Theorem 2.2 may be stated in two equivalent forms.

DEFINITION 3.6 *We say that given numbers* $B_1, \ldots, B_r$ $(B_1 > 0)$ *satisfy the* **contractive substitution condition** *if the polynomial* $B(z) = B_1 z + \cdots + B_r z^r$ *has one and hence both of these properties:*
(i) The inequality

$$(13) \qquad\qquad \langle f(B(z)), f(B(z)) \rangle_{\mathcal{G}_\sigma^\nu} \;\le\; \langle f(z), f(z) \rangle_{\mathcal{G}_\sigma^\nu}$$

holds for every $f(z)$ *in* $\mathcal{G}_\sigma^\nu$ *whenever* ν *is real,* $\sigma_1 \ge \cdots \ge \sigma_r$, *and* $\sigma_n = 0$ *for* $n > r$.

(ii) The inequality

$$(14) \qquad \left\langle \frac{B(z)^\nu - B'(0)^\nu z^\nu}{\nu} + g(B(z)), \frac{B(z)^\nu - B'(0)^\nu z^\nu}{\nu} + g(B(z)) \right\rangle_{\mathcal{G}_\sigma^\nu} - \langle g(z), g(z) \rangle_{\mathcal{G}_\sigma^\nu}$$

$$\leq \frac{1 - B'(0)^{2\nu}}{\nu} \sigma_1$$

holds for every $g(z)$ in $\mathcal{G}_\sigma^\nu$ whenever ν is real, $\sigma_1 \geq \cdots \geq \sigma_{r-1}$, and $\sigma_n = 0$ for $n > r - 1$.

A simpler form of the conditions may be given.

THEOREM 3.7 *For given numbers $B_1, \ldots, B_r$ $(B_1 > 0)$ to satisfy the contractive substitution condition, it is sufficient that substitution by $B(z) = B_1 z + \cdots + B_r z^r$ is contractive in $\mathcal{D}_\nu^r$ for all $\nu \geq -\frac{1}{2}(r + 1)$.*

PROOF. Let $T : f(z) \mapsto f(B(z))$ be contractive in $\mathcal{D}_\nu^r$ whenever $\nu \geq -\frac{1}{2}(r + 1)$. Then Theorem 2.5 implies that T is contractive in $\mathcal{D}_\nu^r$ for all real ν (see the argument in the variant on the proof of Theorem 2.2 at the end of §2). We show next that the operator $T_k : f(z) \mapsto f(B(z))$ is contractive in $\mathcal{D}_\nu^k$ for any real ν and $k = 1, \ldots, r$. If we view $\mathcal{D}_\nu^k$ as a subspace of $\mathcal{D}_\nu^r$ with projection P_k, then $T_k = P_k T|_{\mathcal{D}_\nu^k}$ and $T_k^* = T^*|_{\mathcal{D}_\nu^k}$. By Theorem 2.1, T^* is a contraction. Therefore T_k^* and T_k are contractions. Finally, using the summation by parts identity (5), we verify condition (i) in Definition 3.1.

$$\square$$

Inequalities similar to (13) and (14) appear in the proof of Milin's conjecture [5] and its generalizations [6,15]. Let ν be any real number, and let $\mathcal{G}_{\sigma(t)}^\nu$, $t \geq 1$, be a family of Grunsky spaces such that $\sigma_1(t), \sigma_2(t), \ldots$ are absolutely continuous functions of $t \geq 1$ and $\sigma_n(t)$ vanishes identically for all sufficiently large n. We say that the family $\mathcal{G}_{\sigma(t)}^\nu$, $t \geq 1$, is *admissible* if for all n,

$$(15) \qquad (2\nu + n)\sigma_n(t) + t\sigma_n'(t) = (2\nu + n)\sigma_{n+1}(t) - \frac{2\nu + n}{n+1} t\sigma_{n+1}'(t)$$

a.e. for $t \geq 1$ and the function

$$(16) \qquad \tau_n(t) = (\nu + n)\left[\sigma_n(t) - \sigma_{n+1}(t) + \frac{1}{n+1} t\sigma_n'(t)\right]$$

is nonnegative a.e. for $t \geq 1$. For an example, define $\sigma(t) = \{\sigma_1(t), \sigma_2(t), \ldots\}$ by

$$(17) \qquad \sigma_n(t) = \sum_{j=0}^\infty (-1)^j \frac{(2\nu + n)_j (2\nu + 2n + j + 1)_j}{j!(n+1)_j} \Delta_{n+j} t^{-2\nu - n - j},$$

$n = 1, 2, \ldots$, where

$$(18) \qquad \Delta_k = \frac{4^{-k} \Gamma(k+1)(2\nu + 2\lambda + r + 2)_{k-1}}{\Gamma(\nu + k + 1)\Gamma(\nu + \lambda + k + 1)\Gamma(2\nu + k + 1)\Gamma(r + 1 - k)},$$

$k = 1, 2, \ldots.$ Here r is a fixed positive integer and $\lambda \geq 0$. We view $1/\Gamma(z)$ as an entire function vanishing at the nonpositive integers, and so $\Delta_k = 0, k > r$, and $\sigma_n(t) \equiv 0$ for $n > r$. The family is admissible if either $\nu > -\frac{3}{2}$ or 2ν is a negative integer [6,15]. As we shall see, there are examples of admissible families for arbitrary real ν.

For any admissible family $\mathcal{G}^\nu_{\sigma(t)}$, $t \geq 1$, and any normalized Riemann mapping $B(z)$ of the unit disk into itself, the inequality [6,15]

$$
(19) \quad \left\langle \frac{B(z)^\nu - B'(0)^\nu z^\nu}{\nu} + g(B(z)), \frac{B(z)^\nu - B'(0)^\nu z^\nu}{\nu} + g(B(z)) \right\rangle_{\mathcal{G}^\nu_{\sigma(a)}} - \langle g(z), g(z) \rangle_{\mathcal{G}^\nu_{\sigma(b)}}
$$

$$
\leq 4b^{-2\nu} \sum_{n=1}^\infty (\nu + n) \left[\frac{(2\nu + 1)_{n-1}}{n!} \right]^2 \left[a^{2\nu} \sigma_n(a) - b^{2\nu} \sigma_n(b) \right]
$$

holds whenever

$$
(20) \qquad a \geq 1, \qquad b \geq 1, \qquad a/b = B'(0), \qquad g(z) = \sum_{n=1}^\infty c_n z^{\nu+n}.
$$

Milin's conjecture is proved by choosing $\nu = 0$ with the example of an admissible family given above. With the same example but other values of ν, the method yields coefficient estimates for powers of Riemann mappings.

The inequality (19) may be viewed as a more delicate version of (14), and it can be used in a similar way to impose conditions on coefficients.

DEFINITION 3.8 *We say that given numbers $B_1, \ldots, B_r$ $(B_1 > 0)$ satisfy the* **admissible families condition** *if the polynomial $B(z) = B_1 z + \cdots + B_r z^r$ satisfies (19) for every admissible family $\mathcal{G}^\nu_{\sigma(t)}$, $t \geq 1$, such that $\sigma_n(t) \equiv 0$ for $n > r - 1$.*

In the definition, it is understood that the quantities a, b, and $g(z)$ appearing in (19) are arbitrary except for the restrictions (20). The contractive substitution condition and admissible families condition give the same result when $r = 2$.

THEOREM 3.9 *If B_1, B_2 $(B_1 > 0)$ are given numbers, the following conditions are equivalent :*
(i) B_1, B_2 satisfy the contractive substitution condition,
(ii) B_1, B_2 satisfy admissible families condition,
(iii) $B_1 \leq 1$ and $|B_2| \leq 2B_1(1 - B_1)$,
(iv) B_1, B_2 are the first two coefficients of a normalized Riemann mapping of the unit disk into itself.

PROOF. The equivalence of (i),(iii), and (iv) is proved in [11, Theorem 2]. Also (iii) implies (ii), because (iii) implies (iv) and (19) holds for any normalized Riemann mapping $B(z)$ of the unit disk into itself and any admissible family $\mathcal{G}^\nu_{\sigma(t)}$, $t \geq 1$. It remains to show that (ii) implies (iii).

Assume (ii). If $g(z) = \sum_{n=1}^{\infty} c_n z^{\nu+n}$, then

$$\frac{B(z)^\nu - B'(0)^\nu z^\nu}{\nu} + g(B(z)) = \left[\frac{1}{\nu}B_1(\nu) + B_0(\nu+1)c_1\right] z^{\nu+1} + \cdots.$$

For any real ν, an admissible family $\mathcal{G}^\nu_{\sigma(t)}$, $t \geq 1$, is obtained with $\sigma_1(t) = (\nu+1)t^{-2\nu-1}$ and $\sigma_n(t)$ identically zero otherwise. Then if $a \geq 1, b \geq 1$, and $a/b = B_1$, the inequality (19) asserts that

$$(\nu+1)\sigma_1(a)\left|\frac{1}{\nu}B_1(\nu) + B_0(\nu+1)c_1\right|^2 - (\nu+1)\sigma_1(b)|c_1|^2$$
$$\leq 4b^{-2\nu}(\nu+1)[a^{2\nu}\sigma_1(a) - b^{2\nu}\sigma_1(b)].$$

Using (2), we simplify the conditions on B_1, B_2 to the inequality

$$(21) \qquad \left|B_1^{\frac{1}{2}}c_1 + B_1^{-\frac{3}{2}}B_2\right|^2 - |c_1|^2 \leq 4(B_1^{-1} - 1).$$

Now if p and q are complex numbers, the supremum $\sup\left[|pc + q|^2 - |c|^2\right]$ over all complex numbers c is finite if and only if either $|p| < 1$ or $|p| = 1$ and $q = 0$. When $|p| < 1$ the value of the supremum is $|q|^2/(1 - |p|^2)$. Applied to (21), this yields (iii).

$\qquad \square$

4   Failure of the admissible families condition

We exhibit numbers B_1, B_2, B_3 ($B_1 > 0$) which satisfy the admissible families condition but are not the first three coefficients of a normalized Riemann mapping of the unit disk into itself. The example takes into account all admissible families $\mathcal{G}^\nu_{\sigma(t)}$, $t \geq 1$, with ν real and $\sigma(t)$ of the form $\{\sigma_1(t), \sigma_2(t), 0, 0, \ldots\}$. For any real ν, such a family is obtained with

$$(22) \qquad \begin{cases} \sigma_1(t) &= (\nu+1)t^{-2\nu-1}, \\ \sigma_2(t) &= 0, \end{cases}$$

or

$$(23) \qquad \begin{cases} \sigma_1(t) &= (\nu+2)^2(2\nu+2)t^{-2\nu-1} - (\nu+2)^2(2\nu+1)t^{-2\nu-2}, \\ \sigma_2(t) &= (\nu+2)t^{-2\nu-2}. \end{cases}$$

Apart from constant multiples, these sequences agree with those determined by (17) and (18) with $r = 1, 2$.

THEOREM 4.10 *In order that B_1, B_2, B_3 ($B_1 > 0$) satisfy the admissible families condition it is sufficient that $B(z) = B_1 z + B_2 z^2 + B_3 z^3$ satisfy (19) for the special families (22) and (23) in the case $a = 1$ and $b = 1/B'(0)$.*

PROOF. Assume that $B(z)$ satisfies (19) for the special families (22) and (23) with $a = 1$ and $b = 1/B'(0)$. It must be shown that $B(z)$ satisfies (19) in the general case. We exhibit all admissible families $\mathcal{G}^{\nu}_{\sigma(t)}$, $t \geq 1$, with ν real and $\sigma(t)$ of the form $\{\sigma_1(t), \sigma_2(t), 0, 0, \ldots\}$. For any real ν, an example is obtained with

$$(24) \qquad \begin{cases} \sigma_1(t) &= (\nu+2)^2(2\nu+2)Dt^{-2\nu-1} - (\nu+2)^2(2\nu+1)Ct^{-2\nu-2}, \\ \sigma_2(t) &= (\nu+2)Ct^{-2\nu-2}, \end{cases}$$

where $D \geq C \geq 0$. For $\nu = -1$, another example is

$$(25) \qquad \sigma_1(t) = A + Bt \quad \text{and} \quad \sigma_2(t) = A,$$

where $A \geq 0$. For $\nu = -2$, another example is

$$(26) \qquad \sigma_1(t) = Bt^3 \quad \text{and} \quad \sigma_2(t) = At^2,$$

where $B \leq 0$. This list exhausts all possibilities, because the definition of an admissible family requires that

$$\begin{aligned} (2\nu+1)\sigma_1(t) + t\sigma_1'(t) &= (2\nu+1)\sigma_2(t) - \frac{2\nu+1}{2}t\sigma_2'(t), \\ (2\nu+2)\sigma_2(t) + t\sigma_2'(t) &= 0 \end{aligned}$$

and

$$\begin{aligned} \tau_1(t) &= (\nu+1)[\sigma_1(t) - \sigma_2(t) + \frac{1}{2}t\sigma_2'(t)] \geq 0, \\ \tau_2(t) &= (\nu+2)\sigma_2(t) \geq 0 \end{aligned}$$

for $t \geq 1$. The differential equations have the general solution

$$\begin{aligned} \sigma_1(t) &= Bt^{-2\nu-1} - (\nu+2)(2\nu+1)At^{-2\nu-2}, \\ \sigma_2(t) &= At^{-2\nu-2}, \end{aligned}$$

where A, B are real constants. If $\nu \neq -1, -2$, we may write

$$\begin{aligned} A &= (\nu+2)C, \\ B &= (\nu+2)^2(2\nu+2)D, \end{aligned}$$

and the solutions has the form (24). The requirement that

$$\begin{aligned} \tau_1(t) &= 2(\nu+1)^2(\nu+2)^2t^{-2\nu-2}(Dt - C) \geq 0, \\ \tau_2(t) &= (\nu+2)^2Ct^{-2\nu-2} \geq 0 \end{aligned}$$

for $t \geq 1$ implies $D \geq C \geq 0$. In a similar way, the solution reduces to (25) and (26) when $\nu = -1$ and -2.

If $g(z) = \sum_{n=1}^{\infty} c_n z^{\nu+n}$, then

$$\frac{B(z)^{\nu} - B'(0)^{\nu}z^{\nu}}{\nu} + g(B(z)) = \left[\frac{1}{\nu}B_1(\nu) + B_0(\nu+1)c_1\right]z^{\nu+1}$$
$$+ \left[\frac{1}{\nu}B_2(\nu) + B_1(\nu+1)c_1 + B_0(\nu+2)c_2\right]z^{\nu+2} + \cdots.$$

Thus (19) asserts

$$(27) \qquad (\nu+1)\sigma_1(a)\left|\frac{1}{\nu}B_1(\nu)+B_0(\nu+1)c_1\right|^2$$

$$+(\nu+2)\sigma_2(a)\left|\frac{1}{\nu}B_2(\nu)+B_1(\nu+1)c_1+B_0(\nu+2)c_2\right|^2$$

$$-(\nu+1)\sigma_1(aB_1^{-1})|c_1|^2-(\nu+2)\sigma_2(aB_1^{-1})|c_2|^2$$

$$\leq\ 4(\nu+1)[B_1^{2\nu}\sigma_1(a)-\sigma_1(aB_1^{-1})]+4(\nu+2)\left(\frac{2\nu+1}{2}\right)^2[B_1^{2\nu}\sigma_2(a)-\sigma_2(aB_1^{-1})].$$

By hypothesis, (27) holds for the families (22) and (23) with $a=1$. We show that (27) holds for the families (22) and (23) when $a\geq 1$. The case of (22) is clear since we need only multiply the inequality for $a=1$ by a constant. In the case of (23), (27) becomes

$$4(\nu+1)[B_1^{2\nu}(2\nu+2)a-B_1^{2\nu}(2\nu+1)-(2\nu+2)aB_1^{2\nu+1}+(2\nu+1)B_1^{2\nu+2}]$$

$$-(\nu+1)[(2\nu+2)a-(2\nu+1)]\left|\frac{1}{\nu}B_1(\nu)+B_0(\nu+1)c_1\right|^2$$

$$+(\nu+1)[(2\nu+2)aB_1^{2\nu+1}-(2\nu+1)B_1^{2\nu+2}]|c_1|^2+(2\nu+1)^2[B_1^{2\nu}-B_1^{2\nu+2}]$$

$$+\left|\frac{1}{\nu}B_2(\nu)+B_1(\nu+1)c_1+B_0(\nu+2)c_2\right|^2+B_1^{2\nu+2}|c_2|^2\geq 0$$

Since this holds for $a=1$, it is sufficient to show that the coefficient of a on the left side is nonnegative. The coefficient of a is nonnegative by (27) taken with the family (22), and so (27) holds for the families (22) and (23) and all $a\geq 1$.

It remains to verify (27) for the families (24)-(26). The functions (24) are a linear combination with nonnegative coefficients of those in (22) and (23). Therefore (27) holds for the family (24). Choosing $\nu=-1$ in the inequality, we obtained an inequality equivalent to (27) with the family (25). Choosing instead $D=-B$ and $C=0$, we obtain as a limiting case for $\nu=-2$ an inequality equivalent to (27) with the family (26).

$\square$

THEOREM 4.11 *Given numbers* B_1, B_2, B_3 $(B_1>0)$ *which are not* $1,0,0$ *satisfy the admissible families condition if and only if* $B_1<1$, $|B_2|\leq 2B_1(1-B_1)$, *and*

$$(28) \qquad \left|B_3+\frac{1}{2}\left(\frac{\nu-1}{B_1}+\frac{1}{1-B_1}\right)B_2^2\right|^2\leq\frac{4B_1^2(1-B_1)^2\frac{1}{2}(B_1^{-1}+1)-|B_2|^2}{2B_1(1-B_1)}$$

$$\times\left\{8(\nu+1)^2B_1^2(1-B_1)-(2\nu+3)(2\nu+1)B_1^2(1-B_1^2)-\frac{2\nu+2-(2\nu+1)B_1}{2(1-B_1)}|B_2|^2\right\}$$

for all real ν.

PROOF. We use the supremum [7,9]

$$(29) \qquad \sup_{g\in\mathcal{H}}[\langle f+Tg,f+Tg\rangle_{\mathcal{K}}-\langle g,g\rangle_{\mathcal{H}}],$$

in the case where $\mathcal{H}$ and $\mathcal{K}$ are 2-dimensional Kreĭn spaces, T is a contraction on $\mathcal{H}$ to $\mathcal{K}$, and f is a vector in $\mathcal{K}$. The supremum may be evaluated in terms of any factorization

$$(30) \qquad\qquad 1 - TT^* = DD^*,$$

where D is an operator on a Kreĭn space $\mathcal{D}$ to $\mathcal{K}$ such that $\ker D = \{0\}$. The supremum (29) is finite if and only if f is in the range of D, and if $f = Dh$ its value is

$$(31) \qquad\qquad \langle h, h \rangle_{\mathcal{D}}.$$

We show that the conditions are necessary. Assume that B_1, B_2, B_3 satisfy the admissible families condition, and set $B(z) = B_1 z + B_2 z^2 + B_3 z^3$. By Theorem 3.4, $B_1 \leq 1$ and $|B_2| \leq 2B_1(1 - B_1)$. If $B_1 = 1$, then $B_2 = 0$ and $B_3 = 0$ (apply (27) with the family (23)). Since the numbers are not $1, 0, 0$, $B_1 < 1$. Notice that it is enough to prove (28) except for isolated values of ν, since then by continuity the inequality holds for all real ν.

Consider the supremum formed from the left side of (19),

$$\sup \left\{ \left\langle \frac{B(z)^\nu - B'(0)^\nu z^\nu}{\nu} + g(B(z)), \frac{B(z)^\nu - B'(0)^\nu z^\nu}{\nu} + g(B(z)) \right\rangle_{\mathcal{G}^\nu_{\sigma(a)}} - \langle g(z), g(z) \rangle_{\mathcal{G}^\nu_{\sigma(b)}} \right\},$$

where for the admissible family we use (23). The supremum is over all generalized power series $g(z) = \sum_{n=1}^\infty c_n z^{\nu+n}$. It is not hard to see that $T : g(z) \mapsto g(B(z))$ is a contraction on $\mathcal{G}^\nu_{\sigma(b)}$ to $\mathcal{G}^\nu_{\sigma(a)}$. For if we apply (19) with $g(z)$ replaced by $\epsilon^{-1} g(z)$, multiply by ϵ^2, then let $\epsilon \to 0$, we get

$$\langle g(B(z)), g(B(z)) \rangle_{\mathcal{G}^\nu_{\sigma(a)}} \leq \langle g(z), g(z) \rangle_{\mathcal{G}^\nu_{\sigma(b)}}$$

for any $g(z)$ in $\mathcal{G}^\nu_{\sigma(b)}$. The supremum is thus of the form (29). We calculate a factorization (30). Let C^2 be 2-dimensional Euclidean space with elements written as column vectors $c = [c_1, c_2]^t$. Set

$$D(t) = \left[\begin{array}{cc} (\nu + 1)\sigma_1(t) & 0 \\ 0 & (\nu + 2)\sigma_2(t) \end{array} \right]$$

for $t \geq 1$. Let $\mathcal{H}$ be C^2 in the scalar product $\langle c, c \rangle_{\mathcal{H}} = \langle D(b)c, c \rangle_{C^2}$, and let $\mathcal{K}$ be C^2 in the scalar product $\langle c, c \rangle_{\mathcal{K}} = \langle D(a)c, c \rangle_{C^2}$. Then $\mathcal{H}$ and $\mathcal{K}$ are Kreĭn spaces except for the isolated values of ν where $D(a)$ or $D(b)$ is singular. In an obvious way we may identify $\mathcal{G}^\nu_{\sigma(b)}$ with $\mathcal{H}$ and $\mathcal{G}^\nu_{\sigma(a)}$ with $\mathcal{K}$ by using the coefficients of generalized power series as entries of column vectors. The operator T acts as multiplication by the matrix

$$M = \left[\begin{array}{cc} B_0(\nu + 1) & 0 \\ B_1(\nu + 1) & B_0(\nu + 2) \end{array} \right]$$

on $\mathcal{H}$ to $\mathcal{K}$. Making the usual identifications of operators and matrices, we have

$$T = M, \quad T^* = D(b)^{-1} \overline{M} D(a),$$

where the bar of a matrix denotes its conjugate transpose. By matrix calculus there exists a matrix Q and a diagonal matrix J with diagonal entries either 1 or -1 such that

$$(32) \qquad 1 - TT^* = 1 - MD(b)^{-1} \overline{M} D(a) = D(a)^{-1} Q J \overline{Q},$$

Define $\mathcal{D}$ to be C^2 in the scalar product $\langle c, c\rangle_{\mathcal{D}} = \langle Jc, c\rangle_{C^2}$. Viewing $D = D(a)^{-1}QJ$ as an operator on $\mathcal{D}$ to $\mathcal{K}$, one checks that $D^* = \overline{Q}$, and so (30) holds. From (29)-(31) we get

$$(33) \qquad \sup \left\{ \left\langle \frac{B(z)^\nu - B'(0)^\nu z^\nu}{\nu} + g(B(z)), \frac{B(z)^\nu - B'(0)^\nu z^\nu}{\nu} + g(B(z)) \right\rangle_{\mathcal{G}^\nu_{\sigma(a)}} \right.$$
$$\left. - \langle g(z), g(z)\rangle_{\mathcal{G}^\nu_{\sigma(b)}} \right\}$$

$$= \left\langle D^{-1}\begin{bmatrix} B_1(\nu)/\nu \\ B_2(\nu)/\nu \end{bmatrix}, D^{-1}\begin{bmatrix} B_1(\nu)/\nu \\ B_2(\nu)/\nu \end{bmatrix} \right\rangle_{\mathcal{D}}$$
$$= \left\langle JQ^{-1}D(a)\begin{bmatrix} B_1(\nu)/\nu \\ B_2(\nu)/\nu \end{bmatrix}, Q^{-1}D(a)\begin{bmatrix} B_1(\nu)/\nu \\ B_2(\nu)/\nu \end{bmatrix} \right\rangle_{C^2}.$$

To make the calculation, we must exhibit matrices J and Q satisfying (32). Set

$$D(a) - D(a)MD(b)^{-1}\overline{M}D(a) = \begin{bmatrix} \alpha & \beta \\ \overline{\beta} & \gamma \end{bmatrix}.$$

Then

$$\alpha = (\nu + 1)\sigma_1(a)\left[1 - |B_0(\nu + 1)|^2 \frac{\sigma_1(a)}{\sigma_1(b)}\right]$$
$$= 2(\nu + 1)^2(\nu + 2)^2 a^{-2\nu-1}\frac{(2\nu + 2)a - (2\nu + 1)}{(2\nu + 2)a - (2\nu + 1)B_1}(1 - B_1),$$
$$\beta = -B_0(\nu + 1)\overline{B}_1(\nu + 1) \cdot (\nu + 2)\frac{\sigma_1(a)\sigma_2(a)}{\sigma_1(b)}$$
$$= -(\nu + 1)(\nu + 2)^2\overline{B}_2\, a^{-2\nu-2}\frac{(2\nu + 2)a - (2\nu + 1)}{(2\nu + 2)a - (2\nu + 1)B_1},$$
$$\gamma = (\nu + 2)\sigma_2(a)\left[1 - |B_1(\nu + 1)|^2\frac{\nu + 2}{\nu + 1}\frac{\sigma_2(a)}{\sigma_1(b)} - |B_0(\nu + 2)|^2\frac{\sigma_2(a)}{\sigma_2(b)}\right]$$
$$= (\nu + 2)^2 a^{-2\nu-2}\left[1 - |B_2|^2 B_1^{-1}\frac{\nu + 1}{(2\nu + 2)a - (2\nu + 1)B_1} - B_1^2\right].$$

The determinant $\delta = \alpha\gamma - |\beta|^2$ is given by

$$\delta = (\nu + 1)^2(\nu + 2)^4 a^{-4\nu-4}B_1^{-1}\frac{(2\nu + 2)a - (2\nu + 1)}{(2\nu + 2)a - (2\nu + 1)B_1}\left[2B_1^2(1 - B_1)^2(B_1^{-1} + 1)a - |B_2|^2\right].$$

When $\delta > 0$, (32) holds with

$$J = \begin{bmatrix} 1 & 0 \\ 0 & 1 \end{bmatrix} \quad \text{and} \quad Q = \alpha^{-\frac{1}{2}}\begin{bmatrix} \alpha & 0 \\ \overline{\beta} & \delta^{\frac{1}{2}} \end{bmatrix}.$$

160   K. Y. Li and J. Rovnyak

When $\delta < 0$, we obtain (32) with

$$J = \begin{bmatrix} -1 & 0 \\ 0 & 1 \end{bmatrix} \quad \text{and} \quad Q = |\alpha|^{-\frac{1}{2}} \begin{bmatrix} \alpha & 0 \\ \overline{\beta} & -|\delta|^{\frac{1}{2}} \end{bmatrix}.$$

Formula (33) yields the same result in both cases, namely

$$\sup \left[\left\langle \frac{B(z)^\nu - B'(0)^\nu z^\nu}{\nu} + g(B(z)), \frac{B(z)^\nu - B'(0)^\nu z^\nu}{\nu} + g(B(z)) \right\rangle_{\mathcal{G}^\nu_{\sigma(a)}} - \langle g(z), g(z) \rangle_{\mathcal{G}^\nu_{\sigma(b)}} \right]$$

$$= \frac{1}{\alpha\delta} \left[\delta(\nu+1)^2 \sigma_1(a)^2 \left| \frac{B_1(\nu)}{\nu} \right|^2 + \left| -\overline{\beta}(\nu+1)\sigma_1(a)\frac{B_1(\nu)}{\nu} + \alpha(\nu+2)\sigma_2(a)\frac{B_2(\nu)}{\nu} \right|^2 \right].$$

Therefore (19) gives

$$\frac{1}{\alpha\delta} \left[\delta(\nu+1)^2 \sigma_1(a)^2 \left| \frac{B_1(\nu)}{\nu} \right|^2 + \left| -\overline{\beta}(\nu+1)\sigma_1(a)\frac{B_1(\nu)}{\nu} + \alpha(\nu+2)\sigma_2(a)\frac{B_2(\nu)}{\nu} \right|^2 \right]$$

$$\leq 4b^{-2\nu}(\nu+1) \left[a^{2\nu}\sigma_1(a) - b^{2\nu}\sigma_1(b) \right] + 4b^{-2\nu}(\nu+2)\frac{(2\nu+1)^2}{4} \left[a^{2\nu}\sigma_2(a) - b^{2\nu}\sigma_2(b) \right].$$

When $a = 1$ and $b = B_1^{-1}$, this simplifies to (28). We have shown that the conditions in the theorem are necessary.

      Conversely, assume that $B_1 < 1$, $|B_2| \leq 2B_1(1 - B_1)$, and that (28) holds for all real ν. By Theorem 3.4, $B(z) = B_1 z + B_2 z^2 + B_3 z^3$ satisfies (19) with the family (22). We can reverse the preceding steps to show that $B(z)$ satisfies (19) for the family (23) in the special case $a = 1$ and $b = 1/B'(0)$. Then by Theorem 4.1, the numbers B_1, B_2, B_3 satisfy the admissible families condition.

$$\square$$

      We can now give an example of three numbers which satisfy the admissible families condition but which are not the first three coefficients of a normalized Riemann mapping of the unit disk into itself. They are

$$(34) \qquad\qquad B_1 = \frac{1}{3}, \quad B_2 = \frac{1}{4}e^{i\pi/4}, \quad B_3 = \frac{1}{3}i.$$

Clearly $B_1 < 1$ and $|B_2| < 2B_1(1 - B_1)$. The inequality (28) has the form $uv - w \geq 0$, where

$$u = \frac{4B_1^2(1 - B_1)^2 \frac{1}{2}(B_1^{-1} + 1) - |B_2|^2}{2B_1(1 - B_1)},$$

$$v = 8(\nu+1)^2 B_1^2(1 - B_1) - (2\nu+3)(2\nu+1)B_1^2(1 - B_1^2) - \frac{2\nu + 2 - (2\nu+1)B_1}{2(1 - B_1)}|B_2|^2,$$

$$w = \left| B_3 + \frac{1}{2}\left(\frac{\nu-1}{B_1} + \frac{1}{1 - B_1}\right)B_2^2 \right|^2.$$

We find

$$9 \cdot 81(64)^2(uv - w) = p\nu^2 + q\nu + r,$$

$$p = 415100, \quad q = 582664, \quad r = 242436.$$

Since $q^2 - 4pr = -63043397504$ is negative, the inequality (28) holds for all real ν. Therefore B_1, B_2, B_3 satisfy the admissible families condition by Theorem 4.2.

We show that the numbers (34) do not satisfy the contractive substitution condition, and therefore they are not the first three coefficients of a normalized Riemann mapping of the unit disk into itself. Set $B(z) = B_1 z + B_2 z^2 + B_3 z^3$. The inequality

$$\langle g(B(z)), g(B(z)) \rangle_{\mathcal{D}_\nu^3} \leq \langle g(z), g(z) \rangle_{\mathcal{D}_\nu^3}$$

holds for all $g(z)$ in $\mathcal{D}_\nu^3$ if and only if $D - \overline{M}DM \geq 0$, where

$$D = \begin{bmatrix} \nu+1 & 0 & 0 \\ 0 & \nu+2 & 0 \\ 0 & 0 & \nu+3 \end{bmatrix},$$

$$M = \begin{bmatrix} B_0(\nu+1) & 0 & 0 \\ B_1(\nu+1) & B_0(\nu+2) & 0 \\ B_2(\nu+1) & B_1(\nu+2) & B_0(\nu+3) \end{bmatrix}.$$

The inequality fails when $\nu = -\frac{3}{2}$, because then $\det(D - \overline{M}DM) = -59363/1769472$.

The preceding calculations were carried out using MACSYMA and confirmed with a hand calculator. The example was found from numerical experiments. The experiments showed that, in fact, the contractive substitution condition and the admissible families condition each do a good job in constraining the first three coefficients. This was tested against large random sets of coefficients generated by composing Koebe mappings. The first numerical examples of the failure of the admissible families condition were dependent on the computer programs to verify, but with some adjustment of parameters we were able to obtain the simple numbers (34) for which the verification can be made by any method to do rational arithmetic.

Conclusions should not be drawn hastily. The admissible families method produces sharp coefficient estimates, and it could yet be used in the problem of characterizing initial segments of coefficients of normalized Riemann mappings of the unit disk into itself. Some modification of the method would, however, be required for this purpose.

Added in Proof. We thank the referee for pointing out that in a sequal to [13] titled "Operator-valued measures and coefficients of univalent functions", *Algebra i Analiz* 3 (6) (1991), 1-75; English translation, *St. Petersburg Math. J.* 3 (6) (1992), Vasyunin and Nikol'skiĭ have shown that a stronger condition than the "admissible families" condition is not sufficient for characterizing the initial segments of bounded univalent functions, and that this can be seen in other ways. Also, making use of an additional free parameter, which they called the *isometric trajectory*, they obtained more inequalities, which include inequality (28) in Theorem 4.2 as a special case. Unfortunately, the authors have not seen the key portions of this work and have only been able to refer to [13].

In a joint paper with Gene Christner, the authors have explicitly identified the inequalities corresponding to the contractive substitution condition. In a separate joint paper with Daniel Dreibelbis, making use of these inequalities, the authors have shown that the contractive substitution condition and the admissible families condition together are not sufficient to characterize the initial segments of the coefficients of bounded univalent functions on the unit disk. The first of these papers has been submitted to *Operator Theory: Advances and Applications*, and the second to the *Proceedings of the International Conference and Special Year on Complex Analysis (Nankai Institute of Mathematics, Tianjin, China, 1992)*.

REFERENCES

[1] J. Bognár, *Indefinite Inner Product Spaces*, Springer-Verlag, Berlin-New York, 1974.

[2] L. de Branges, *Square Summable Power Series*, Springer-Verlag, in preparation.

[3] L. de Branges, "Unitary linear systems whose transfer functions are Riemann mapping functions," *Integral Equations and Operator Theory* 19 (1986), 105-124.

[4] L. de Branges, "Underlying concepts in the proof of the Bieberbach conjecture," *Proceedings of the International Congress of Mathematicians* 1986, pp.25-42, Berkeley, California, 1986.

[5] L. de Branges, "A proof of the Bieberbach conjecture," *Acta Math.* 154 (1985), 137-152.

[6] L. de Branges, "Powers of Riemann mapping functions," *The Bieberbach Conjecture (Proceedings of the Symposium on the Occasion of its Proof, Purdue University, 1985)*, A. Baernstein II, D. Drasin, P. Duren and A. Marden, eds., pp. 51-67, Amer. Math. Soc., Providence, 1986.

[7] L. de Branges, "Complementation in Krein spaces," *Trans. Amer. Math. Soc.* 305 (1988), 277-291.

[8] M. Dritschel and J. Rovnyak, "Extension theorems for contraction operators on Kreĭn spaces," *Oper. Theory: Adv. Appl.*, vol. 47, pp.221-305, Birkhauser, Basel-Boston, 1990.

[9] M. Dritschel and J. Rovnyak, "Julia operators and complementation in Kreĭn spaces," Indiana Univ. Math. J. 47 (1991), 885-901.

[10] K. Y. Li, *Applications of de Branges' Theory of Contractively Contained Spaces*, Dissertation, University of California, Berkeley, 1989.

[11] K. Y. Li, "De Branges' conjecture on bounded Riemann mapping coefficients," *J. Analyse Math.* 54 (1990), 289-296.

[12] Z. Nehari, *Conformal Mapping*, Dover, New York, 1975.

[13] V. I. Vasyunin and N. K. Nikol'skiĭ, "Quasiorthogonal decompositions in complementary metrics and estimates of univalent functions," *Algebra and Analysis* 2 (4) (1990), 1-81.

[14] V. P. Potapov, "The multiplicative structure of J-contractive matrix functions," *Trudy Moskov. Mat. Obshch.* 4 (1955), 125-136; *Amer. Math. Soc. Transl.* (2) 15 (1960), 131-243.

[15] J. Rovnyak, "Coefficient estimates for Riemann mapping functions," *J. Analyse Math.* 52 (1989), 53-93.

Department of Mathematics
Hong Kong University of Science and Technology
Clear Water Bay
Kowloon
Hong Kong

Department of Mathematics
Mathematics-Astronomy Building
University of Virginia
Charlottesville, Virginia 22903
U. S. A.

AMS Subject Classification: Primary 47A57, 30C50, Secondary 47B50

Operator Theory:
Advances and Applications, Vol. 62
© 1993 Birkhäuser Verlag Basel

WEAK-STAR LIMITS OF POLYNOMIALS AND THEIR DERIVATIVES

WILLIAM T. ROSS AND JOSEPH A. BALL

Dedicated to T. Ando

ABSTRACT. Let μ and ν be regular finite Borel measures with compact support in the real line $\mathbb{R}$ and define the differential operator $D : L^\infty(\mu) \to L^\infty(\nu)$ with domain equal to the polynomials $\mathcal{P}$ by $Dp = p'$. In this paper we will characterize the weak-star closure of the graph of D

$$\mathcal{G} = \{p \oplus Dp : p \in \mathcal{P}\}$$

in $L^\infty(\mu) \oplus L^\infty(\nu)$. As a consequence we will characterize when D is closable (i.e. the weak-star closure of $\mathcal{G}$ contains no non-zero elements of the form $0 \oplus g$) and when $\mathcal{G}$ is weak-star dense in $L^\infty(\mu) \oplus L^\infty(\nu)$. We will also consider the same problem where μ and ν are measures supported on the unit circle $\mathbb{T}$.

1. INTRODUCTION

Let μ and ν be compactly supported regular finite Borel measures on the real line $\mathbb{R}$ and define the differential operator D from $L^\infty(\mu)$ into $L^\infty(\nu)$ with dense (in the weak-star sense) domain equal to the polynomials $\mathcal{P}$ by $Dp = p'$. In this paper we will give a complete characterization of the weak-star closure of the graph of D

$$(1.1) \qquad \mathcal{G} = \{p \oplus Dp : p \in \mathcal{P}\}$$

in $L^\infty(\mu) \oplus L^\infty(\nu)$, which we denote by $\mathcal{G}^-(\mu,\nu)$. As a consequence of this characterization, we also obtain characterizations of when D is closable (i.e. $\mathcal{G}^-(\mu,\nu)$ contains no non-zero elements of the form $0 \oplus g$) and when $\mathcal{G}$ is weak-star dense in $L^\infty(\mu) \oplus L^\infty(\nu)$. This will continue work of [1] and [2] where the authors look at the closure of the graph of D in $L^p(\mu) \oplus L^p(\nu)$, $1 \le p < \infty$, and $C(E) \oplus C(E)$. Here $C(E)$ is the space of continuous functions on a compact set $E \subset \mathbb{R}$ with the uniform topology. Many of the same techniques of [1] and [2] will be used here and we will refer the reader to these papers for some of the needed technical lemmas.

To state our main results, we first make some definitions. Let m be Lebesgue measure on $\mathbb{R}$ and let $\nu = \nu_a \oplus \nu_s$ be the Lebesgue decomposition of the measure ν with respect to m, where $\nu_a \ll m$ and $\nu_s \perp m$. Note that if $k \in L^\infty(\nu)$ then $k = k_a \oplus k_s$, where $k_a \in L^\infty(\nu_a)$ and $k_s \in L^\infty(\nu_s)$, i.e. $L^\infty(\nu)$ splits as $L^\infty(\nu_a) \oplus L^\infty(\nu_s)$. Our first result will be that $\mathcal{G}^-(\mu,\nu)$ has the following splitting property.

Theorem 1.1. $\mathcal{G}^-(\mu,\nu) = \mathcal{G}^-(\mu,\nu_a) \oplus [0 \oplus L^\infty(\nu_s)]$.

This splitting allows us to focus our attention on $\mathcal{G}^-(\mu, \nu_a)$ for which we make the following definition.

Definition: We say a closed interval I is an *interval of absolute continuity* if

$$m|_I \ll \nu_a|_I,$$

and we say I is a *maximum interval of absolute continuity* (abbreviated MIAC) if I is maximal with respect to this property. (Note that if $w = d\nu/dm$ then a MIAC is the same as a maximal interval for which $w > 0$ m-a.e.)

We now state our main theorem which characterizes $\mathcal{G}^-(\mu, \nu_a)$

Theorem 1.2. *A function $h \oplus k \in L^\infty(\mu) \oplus L^\infty(\nu_a)$ belongs to $\mathcal{G}^-(\mu, \nu_a)$ if and only if for each MIAC $I = [a, b]$ there is a function $\tilde{h}$ satisfying the following three conditions:*

(i) $\tilde{h}$ *is absolutely continuous on* I
(ii) $\tilde{h}|_I = h|_I$ $\mu - a.e.$
(iii) $\tilde{h}(x) = \tilde{h}(a) + \int_a^x k(t)dt$ *for all* $x \in I$

As a consequence of Theorem 1.1 and Theorem 1.2 we can give a complete description of when D is closable. For simplicity we assume, for the rest of the paper, that μ and ν are supported in the unit interval $[0, 1]$.

Theorem 1.3. *The operator $D : L^\infty(\mu) \to L^\infty(\nu)$ is closable if and only if the following three conditions are satisfied:*

(i) $\nu \ll m$
(ii) *The complement of the union of all the MIAC's has ν measure zero.*
(iii) $[0, 1]\backslash supp t(\mu)$ *contains no intervals in a MIAC.*

We also describe when $\mathcal{G}$ is weak-star dense in $L^\infty(\mu) \oplus L^\infty(\nu)$.

Theorem 1.4. $\mathcal{G}^-(\mu, \nu) = L^\infty(\mu) \oplus L^\infty(\nu)$ *if and only if for each MIAC I, $\mu|_I$ is either zero or a point mass.*

When looking at the closure of $\mathcal{G}$ in $L^p(\mu) \oplus L^p(\nu)$, $1 \leq p < \infty$ one has analogs for Theorems 1.1 - 1.4, see [1] and [6]. For $1 < p < \infty$ the MIAC's are replaced by MILI's for w^{1-q}. Here $w = d\nu/dm$, q is the conjugate index for p, and a MILI is a *maximum interval of local integrability*. For $p = 1$, one looks at maximum intervals on which w^{-1} is essentially bounded. In characterizing the uniform closure of $\mathcal{G}$ in $C(E) \oplus C(E)$, see [2], one replaces the MIAC's with maximum intervals in the interior of E. More specifically, write the interior of E as a countable disjoint union of open intervals and let $\hat{E}$ be the union of the closure of these intervals. The intervals in $\hat{E}$ are the analogs for the MIAC's.

As in [1] and [2] we let $\hat{\mathcal{G}}(\mu, \nu_a)$ be the linear manifold of functions $h \oplus k \in L^\infty(\mu) \oplus L^\infty(\nu_a)$ satsfying the conditions (i), (ii), and (iii) of Theorem 1.2 and notice our aim is to prove

$$\mathcal{G}^-(\mu, \nu) = \hat{\mathcal{G}}(\mu, \nu_a) \oplus [0 \oplus L^\infty(\nu_s)].$$

We first establish the splitting in Theorem 1.1 which reduces the problem to showing $\mathcal{G}^-(\mu, \nu_a) = \hat{\mathcal{G}}(\mu, \nu_a)$. The inclusion $\mathcal{G}^-(\mu, \nu_a) \subset \hat{\mathcal{G}}(\mu, \nu_a)$ is reasonably straightforward and for the reverse inclusion, we use the the Hahn-Banach theorem to write $\mathcal{G}^-(\mu, \nu_a) = (^\perp \mathcal{G})^\perp$ and obtain an explicit characterization of the pre-annihilator, $^\perp \mathcal{G}$, of $\mathcal{G}$, in $L^1(\mu) \oplus L^1(\nu_a)$.

Here, for two measures α and β on $[0,1]$, we identify the pre-dual of $L^\infty(\alpha) \oplus L^\infty(\beta)$ with $L^1(\alpha) \oplus L^1(\beta)$ with respect to the usual pairing

$$< h \oplus k, f \oplus g >= \int_{[0,1]} hf d\alpha + \int_{[0,1]} kg d\beta,$$

where $h \oplus k \in L^\infty(\alpha) \oplus L^\infty(\beta)$ and $f \oplus g \in L^1(\alpha) \oplus L^1(\beta)$.

An open-ended problem is to obtain similar results for the case where μ and ν are compactly supported measures in the complex plane $\mathbb{C}$. For the case where $\nu \equiv 0$, a complete characterization of $\mathcal{G}^-(\mu, 0)$ (the weak-star closure of the polynomials) has been obtained by Sarason [8] and so this problem can be thought of as a generalization of Sarason's theorem with derivatives. Conway and Olin [5] have applied this result to operator theory; specifically, they used Sarason's characterization of $\mathcal{G}^-(\mu, 0)$ to develop a functional calculus for subnormal operators. In the last section of the paper we obtain some partial results on $\mathcal{G}^-(\mu, \nu)$ for the case where μ and ν are two measures supported on the unit circle. Conceivably these results will have operator-theoretic applications to the development of a functional calculus for subjordan operators (see [1], [2], and [6]) analogous to those mentioned above for subnormal operators.

2. The Weak-star Closure of $\mathcal{G}$

Our first aim is to establish the splitting property of Theorem 1.1 and for this we need an explicit description of the pre-annihilator, $^\perp\mathcal{G}$, of $\mathcal{G}$ in $L^1(\mu) \oplus L^1(\nu)$. To do this we recall that $w = d\nu/dm$ and use the same proof as in [1], p. 101, (except replace $L^2(\mu) \oplus L^2(\nu)$ with $L^1(\mu) \oplus L^1(\nu)$). Also recall that we are assuming, without loss of generality, that μ and ν are supported on $[0,1]$.

Lemma 2.1. *A pair $f \oplus g \in L^1(\mu) \oplus L^1(\nu)$ belongs to $^\perp\mathcal{G}$ if and only if the following three conditions are satisfied:*

(i) $gd\nu = gd\nu_a = gw dm$
(ii) $\int_{[0,1]} f(x) d\mu(x) = 0$
(iii) $g(w)w(x) = - \int_{(x,1]} f(t) d\mu(t) \ \ m-a.e.$

As a consequence of Lemma 2.1 we have this following corollary which is Theorem 1.1 and states that there is no restriction on k_s for a pair $h \oplus k \in L^\infty(\mu) \oplus L^\infty(\nu)$ to belong to $\mathcal{G}^-(\mu, \nu)$.

Corollary 2.2. $\mathcal{G}^-(\mu, \nu) = \mathcal{G}^-(\mu, \nu_a) \oplus [0 \oplus L^\infty(\nu_s)]$.

Proof. For $k \in L^\infty(\nu)$, let $k = k_a \oplus k_s$ be the Lebesgue decomposition of k with respect to $\nu_a \oplus \nu_s$. Using Lemma 2.1 we see that $0 \oplus k_s \in (^\perp\mathcal{G})^\perp = \mathcal{G}^-(\mu, \nu)$. $\square$

Remark: By our splitting property in Theorem 1.1, we are able to characterize $\mathcal{G}^-(\mu, \nu)$ by characterizing $\mathcal{G}^-(\mu, \nu_a)$. We also note that for the operator D to be closable (i.e. $\mathcal{G}^-(\mu, \nu)$ contains no non-zero elements of the form $0 \oplus g$) we see that we must have $\nu \ll m$. With regards to the question of density, we see from our splitting that $\mathcal{G}^-(\mu, \nu) = L^\infty(\mu) \oplus L^\infty(\nu)$ if and only if $\mathcal{G}^-(\mu, \nu_a) = L^\infty(\mu) \oplus L^\infty(\nu_a)$. With these observations, we see that the answers to our questions depend on ν_a, not ν_s, and thus the following assumption will be in force for the rest of the paper (unless otherwise noted):

Assumption: $\nu = \nu_a$.

Recall that $\hat{\mathcal{G}}(\mu, \nu)$ is the linear manifold of functions $h \oplus k \in L^\infty(\mu) \oplus L^\infty(\nu)$ satisfying (i) - (iii) of Theorem 1.2. To prove Theorem 1.2 we must show

$$\mathcal{G}^-(\mu, \nu) = \hat{\mathcal{G}}(\mu, \nu).$$

Clearly $\mathcal{G} \subset \hat{\mathcal{G}}(\mu, \nu)$ and we now show $\mathcal{G}^-(\mu, \nu) \subset \hat{\mathcal{G}}(\mu, \nu)$.

Proposition 2.3. $\hat{\mathcal{G}}(\mu, \nu)$ *is weak-star closed and hence* $\mathcal{G}^-(\mu, \nu) \subset \hat{\mathcal{G}}(\mu, \nu)$.

Proof. Since μ and $\nu = wdm$ are regular Borel measures, then $L^1(\mu)$ and $L^1(\nu)$ are separable and thus, by a consequence the Krein-Smulian theorem [4], p. 165, to show the linear manifold $\hat{\mathcal{G}}(\mu, \nu)$ is weak-star closed it suffices to show $\hat{\mathcal{G}}(\mu, \nu)$ is weak-star sequentially closed. To this end, let $\{h_n \oplus k_n\}$ be a sequence in $\hat{\mathcal{G}}(\mu, \nu)$ with $h_n \oplus k_n \to h \oplus k$ weak-star in $L^\infty(\mu) \oplus L^\infty(\nu)$. Let $I = [a, b]$ be a MIAC and note that $dm|_I = w^{-1}d\nu|_I$. In particular $w^{-1}|_I \in L^1(d\nu|_I)$. Since $\{h_n \oplus k_n\} \subset \hat{\mathcal{G}}(\mu, \nu)$, then for μ-a.e. all $x, y \in I$ we have

$$(2.1) \qquad h_n(y) - h_n(x) = \int_{[x,y]} k_n(s)ds = \int_{[x,y]} k_n(s)w(s)^{-1}d\nu(s)$$

and the right hand side of (2.1) converges to

$$\int_{[x,y]} k(s)w(s)^{-1}d\nu(s) = \int_{[x,y]} k(s)ds.$$

Thus h_n converges pointwise μ-a.e. on I to the absolutely continuous function

$$\tilde{h}(x) = \tilde{h}(a) + \int_{[a,x]} k(s)ds.$$

We will complete the proof by showing $h|_I = \tilde{h}|_I$ μ-a.e. We do this by showing $h_n|_I \to \tilde{h}|_I$ weak-star in $L^\infty(\mu|_I)$.

For μ-a.e. $x \in I$ we have

$$(2.2) \quad |h_n(x) - \tilde{h}(x)| \le |\int_{[a,x]} (k_n(s) - k(s))w(s)^{-1}d\nu(s)| \le \|k_n - k\|_{L^\infty(\nu)}\|w^{-1}\|_{L^1(\nu|_I)}.$$

Note that $\|k_n - k\|_{L^\infty(\nu)}$ is uniformly bounded in n (since $\{k_n\}$ is weak-star convergent in $L^\infty(\nu)$). Using this fact, (2.2), and the Lebesgue dominated convergence theorem we have $h_n|_I \to \tilde{h}|_I$ weak-star in $L^\infty(\mu|_I)$. $\square$

Recalling the definition of a MIAC, one sees that the set of MIAC's is a disjoint collection of closed intervals (each of positive length) hence countable. Letting $\mathcal{M}$ be the union of all the MIAC's, these next two lemmas will show that the functions in $^\perp\mathcal{G}$ vanish on the set

$$(2.3) \qquad\qquad\qquad K \equiv [0, 1]\backslash\mathcal{M}$$

Lemma 2.4. *If* $x \in K$ *and* $f \oplus g \in {}^\perp\mathcal{G}$ *then*

$$\int_{[x,1]} f(t)d\mu(t) = \int_{(x,1]} f(t)d\mu(t) = 0.$$

Proof. Since the measure $\lambda(A) = \int_A f d\mu$ is countably additive, then

$$\lim_{\delta \to 0} \int_{(x+\delta, 1]} f d\mu = \int_{(x, 1]} f d\mu$$

for all $x \in [0, 1]$. Now assume $x \in K$ and the above limit is not zero. Then there is an $\varepsilon > 0$ and an interval J with left-hand endpoint x such that for all $y \in J$, $|\int_{(y, 1]} f d\mu| > \varepsilon$. By Lemma 2.1 (iii)

$$|g(y)w(y)| = |\int_{(y, 1]} f(t) d\mu(t)| > \varepsilon$$

m-a.e. on J. Letting A be a subset of J with $\nu(A) = 0$, we have

$$0 = \int_A |g| d\nu = \int_A |g| w \, dm \geq \varepsilon m(A).$$

From this we obtain $m(A) = 0$, hence $m|_J \ll \nu|_J$, implying that $x \notin K$, a contradiction. A similar argument shows $\int_{[x, 1]} f d\mu = 0$. $\square$

Lemma 2.5. *If $f \oplus g \in {}^{\perp}\mathcal{G}$ then $f = 0$ μ-a.e. on K and $g = 0$ ν-a.e on K.*

Proof. This is a direct analog of a proof in [1], p. 106, of an analogous result for $L^2(\mu) \oplus L^2(\nu)$. Let λ be the measure $\lambda(A) = \int_A f d\mu$. Applying Lemma 2.1 and Lemma 2.4 we see that for m-a.e. $x \in K$, $g(x)w(x) = -\lambda((x, 1]) = 0$. That is $gw = 0$ m-a.e. (hence by assumption $g = 0$ ν-a.e) on K. To complete the proof we need only show that $\lambda(A) = 0$ for all $A \subset K$, since this will show $f = 0$ μ-a.e.

Since $\{x\} = [x, 1] \backslash (x, 1]$, one can use Lemma 2.4 and the additivity of λ to get $\lambda(\{x\}) = 0$ for all $x \in K$, i.e. λ has no point masses in K. Now let $A \subset K$. By regularity of the total variation measure $|\lambda|$, we may assume A is closed. Let $\varepsilon > 0$ be given and choose an open set $U \supset A$ so that $|\lambda|(U \backslash A) < \varepsilon$. Write

$$U = \bigcup_{n \in \mathbb{Z}} (a_n, b_n)$$

as a countable disjoint union of intervals. We construct a new open set U' as follows: Let

$$\mathbb{Z}' = \{n \in \mathbb{Z} : (a_n, b_n) \cap A \neq \emptyset\}$$

and for each $n \in \mathbb{Z}$ let

$$a'_n = \inf\{x : x \in (a_n, b_n) \cap A\}$$
$$b'_n = \sup\{x : x \in (a_n, b_n) \cap A\}$$

and let

$$U' = \bigcup_{n \in \mathbb{Z}'} (a'_n, b'_n).$$

Since A is closed, we have $a'_n, b'_n \in A$ for all n, and $A \subset U'$ except for the endpoints a'_n and b'_n. But since these points are countable and λ has no point masses in K, this set has λ-measure zero. Furthermore since each $a'_n, b'_n \in K$ we have (by Lemma 2.4)

$$\lambda((a'_n, b'_n)) = \lambda((a'_n, 1]) - \lambda([b'_n, 1]) = 0.$$

Thus $\lambda(U') = 0$ and so

$$|\lambda(A)| = |\lambda(U') - \lambda(A)| = |\lambda(U' \backslash A)| \leq |\lambda|(U' \backslash A) \leq |\lambda|(U \backslash A) \leq \varepsilon. \quad \square$$

Lemma 2.6. *For each MIAC $I = [a, b]$ we have*

$$\int_{[a,1]} f(t)d\mu(t) = \int_{(b,1]} f(t)d\mu(t) = 0$$

for all $f \oplus g \in {}^{\perp}\mathcal{G}$.

Proof. By the maximality of I, there is a sequence $\{x_n\} \subset K$ with $x_n \nearrow a$ and hence

$$\int_{[a,1]} f(t)d\mu(t) = \lim_{n\to\infty} \int_{(x_n,1]} f(t)d\mu(t),$$

which by Lemma 2.4 is equal to zero. A similar argument shows $\int_{(b,1]} f(t)d\mu(t)$ is equal to zero. $\square$

Lemma 2.7. *If I is a MIAC and $h \oplus k \in \hat{\mathcal{G}}(\mu, \nu)$ then*

$$\int_I hf d\mu + \int_I kg d\nu = 0$$

for all $f \oplus g \in^{\perp} \mathcal{G}$.

Proof. Let $I = [a, b]$. Using Lemma 2.1 we have

$$\int_{[a,b]} k(x)g(x)d\nu(x) = -\int_{[a,b]} k(x) \left(\int_{(x,1]} f(t)d\mu(t) \right) dx$$

$$= -\int_{[a,b]} \left(\int_{[a,x]} k(t)dt \right) f(x)d\mu(x) - \int_{(b,1]} \left(\int_{[a,b]} k(t)dt \right) f(x)d\mu(x)$$

by reversing the order of integration. By definition of $\hat{\mathcal{G}}(\mu, \nu)$, there is a $\tilde{h}$ so that $h|_I = \tilde{h}|_I$ μ-a.e. and for all $x, y \in I$,

$$\tilde{h}(y) - \tilde{h}(x) = \int_{[x,y]} k(t)dt.$$

Thus

$$\int_{[a,b]} k(x)g(x)d\nu(x) = -\int_{[a,b]} h(x)f(x)d\mu(x) + \tilde{h}(a) \int_{[a,b]} f(t)d\mu(t)$$

$$-\tilde{h}(b) \int_{(b,1]} f(t)d\mu(t) + \tilde{h}(a) \int_{(b,1]} f(t)d\mu(t)$$

$$= -\int_{[a,b]} h(x)f(x)d\mu(x) + \tilde{h}(a) \int_{[a,1]} f(t)d\mu(t) - \tilde{h}(b) \int_{(b,1]} f(t)d\mu(t).$$

By Lemma 2.6 the last two integrals are zero, hence

$$\int_I hf d\mu + \int_I kg d\nu = \int_I hf d\mu - \int_I hf d\mu = 0. \ \square$$

Proof of Theorem 1.2

Proof. By Proposition 2.3, $\mathcal{G}^{-}(\mu, \nu) \subset \hat{\mathcal{G}}(\mu, \nu)$. For the other direction, we let $h \oplus k \in \hat{\mathcal{G}}(\mu, \nu)$ and show that $h \oplus k \in ({}^{\perp}\mathcal{G})^{\perp} = \mathcal{G}^{-}(\mu, \nu)$, by showing

$$(2.4) \qquad \int_{[0,1]} fh d\mu + \int_{[0,1]} kg d\nu = 0$$

for all $f \oplus g \in {}^{\perp}\mathcal{G}$. By Lemma 2.5, the left-hand side of (2.4) is equal to

$$\int_{\mathcal{M}} fh d\mu + \int_{\mathcal{M}} kg d\nu,$$

which be Lemma 2.7 and countable additivity is equal to zero. $\square$

3. Closability of D

To characterize when $D : L^\infty(\mu) \to L^\infty(\nu)$ is closable we will closely follow a procedure used in [1] to characterize when $D : L^2(\mu) \to L^2(\nu)$ is closable. Before we prove Theorem 1.3, we note that

$$\mathcal{G}^-(\mu, \nu) = \mathcal{G}^-(\mu, \nu_a) \oplus [0 \oplus L^\infty(\nu_s)],$$

hence if $\nu_s \neq 0$, then we have the following:

Lemma 3.1. *If $\nu_s \neq 0$, then D is not closable.*

Proof of Theorem 1.3

Proof. Suppose $\nu \ll m$, $\nu(K) = 0$ and the complement of the $\mathrm{suppt}(\mu)$ contains no intervals in a MIAC. We want to show that if $0 \oplus k \in \mathcal{G}^-(\mu, \nu)$, then $k = 0$ ν-a.e. Since $\nu(K) = 0$, it suffices to show $k|_I = 0$ for each MIAC I. Since $\mathcal{G}^-(\mu, \nu) = \hat{\mathcal{G}}(\mu, \nu)$ we know that there is a $\tilde{h}$ which is absolutely continuous on I with $\tilde{h} = 0$ μ-a.e. and

$$\tilde{h}(x) = \tilde{h}(a) + \int_a^x k(t)dt$$

on I. Now $\mathrm{suppt}(\mu) \cap I$ is dense in I, by assumption, so $\tilde{h} \equiv 0$ on I. Thus $\int_x^y k(t)dt = 0$ for all $x, y \in I$ which implies that $k = 0$ m-a.e. on I. But since $\nu \ll m$ we have $k = 0$ ν-a.e. on I. Thus D is closable.

Conversely, if ν is not absolutely continuous with respect to m we use Lemma 3.1 to conclude D is not closable.

If $\nu \ll m$, $\nu(K) > 0$, consider the non-zero element of $L^\infty(\mu) \oplus L^\infty(\nu)$, $0 \oplus \chi_K$. By Lemma 2.5, this element annihilates $^\perp\mathcal{G}$ and thus D is not closable.

Finally suppose $\nu \ll m$, $\nu(K) = 0$, but there is an interval $(c, d) \subset I \backslash \mathrm{suppt}(\mu)$ where I is some MIAC. Choose $c < c' < d' < d$ and set

$$k(t) = \chi_{[c',r]}(t) - \chi_{[r,d']}(t),$$

where $r = c' + \frac{1}{2}(d' - c')$. Then $\int_{[c',d']} k(t)dt = 0$ and $\mathrm{suppt}(k) \subset [c', d']$. If we define h_1 on $[c, d]$ by

$$h_1(x) = \int_{c'}^x k(t)dt$$

and note that h_1 is absolutely continuous on I with support in $[c', d']$. Hence $h_1 = 0$ μ-a.e. and so $0 \oplus k \in \hat{\mathcal{G}}(\mu, \nu) = \mathcal{G}^-(\mu, \nu)$, thus D is not closable. $\square$

4. Density of $\mathcal{G}$

We finally consider the most degenerate case of when $\mathcal{G}^-(\mu, \nu) = L^\infty(\mu) \oplus L^\infty(\nu)$ but first we need the following localization theorem whose proof can be found in [1], p. 123. (Just replace $L^2(\mu) \oplus L^2(\nu)$ with $L^1(\mu) \oplus L^1(\nu)$.)

Lemma 4.1. *Suppose I is a MIAC and $f \oplus g \in L^1(\mu) \oplus L^1(\nu)$ has support in I. Then $f \oplus g \in {}^\perp\mathcal{G}$ if and only if the following three conditions are satisfied:*

    (i) $g d\nu = g d\nu_a = g w dm$ *on I*
    (ii) $\int_I f d\mu = 0$
    (iii) *For m-a.e. $x \in I$, $g(x)w(x) = -\int_{(x,1]\cap I} f(t)d\mu(t)$.*

172   W. T. Ross and J. A. Ball

Remark: We now prove our main result about the weak-star density of $\mathcal{G}$ in $L^\infty(\mu)\oplus L^\infty(\nu)$ and again we closely follow [1]. But first we observe that by Theorem 1.1 $\mathcal{G}^-(\mu,\nu) = L^\infty(\mu)\oplus L^\infty(\nu)$ if and only if $\mathcal{G}^-(\mu,\nu_a) = L^\infty(\mu)\oplus L^\infty(\nu_a)$. Thus we remind the reader that we are assuming without loss of generality that $\nu = \nu_a = wdm$. (And thus condition (i) Lemma 4.1 is vacuous.)

Proof of Theorem 1.4

Proof. Suppose that for some MIAC $I = [a,b]$, $\mu|_I$ is non-zero and not a single point mass. Then there is a set $S \subset \mathrm{suppt}(\mu|_I)$ so that $0 < \mu(S) < \mu(\mathrm{suppt}(\mu|_I))$. Let

$$f(x) = -\frac{\mu(I\backslash S)}{\mu(S)}\chi_S(x) + \chi_{I\backslash S}(x)$$

if $x \in I$ and zero otherwise. Then $f \in L^1(\mu)$ and $f \not\equiv 0$. Define

$$g(x) = -\frac{1}{w(x)}\int_{[x,b]} f(t)d\mu(t)$$

if $w(x) \neq 0$ and zero otherwise. Note that since I is a MIAC then $w(x) > 0$ m-a.e. on I, so $g \in L^1(\nu)$ and satisfies (iii) of Lemma 4.1 (Condition (i) is satisfied by assumption.). Finally we see that (ii) of Lemma 4.1 is satisfied, since

$$\int_I f(t)d\mu(t) = -\frac{\mu(I\backslash S)}{\mu(S)}\int_I \chi_S(t)d\mu + \int_I \chi_{I\backslash S}(t)d\mu$$

$$= -\frac{\mu(I\backslash S)}{\mu(S)}\mu(S) + \mu(I\backslash S) = 0.$$

Hence $^\perp\mathcal{G} \neq (0)$, so $\mathcal{G}^-(\mu,\nu) \neq L^\infty(\mu)\oplus L^\infty(\nu)$.

To finish, we must show that if either $\mu|_I = 0$ or $\mu|_I$ is a single point mass for every MIAC I, then $\mathcal{G}^-(\mu,\nu) = L^\infty(\mu)\oplus L^\infty(\nu)$. First suppose $\mu|_I = 0$. Then $L^1(\mu|_I) = (0)$. If $g \in L^1(\nu)$ is such that $0\oplus g \in {}^\perp\mathcal{G}$, then by Lemma 4.1 $gw = 0$ m-a.e. (and by assumption $g = 0$ ν-a.e) on I. Thus $^\perp\mathcal{G}|_I = 0$. Now suppose $\mu|_I$ is a single point mass at $x_0 \in I$ and let $f\oplus g \in {}^\perp\mathcal{G}$. Then by Lemma 4.1, $f(x_0)\mu(\{x_0\}) = 0$. Thus $f = 0$ μ-a.e. on I. Hence by Lemma 4.1 $gw = 0$ m-a.e. (and hence $g = 0$ ν a.e.) on I and thus $^\perp\mathcal{G}|_I = (0)$.

Combining these two cases we get $^\perp\mathcal{G}|_I = (0)$ for all MIAC's I. By Lemma 2.5 $^\perp\mathcal{G}|_K = (0)$ and so $^\perp\mathcal{G} = (0)$. $\square$

5. Measures on the Circle

For two finite regular Borel measures μ and ν with compact support in the complex plane, one might consider the more general problem of characterizing, $\mathcal{G}^-(\mu,\nu)$, the closure of the graph

$$\mathcal{G} = \{p\oplus Dp,\ p \in \mathcal{P}\}$$

of $D = \frac{d}{dz}$ in the weak-star topology of $L^\infty(\mu)\oplus L^\infty(\nu)$. (Here $\mathcal{P}$ is the set of polynomials in the complex variable z.) The complete description of $\mathcal{G}^-(\mu,\nu)$ seems to be very complicated and is yet unknown. For example if $\nu \equiv 0$, then $\mathcal{G}^-(\mu,0)$, the weak-star closure of the polynomials in $L^\infty(\mu)$ has been characterized by Sarason [8]. Thus this problem can be viewed as a generalization of Sarason's theorem with derivatives. In this section, we consider the special case where μ and ν are measures supported on the unit circle $\mathbb{T}$ and generalize results of [3] (which look at the closure of $\mathcal{G}$ in $L^2(\mu)\oplus L^2(\nu)$) to the weak-star case. Again our main tool will be the use of the Hahn-Banach theorem to write $\mathcal{G}^-(\mu,\nu) = (^\perp\mathcal{G})^\perp$ and then use the F. and M. Riesz theorem to obtain an explicit characterization of $^\perp\mathcal{G}$.

To establish some notation, we let $\nu = w|dz| + \nu_s$, $w = d\nu/|dz|$, $\nu_s \perp |dz|$, be the Lebesgue decomposition of ν with respect to arc length measure $|dz|$ on the unit circle $\mathbb{T}$. We also let H_0^1 denote the Hardy subspace of $L^1(|dz|)$ of functions with vanishing non-positive Fourier coefficients. We now refer the reader to [3] for the proof of this next result which is similar to Lemma 2.1 and uses the F. and M. Riesz theorem to characterize $^\perp\mathcal{G}$.

Proposition 5.1. *A function pair $f \oplus g \in L^1(\mu) \oplus L^1(\nu)$ belongs to $^\perp\mathcal{G}$ if and only if the following conditions are satisfied:*

   (i) $gd\nu = gw|dz|$
   (ii) $\int_\mathbb{T} f(z)d\mu(z) = 0$
   (iii) *The function*

$$(5.1) \qquad \frac{dz}{|dz|} \int_z^{z_0} f(\zeta)d\mu(\zeta) + g(z)w(z)$$

belongs to H_0^1 for all $z_0 \in \mathbb{T}$.

As a corollary of Proposition 5.1 we have the following splitting property which is analogous to Corollary 2.2 and whose proof is the same.

Corollary 5.2. $\mathcal{G}^-(\mu,\nu) = \mathcal{G}^-(\mu,\nu_a) \oplus [0 \oplus L^\infty(\nu_s)]$

As in section 1, Corollary 5.2 allows us to focus our attention on $\mathcal{G}(\mu,\nu_a)$ and so as before we make the following assumption:

Assumption: $\nu = \nu_a$

This next corollary is a refinement of Proposition 5.1 to the case when $\mathrm{suppt}(\mu)$ and $\mathrm{suppt}(\nu)$ are contained in some proper subset of $\mathbb{T}$ and will be the key step in our characterization of $\mathcal{G}^-(\mu,\nu)$ in this case.

Proposition 5.3. *If $\mathrm{suppt}(\mu)$ and $\mathrm{suppt}(\nu)$ are contained in some proper subset of $\mathbb{T}$ then the function pair $f \oplus g \in L^1(\mu) \oplus L^1(\nu)$ belongs to $^\perp\mathcal{G}$ if and only if (i) and (ii) of Proposition 5.1 hold in addition to*

$$(5.2) \qquad \frac{dz}{|dz|} \int_z^{z_0} f(\zeta)d\mu(\zeta) + g(z)w(z) = 0 \ \forall \, z_0 \in \mathbb{T}$$

Proof. By hypothesis, there is an arc J which is disjoint from the supports of μ and ν. Thus if $z_0 \in J$, then by Proposition 5.1 the function in (5.1) is an H^1 function which vanishes on a set of positive Lebesgue measure, which, by [7], p. 52, vanishes identically. $\square$

If μ and ν have support in some proper subset of $\mathbb{T}$, then the structure of $\mathcal{G}^-(\mu,\nu)$ is similar to the case where μ and ν are supported on the real line and one can readily check that all the results concerning $\mathcal{G}^-(\mu,\nu)$, the closablilty of $D = \frac{d}{dz}$, and the density of $\mathcal{G}$ in $L^\infty(\mu) \oplus L^\infty(\nu)$ are analogous to those for the real line case except that we replace m with $|dz|$ and MIAC with MAAC (*maximum arc of absolute continuity*). Here a closed arc $I \subset \mathbb{T}$ is an *arc of absolute continuity* if

$$|dz||_I \ll \nu|_I$$

and a *maximum arc of absolute continuity* if I is maximal with respect to this property. With these definitions one checks that our results are the same once Lemma 2.1 is replaced with Proposition 5.3.

When $\text{suppt}(\mu) = \mathbb{T}$ or $\text{suppt}(\nu) = \mathbb{T}$, the situation becomes more complicated in the sense that the polynomials (the domain of D) may fail to be weak-star sense in $L^\infty(\mu)$. For example, if the support of μ is properly contained in $\mathbb{T}$, then the polynomials are weak-star dense in $L^\infty(\mu)$. If the support of μ is all of $\mathbb{T}$ then, by the F. and M. Riesz theorem, the polynomials fail to be weak-star dense in $L^\infty(\mu)$ if any only if Lebesgue measure on the unit circle is absolutely continuous with respect to μ. In the special case where $\nu = w|dz|$ and $\text{suppt}(\nu) = \mathbb{T}$ (i.e. $L^\infty(|dz|) = L^\infty(w|dz|)$) we are able to give a complete description of $\mathcal{G}^-(\mu, \nu)$.

Theorem 5.4. *Let μ and ν be finite positive Borel measures on $\mathbb{T}$ with $\nu = w|dz|$ and $\text{suppt}(\nu) = \mathbb{T}$. Then $h \oplus k \in L^\infty(\mu) \oplus L^\infty(\nu)$ belongs to $\mathcal{G}^-(\mu, \nu)$ if and only if there is a function $\tilde{h}$ such that*

    (i) *$\tilde{h}$ is absolutely continuous on $\mathbb{T}$ and $h = \tilde{h}$ μ-a.e.*
    (ii) *$k \in H^\infty$*
    (iii) *$\tilde{h}' = k$ $|dz|$-a.e.*

Proof. Let $\hat{\mathcal{G}}(\mu, \nu)$ be the linear manifold of $h \oplus k$ in $L^\infty(\mu) \oplus L^\infty(\nu)$ satisfying (i) - (iii) of Theorem 5.4. We now show that $\hat{\mathcal{G}}(\mu, \nu) = \mathcal{G}^-(\mu, \nu)$. To show $\mathcal{G}^-(\mu, \nu) \subset \hat{\mathcal{G}}(\mu, \nu)$ we follow the proof of Proposition 5.1 and show that $\hat{\mathcal{G}}(\mu, \nu)$ is weak-star closed; which by the Krein-Smulian theorem it suffices to show $\hat{\mathcal{G}}(\mu, \nu)$ is weak-star sequentially closed. So let $\{h_n \oplus k_n\}$ be a sequence in $\hat{\mathcal{G}}(\mu, \nu)$ with $h_n \oplus k_n \to h \oplus k$ weak-star in $L^\infty(\mu) \oplus L^\infty(\nu)$. Since $L^\infty(|dz|) = L^\infty(w|dz|)$ and H^∞ is weak-star closed in $L^\infty(|dz|)$, then $k \in H^\infty$, so (ii) holds. Now write, for μ-a.e. $z_1, z_2 \in \mathbb{T}$,

$$h_n(z_2) - h_n(z_1) = \int_{z_1}^{z_2} k_n(\zeta) d\zeta$$

and proceed as in the proof of Proposition 2.3 to show (i) and (iii) of Theorem 5.4. Thus $\mathcal{G}^-(\mu, \nu) \subset \hat{\mathcal{G}}(\mu, \nu)$.

For the other direction, suppose $h \oplus k \in \hat{\mathcal{G}}(\mu, \nu)$. To finish, we need to show $h \oplus k$ annihilates every $f \oplus g \in {}^\perp\mathcal{G}$, i.e.

$$\int_\mathbb{T} hf d\mu + \int_\mathbb{T} gk d\nu = 0$$

for all $f \oplus g$ satisfying (i) - (iii) of Proposition 5.1. Since $h \oplus k \in \hat{\mathcal{G}}(\mu, \nu)$ then $k \in H^\infty$ and by Proposition 5.1

$$A(z) = g(z) + w(z)^{-1} \frac{dz}{|dz|} \int_z^{z_0} f(\zeta) d\mu(\zeta)$$

belongs to $w^{-1}H_0^1 \cap L^\infty(w|dz|)$. Hence

$$\int_\mathbb{T} kAw|dz| = 0,$$

that is to say

$$\int_\mathbb{T} kgw|dz| = -\int_\mathbb{T} k(z) \left(\int_z^{z_0} f(\zeta) d\mu(\zeta) \right) dz.$$

Since

$$\int_{z_0}^{\zeta} k(z)dz = \tilde{h}(\zeta) - \tilde{h}(z_0),$$

then

$$\int_{\mathbb{T}} kgw|dz| = -\int_{\mathbb{T}} \tilde{h}(\zeta)f(\zeta)d\mu(\zeta) + \tilde{h}(z_0)\int_{\mathbb{T}} f(\zeta)d\mu(\zeta)$$

and the last integral (by Proposition 5.1) above is zero. Hence

$$\int_{\mathbb{T}} hfd\mu + \int_{\mathbb{T}} kgw|dz| = 0 \quad \Box$$

REFERENCES

1. J.A. Ball and T.R. Fanney, *Closability of differential operators and real subjordan operators*, Topics in Operator Theory: Ernst D. Hellinger Memorial Volume, (L. de Branges, I. Gohberg, and J. Rovnyak, eds.), Operator Theory: Advances and Applications, vol. 48, Birkhäuser-Verlag, Basel, 1990, p. 93-156.
2. J.A. Ball and T.R. Fanney, *Uniform limits of sequences of polynomials and their derivatives*, Proc. Amer. Math. Soc. **114** (1992), p. 749 - 755.
3. J.A. Ball and T.R. Fanney, *Pure subjordan operators and simultaneous approximation by a polynomial and its derivative*, to appear, Jour. Op. Thry.
4. J.B. Conway, *A Course in Functional Analysis*, Springer-Verlag, New York-Berlin, 1985.
5. J.B. Conway, and R.F. Olin, *A functional calculus for subnormal operators II*, Mem. Amer. Math. Soc., No. 184, Amer. Math. Soc., Providence, 1970.
6. T.R. Fanney, *Closability of certain differential operators and subjordan operators*, Ph.D. dissertation, Virginia Polytechnic Institute and State University, 1989.
7. K. Hoffman, *Banach Spaces of Analytic Functions*, Prentice-Hall, Englewood Cliff, New Jersey, 1962.
8. D. Sarason, *Weak-star density of polynomials*, J. Reine Angew. Math., **252** (1972), p. 1 - 15.
9. W. Rudin, *Functional Analysis*, McGraw-Hill, New York, 1973.

DEPARTMENT OF MATHEMATICS, UNIVERSITY OF RICHMOND, RICHMOND, VA 23173, USA
E-mail address: rossb@aurora.urich.edu

DEPARTMENT OF MATHEMATICS, VIRGINIA POLYTECHNIC INSTITUTE AND STATE UNIVERSITY, BLACKSBURG, VA 24061, USA
E-mail address: ball@math.vt.edu

AMS subject classification: Primary 41A10, Secondary 47E05

Operator Theory:
Advances and Applications, Vol. 62
© 1993 Birkhäuser Verlag Basel

HAUSDORFF DIMENSION OF SOME FRACTALS
AND PERRON-FROBENIUS THEORY

F. Takeo

Dedicated to Prof. T. Ando on the occasion of his sixtieth birthday.

We shall describe a method to calculate the Hausdorff dimension of a set of sequences of some type, by using Perron-Frobenius theory of non-negative matrices. It is applied to a fractal set which is not self-similar but is a subset of the invariant set with respect to an iterated function system. It is also applied to images of certain function systems which do not satisfy the open set condition.

INTRODUCTION

The Hausdorff dimension plays an important role in various studies. For example, a fractal is characterized by Mandelbrot as a set for which the Hausdorff dimension strictly exceeds the topological dimension [4, p.15]. It is known how to calculate the Hausdorff dimension of some fractals such as self-similar sets and some others. Self-similar sets are the invariant set with respect to an iterated function system of similitudes, which satisfies Moran's open set condition [3,§5.1]. Typical fractals such as the Cantor set, the Koch curve and the Sierpinski gasket are self-similar sets. The Hausdorff dimensions of these self-similar sets are the same with their similarity dimensions, which are easily calculable [5,Theorem III]. However, the Hausdorff dimension of fractal sets which are not self-similar, is not so easy to calculate.

V. Drobot and J. Turner computed the Hausdorff dimension of certain subsets of the unit interval, by using the Perron-Frobenius theory of non-negative matrices in [1]. Those sets are images of function systems which satisfy a kind of Moran's open set condition.

In this paper, we shall describe a method to calculate the Hausdorff dimension of certain sets of sequences on k symbols, endowed with some metric and we apply this result to certain subsets of $\mathbf{R}^d$, which may not be self-similar. These subsets may also be images of function systems which do not satisfy Moran's open set condition. The subsets of the unit interval shown in [1] are special ones among the above sets.

In Section 2, we investigate the Hausdorff dimension of a subset of the space $E_k^{(\omega)}$ of infinite sequences. For a finite subset F of the space $E_k^{(r+1)}$ of finite sequences of length $r+1(r \geq 1)$, we consider the set $\mathcal{A}(F) = \{\underline{x} = (x_j)_{j=1}^{\infty} \in E_k^{(\omega)} \mid (x_n x_{n+1} \cdots x_{n+r}) \in F$ for any $n = 1, 2, \ldots\}$ and define a matrix M induced from F. By introducing the metric $\rho_\delta(0 < \delta < 1)$ in $E_k^{(\omega)}$ and by using the spectral radius λ_0 of M, we show that the Hausdorff dimension of $\mathcal{A}(F)$ in the metric space $(E_k^{(\omega)}, \rho_\delta)$ is $\log_{\frac{1}{\delta}} \lambda_0$ [Proposition 2.5 and Theorem 2.6]. In Section 3 as an application of the above result we calculate the Hausdorff dimension of fractals in $\mathbf{R}^d$ by investigating the condition for the mapping of $\mathcal{A}(F)$ into $\mathbf{R}^d$ to keep the Hausdorff dimension unchanged. Theorem 3.1 shows that the mapping of $\mathcal{A}(F)$ into $\mathbf{R}^d$ keeps the Hausdorff dimension unchanged if a kind of open set condition is satisfied, which is a weaker condition than Moran's open set condition. Theorem 3.2 shows that even if the open set condition (3.1) in Theorem 3.1 is not satisfied, the Hausdorff dimension is kept unchanged under certain condition. Section 4 is devoted to some examples. The Hausdorff dimension of the dragon boundary, which is not self-similar, is calculated by using Theorem 3.1. The fractals shown at Fig.3 are images of function systems which do not satisfy the open set condition in Theorem 3.1 so that $f_1(U)$ and $f_3(U)$ overlap for any open set U since $f_1 f_1$ and $f_3 f_3$ are the same. By using Theorem 3.2 we also calculate the Hausdorff dimension of these fractals.

1. NOTATIONS AND PRELIMINARIES

(1) We will write $E_k^{(\omega)}$ for the set $\{\underline{x} = (x_j)_{j=1}^{\infty} \mid x_j \in \{1, 2, \ldots, k\}\}$ of all infinite sequences on k symbols , $E_k^{(n)}$ for the set $\{\underline{\alpha} = (\alpha_j)_{j=1}^{n} \mid \alpha_j \in \{1, 2 \ldots, k\}\}$ of all sequences of length n and $E_k^{(*)}$ for the set of all finite sequences, that is,

$$E_k^{(*)} = \bigcup_{n \in \mathbf{N}} E_k^{(n)}.$$

For $n \in \mathbf{N} \cup \{0\}$, we define maps $\sigma^n : E_k^{(\omega)} \to E_k^{(\omega)}$ and $P_n : E_k^{(\omega)} \to$

$E_k^{(*)} \cup \{\emptyset\}$ by

$$\sigma^n(x_1 x_2 \ldots \ldots) = (x_{n+1} x_{n+2} \ldots)$$
$$P_n(x_1 x_2 \ldots \ldots) = \begin{cases} (x_1 \ldots x_n) & (n \geq 1) \\ \emptyset & (n = 0). \end{cases}$$

In the same way, $\sigma^n : \bigcup_{p>n} E_k^{(p)} \to E_k^{(*)}$ and $P_n : \bigcup_{p \geq n} E_k^{(p)} \to E_k^{(*)} \cup \{\emptyset\}$ are defined. We also use σ instead of σ^1.

For $\underline{\alpha} \in E_k^{(n)}$ and $\underline{\beta} \in E_k^{(*)} \cup E_k^{(\omega)}$, we will write $\underline{\beta} \succ \underline{\alpha}$ if $\underline{\alpha}$ is an initial segment of $\underline{\beta}$, that is, $P_n \underline{\beta} = \underline{\alpha}$.

For $\underline{\alpha} \in E_k^{(*)}$, we define a cylinder $[\underline{\alpha}]$ to be the set of all infinite sequences that begin with $\underline{\alpha}$, that is,

$$[\underline{\alpha}] = \{\underline{\beta} \in E_k^{(\omega)} \mid \underline{\beta} \succ \underline{\alpha}\}.$$

(2) For $0 < \delta < 1$, we introduce a metric ρ_δ on the space $E_k^{(\omega)}$. For $\underline{x} = (x_j)$, $\underline{y} = (y_j) \in E_k^{(\omega)}$, define

$$\rho_\delta(\underline{x}, \underline{y}) = \begin{cases} \delta^n & \text{if } P_n\underline{x} = P_n\underline{y} \text{ and } x_{n+1} \neq y_{n+1} \text{ for some } n \in \mathbf{N} \cup \{0\} \\ 0 & \text{if } \underline{x} = \underline{y}. \end{cases}$$

(3) Let F be a finite subset $\{\underline{\alpha}^1, \underline{\alpha}^2, \ldots, \underline{\alpha}^l\}$ of $E_k^{(r+1)}$ with $r \geq 1$, $l \geq 1$ and $\underline{\alpha}^i \neq \underline{\alpha}^j (i \neq j)$. Then we define subsets $A(F;p)$ and $A(F)$ of $E_k^{(*)}$ as follows:

$$A(F;p) = \begin{cases} \{\underline{\gamma} \in E_k^{(p)} \mid P_{r+1}\sigma^{j-1}\underline{\gamma} \in F \text{ for } j = 1, 2, \ldots, p-r\} & \text{if } p \geq r+1 \\ \{P_p\underline{\alpha} \mid \underline{\alpha} \in F\} & \text{if } 1 \leq p \leq r \end{cases}$$

and

$$A(F) = \bigcup_{p \in \mathbf{N}} A(F;p).$$

Let $\mathcal{A}_p(F)$ and $\mathcal{A}(F)$ be the subsets of $E_k^{(\omega)}$ as follows:

$$\mathcal{A}_p(F) = \{\underline{x} \in E_k^{(\omega)} \mid \underline{x} \succ \underline{\gamma} \text{ for some } \underline{\gamma} \in A(F;p)\}$$

and

$$\mathcal{A}(F) = \bigcap_{p \in \mathbf{N}} \mathcal{A}_p(F),$$

that is, $\mathcal{A}(F) = \{\underline{x} \in E_k^{(\omega)} \mid P_{r+1}\sigma^{j-1}\underline{x} \in F \text{ for any } j \in \mathbf{N}\}$.

For a subset A of $E_k^{(*)}$, we shall write $\natural(A)$ for the number of elements of A. Let F' be the set $P_r(F) = \{P_r\underline{\alpha} \mid \underline{\alpha} \in F\}$ and $\natural(F') = m$, that is, $F' = \{\underline{\beta}^1, \underline{\beta}^2, \ldots, \underline{\beta}^m\}$.

We introduce an $m \times m$ matrix $M = M(F) = (m_{i,j})$ by putting

$$m_{i,j} = \begin{cases} 1, & \text{if there exists } \underline{\alpha} \in F \text{ such that} \\ & \quad \underline{\alpha} \succ \beta^i \in F' \text{ and } \sigma\underline{\alpha} \succ \beta^j \in F' \\ 0, & \text{otherwise} \end{cases}$$

for $i, j \in \{1, \ldots, m\}$, which we shall call the *induced matrix from* F.

Let Q be an $m \times m$ permutation matrix and M_1 be an $m_0 \times m_0$ matrix $(1 \le m_0 \le m)$ with some $n_1 (0 \le n_1 \le m - m_0)$ satisfying

$$(i, j) \text{ element of } M_1 = (i + n_1, j + n_1) \text{ element of } QMQ^T$$

for any $i, j (1 \le i, j \le m_0)$. Put $v^T = (\overbrace{0, \ldots, 0}^{n_1}, \overbrace{1, \ldots, 1}^{m_0}, \overbrace{0, \ldots, 0}^{m-m_0-n_1})$, $(u_1, \ldots, u_m)^T = Q^T v$ and $J_1 = \{i \in \{1, \ldots, m\} \mid u_i = 1\} = \{i_1, \ldots, i_{m_0}\}$. Put $F_1' = \{\underline{\beta}^{i_1}, \ldots, \underline{\beta}^{i_{m_0}}\}$ and $F_1 = \{\underline{\alpha} \in F \mid P_r\underline{\alpha} \in F_1'\}$. We shall call F_1 [resp. J_1] to be a *subset of* F [resp. *of the set* $\{1, \ldots, m\}$] *corresponding to* M_1.

REMARK. It holds that the induced matrix from F_1 is M_1 and $P_r(F_1) = F_1'$.

(4) For a subset E of a metric space (X, d) and $\varepsilon > 0$, define for each $s \ge 0$,

$$H_\varepsilon^s(E) = \inf\{\textstyle\sum_{j=1}^\infty (\operatorname{diam} A_j)^s \mid E \subset \bigcup_{j \in \mathbf{N}} A_j, \operatorname{diam} A_j \le \varepsilon\},$$

where $\operatorname{diam} A_j$ is the diameter of a set A_j with respect to metric d and the infimum is taken over all countable ε-cover of the set E. It is clear that $H_\varepsilon^s(E)$ gets larger if ε gets smaller. We define

$$H^s(E) = \lim_{\varepsilon \to 0} H_\varepsilon^s(E).$$

There exists an $s_0 \ge 0$ such that

$$H^s(E) = \begin{cases} 0 & \text{if } s > s_0 \\ \infty & \text{if } s < s_0. \end{cases}$$

The number s_0 is called the *Hausdorff dimension of* E and is denoted by $\dim E$.

2. HAUSDORFF DIMENSION OF A SET
OF INFINITE SEQUENCES

Throughout this section we suppose that F is a finite subset of $E_k^{(r+1)}$ $(r \ge 1)$ with $F' = P_r(F) = \{\underline{\beta}^1, \ldots, \underline{\beta}^m\}$ such that $\mathcal{A}(F) = \{\underline{x} = (x_j)_{j=1}^\infty \in E_k^{(\omega)} \mid (x_n x_{n+1} \cdots x_{n+r}) \in F \text{ for any } n \in \mathbf{N}\}$ is a non-empty set, M is the induced $m \times m$ matrix from F, λ_0 is the spectral radius of M and $0 < \delta < 1$.

In this section we shall compute the Hausdorff dimension of $\mathcal{A}(F)$. For this we need a well known theorem of matrix theory. Recall that a square matrix S is called to be *reducible* if there exists a permutation matrix Q such that $QSQ^T = \begin{pmatrix} S_{11} & 0 \\ S_{21} & S_{22} \end{pmatrix}$. If no such permutation matrix exists, it is called to be *irreducible*.

REMARK. If M is an irreducible $m \times m$ matrix with $m \geq 2$, then M is a nonzero matrix.

THEOREM A (Perron-Frobenius Theorem). *Let S be a nonnegative $m \times m$ matrix with $m \geq 2$ and $r(S)$ the spectral radius of S.*

(1) *If there is a nonzero $v \geq 0$ with $Sv \geq \lambda v$, then $r(S) \geq \lambda$.*

(2) *If S is irreducible, then S has a positive real eigenvalue equal to $r(S)$ and there is a strictly positive eigenvector $u = (u_j)$ corrresponding to $r(S)$, i.e.*

$$Su = r(S)u \text{ and } u_j > 0 \text{ for any } 1 \leq j \leq m.$$

Concerning the assumption that $\mathcal{A}(F)$ is non-empty, we have

LEMMA 2.1. *The following are equivalent.*

(1) $\mathcal{A}(F)$ is non-empty.

(2) $\lambda_0 \geq 1$.

PROOF. (1) $\rightarrow$ (2): Since $\mathcal{A}(F)$ is a non-empty set, there exist $i_1, \ldots, i_{m_0} \in \{1, \ldots, m\}$ such that $m_{i_j i_{j+1}} = 1$ for $j = 1, \ldots, m_0 - 1$ and $m_{i_{m_0} i_1} = 1$. Put $v = (v_l)_{l=1}^m$ with $v_l = \begin{cases} 1 & (l = i_j \text{ for some } j \in \{1, \ldots, m_0\}) \\ 0 & \text{(otherwise)}. \end{cases}$
Then $Mv \geq v$ and we have $\lambda_0 \geq 1$ by Theorem A (1).

(2) $\rightarrow$ (1): If M is irreducible, it is clear that $\mathcal{A}(F)$ is non-empty. If M is reducible, there exists a permutation matrix Q satisfying either of the following

(i) $QMQ^T = \begin{pmatrix} M_{11} & 0 \\ M_{21} & M_{22} \end{pmatrix}$ and either of M_{11} or M_{22} is an irreducible $m_0 \times m_0$ matrix $(1 \leq m_0 \leq m - 1)$ and its spectral radius is λ_0,

(ii) $QMQ^T = \begin{pmatrix} M_{11} & 0 & 0 \\ M_{21} & M_{22} & 0 \\ M_{31} & M_{32} & M_{33} \end{pmatrix}$ and M_{22} is an irreducible $m_0 \times m_0$ matrix $(1 \leq m_0 \leq m - 2)$ and $r(M_{22}) = \lambda_0$.

If M_{jj} is irreducible and $r(M_{jj}) = \lambda_0$ with some $j \in \{1,2\}$, let F_1 be the subset of F corresponding to M_{jj}. Since the induced matrix from F_1 is M_{jj} and M_{jj} is a nonzero irreducible matrix, it holds that $\mathcal{A}(F_1)$ is non-empty, which implies that $\mathcal{A}(F)$ is non-empty by using the relation $\mathcal{A}(F) \supset \mathcal{A}(F_1)$. $\square$

For $\underline{\gamma}^o \in A(F;t)$ and $p \geq t$, let $A(F;p,\underline{\gamma}^o)$ be the set $\{\underline{\gamma} \in A(F;p) \mid \underline{\gamma} \succ \underline{\gamma}^o\}$. Then we have

LEMMA 2.2.   For $p \geq t \geq r$ and $\underline{\gamma}^o \in A(F;t)$ satisfying $\sigma^{t-r}\underline{\gamma}^o \succ \underline{\beta}^j$ with some $\underline{\beta}^j \in F'$, we have
$$\natural(A(F;p,\underline{\gamma}^o)) = \langle e_j, M^{p-t}1_m\rangle \ \text{ and } \ \natural(A(F;p)) = \langle 1_m, M^{p-r}1_m\rangle,$$
where the transpose $e_j^T, 1_m^T$ of $e_j, 1_m$ are vectors $(0,\dots,0,\overset{j}{1},0,\dots,0)$, $1_m^T = (1,\dots,1)$ in $\mathbf{R}^m$ and $\langle\cdot,\cdot\rangle$ is the inner product in $\mathbf{R}^m$.

PROOF.   It holds that $\natural(A(F;r,\underline{\beta}^j)) = 1 = \langle e_j, M^0 1_m\rangle$ and
$$\natural(A(F;r+1,\underline{\beta}^j)) = \sum_{i=1}^m \natural\{\underline{\gamma} \in A(F;r+1) \mid \underline{\gamma} \succ \underline{\beta}^j, \sigma\underline{\gamma} \succ \underline{\beta}^i\}$$
$$= \sum_{i=1}^m m_{j,i} = \langle e_j, M1_m\rangle.$$
By induction, we have
$$\natural(A(F;p,\underline{\gamma}^o)) = \natural(A(F;p+r-t,\underline{\beta}^j))$$
$$= \sum_{i=1}^m \natural\{\underline{\gamma} \in A(F;p+r-t) \mid \underline{\gamma} \succ \underline{\beta}^j, \sigma\underline{\gamma} \in A(F;p+r-t-1,\underline{\beta}^i)\}$$
$$= \sum_{i=1}^m m_{j,i}\langle e_i, M^{p-t-1}1_m\rangle = \langle e_j, M^{p-t}1_m\rangle$$
and
$$\natural(A(F;p)) = \sum_{i=1}^m \natural(A(F;p,\underline{\beta}^i))$$
$$= \sum_{i=1}^m \langle e_i, M^{p-r}1_m\rangle = \langle 1_m, M^{p-r}1_m\rangle. \ \square$$

REMARK.   In [1], the authors considered an $l \times l$ matrix $\tilde{M} = (\tilde{m}_{i,j})$, where $\tilde{m}_{i,j}$ are defined from $F = \{\underline{\alpha}^1, \underline{\alpha}^2, \dots, \underline{\alpha}^l\}$ by
$$\tilde{m}_{i,j} = \begin{cases} 1 & \text{if } \ \sigma\underline{\alpha}^i = P_r\underline{\alpha}^j \\ 0 & \text{if } \ \sigma\underline{\alpha}^i \neq P_r\underline{\alpha}^j. \end{cases}$$

In this case, the number $\natural(A(F;p))$ is equal to $\langle 1_l, \tilde{M}^{p-r-1}1_l\rangle$. Since m is smaller than l, $\langle 1_m, M^{p-r}1_m\rangle$ is much easier to calculate than $\langle 1_l, \tilde{M}^{p-r-1}1_l\rangle$.

LEMMA 2.3. *Let E be a compact subset of $(E_k^{(\omega)}, \rho_\delta)$ and let $s \geq 0$ be given. Then* dim $E \geq s$ *holds if there is an $\varepsilon > 0$ with the following property:*

" Given any finite set $\{\gamma^1, \ldots, \gamma^n\}$ of $E_k^{(*)}$ such that

$$\text{diam } [\gamma^j] < \varepsilon \ (j = 1, \ldots, n), \quad [\gamma^i] \cap [\gamma^j] = \emptyset \ (i \neq j)$$

$$\text{and} \quad \sum_{j=1}^n (\text{diam } [\gamma^j])^s \leq 1,$$

then it follows that $\bigcup_{j=1}^n [\gamma^j]$ does not cover E."

PROOF. Suppose, by contradiction, that dim $E < s$. Then for any $\varepsilon (\delta > \varepsilon > 0)$, $H_\varepsilon^s(E) = 0$. So there exists $(A_j)_{j=1}^\infty$ such that diam $A_j < \varepsilon$, $\sum_{i=1}^\infty (\text{diam } A_j)^s \leq \frac{1}{2}$ and $\bigcup_{j \in \mathbf{N}} A_j \supset E$. Then by the definition of diameter, for each A_j, there exists $\gamma^j \in E_k^{(*)}$ such that $A_j \subset [\gamma^j]$ and diam $A_j =$ diam $[\gamma^j]$. Since $[\gamma^j]$ is open and E is compact, there exist a finite set $\{\gamma^{j_1}, \ldots, \gamma^{j_k}\}$ such that $E \subset \bigcup_{l=1}^k [\gamma^{j_l}]$, which contradicts the assumed property. $\square$

Concerning the number of elements of $A(F; p)$ or of $A(F; p, \gamma^o)$, which plays an important role in computation of dim $\mathcal{A}(F)$, we have

LEMMA 2.4. *1) There exists $c_0 > 0$ such that*

$$\sharp(A(F; p)) \leq c_0 (p - r)^m \lambda_0^{p-r} \quad \text{for } p \geq r.$$

2) If the induced matrix M from F is irreducible, then there exists $c > 0 $ such that

$$\lambda_0^{p-r} \leq \sharp(A(F; p)) \leq c \lambda_0^{p-r} \quad \text{for } p \geq r$$

and

$$\sharp(A(F; p, \gamma^o)) \leq c \lambda_0^{p-t} \quad \text{for } \gamma^o \in A(F; t) \text{ with } p \geq t \geq r.$$

PROOF. 1) Since $\mathcal{A}(F)$ is non-empty, $\lambda_0 \geq 1$ holds by Lemma 2.1. There exists a regular matrix V such that $V^{-1}MV = \Lambda$ is a Jordan canonical form

$$\Lambda = \begin{pmatrix} J_1 & 0 & \cdots & 0 \\ 0 & J_2 & \cdots & 0 \\ \vdots & & \ddots & \vdots \\ 0 & \cdots & 0 & J_s \end{pmatrix} \quad \text{with } J_j = \begin{pmatrix} \lambda_j & 1 & 0 & \cdots & 0 \\ 0 & \lambda_j & 1 & \cdots & 0 \\ \vdots & \ddots & \ddots & \ddots & \vdots \\ \vdots & & \ddots & \ddots & 1 \\ 0 & \cdots & \cdots & 0 & \lambda_j \end{pmatrix}.$$

For $\Lambda = (\lambda_{ij})$, put $|\Lambda| = (|\lambda_{ij}|)$.   Since $|\lambda_j| \leq \lambda_0 (j = 1,\ldots,s)$ by Theorem A and $a \leq \lambda_0$ by Lemma 2.1, we have

$$0 \leq |\Lambda|^n 1_m \leq \begin{pmatrix} \lambda_0 & 1 & 0 & \cdots & 0 \\ 0 & \lambda_0 & 1 & \cdots & 0 \\ \vdots & \ddots & \ddots & \ddots & \vdots \\ \vdots & \ddots & \ddots & \ddots & 1 \\ 0 & \cdots & \cdots & 0 & \lambda_0 \end{pmatrix}^n 1_m \leq mn^m \lambda_0^n 1_m.$$

For $V = (v_{ij})$ and $V^{-1} = (v_{ij}^1)$, put $|V| = (|v_{ij}|)$, $|V^{-1}| = (|v_{ij}^1|)$ and $v = \max\{|v_{ij}|, |v_{ij}^1| \mid 1 \leq i,j \leq m\}$. Then

$$0 \leq M^n 1_m = V\Lambda^n V^{-1} 1_m \leq |V||\Lambda|^n |V^{-1}| 1_m \leq m^3 v^2 n^m \lambda_0^n 1_m.$$

Hence by putting $c_0 = m^4 v^2$, we have

$$\natural(A(F;p)) = \langle 1_m, M^{p-r} 1_m \rangle \leq c_0 (p-r)^m \lambda_0^{p-r}$$

by using Lemma 2.2.

2) If m=1, M is a matrix (1) since $\mathcal{A}(F)$ is non-empty. So it is obvious. If $m \geq 2$, there exists an eigenvector u and $c_1 > 0$ such that $Mu = \lambda_0 u$, $\|u\|_\infty = 1$ and $1_m \geq u \geq c_1 1_m$ by Theorem A (2), since M is irreducible. So we have

$$M^p 1_m \geq M^p u = \lambda_0^p u$$

and

$$M^p 1_m \leq \tfrac{1}{c_1} M^p u = \tfrac{1}{c_1} \lambda_0^p u \leq \tfrac{1}{c_1} \lambda_0^p 1_m.$$

Hence

$$\lambda_0^{p-r} \leq \lambda_0^{p-r} \langle 1_m, u \rangle \leq \langle 1_m, M^{p-r} 1_m \rangle \leq \tfrac{1}{c_1} \lambda_0^{p-r} m.$$

By putting $c = \frac{m}{c_1}$ and by using Lemma 2.2, we have

$$\lambda_0^{p-r} \leq \natural(A(F;p)) \leq c\lambda_0^{p-r}.$$

By using Lemma 2.2, we have

$$\natural(A(F;p,\underline{\gamma}^\circ)) = \langle e_j, M^{p-t} 1_m \rangle \leq \langle 1_m, M^{p-t} 1_m \rangle \leq c\lambda_0^{p-t}. \quad \square$$

PROPOSITION 2.5. *If M is irreducible, then*

$$\dim \mathcal{A}(F) = \log_{\frac{1}{\delta}} \lambda_0,$$

where $\mathcal{A}(F)$ is considered as a subset of the metric space $(E_k^{(\omega)}, \rho_\delta)$.

PROOF. Suppose $s > \log_{\frac{1}{\delta}} \lambda_0$. Then $\delta^s \lambda_0 < 1$. For any $\varepsilon > 0$, find $p \in \mathbf{N}$ such that $\delta^p < \varepsilon$. Since $\mathcal{A}(F) \subset \mathcal{A}_p(F) \subset \bigcup_{\underline{\gamma} \in A(F;p)} [\underline{\gamma}]$ holds, $\{[\underline{\gamma}] \mid \underline{\gamma} \in A(F;p)\}$ is an

ε-covering of $\mathcal{A}(F)$. So

$$H^s_\varepsilon(\mathcal{A}(F)) \leq \natural(A(F;p))(\delta^p)^s.$$

By Lemma 2.4 and $\delta^s \lambda_0 < 1$, we have $H^s_\varepsilon(\mathcal{A}(F)) < 1$ for any $\varepsilon > 0$, by taking p so large. So dim $\mathcal{A}(F) \leq s$, which implies dim $\mathcal{A}(F) \leq \log_{\frac{1}{\delta}} \lambda_0$.

So we have dim $\mathcal{A}(F) = \log_{\frac{1}{\delta}} \lambda_0 = 0$, if $\lambda_0 = 1$.

When $\lambda_0 > 1$, suppose $0 < s < \log_{\frac{1}{\delta}} \lambda_0$, so that $\lambda_0^{-1} \delta^{-s} \leq 1$. For any $t \in \mathbf{N}$, there is an $\varepsilon > 0$ such that $\delta^t > \varepsilon$. Let $\mathbf{U} = \{\underline{\gamma}^1, \ldots, \underline{\gamma}^{n_0}\}$ be a finite set of $E_k^{(*)}$ such that diam $[\underline{\gamma}^j] < \varepsilon (j = 1, \ldots, n_0)$, $[\underline{\gamma}^i] \cap [\underline{\gamma}^j] = \emptyset$ $(i \neq j)$ and $\sum_{j=1}^{n_0} (\text{diam}[\underline{\gamma}^j])^s < 1$. Consider the set $\mathbf{U}(p) = \{\underline{\gamma} \in \mathbf{U} \mid \text{diam } [\underline{\gamma}] = \delta^{t+p}\}$ for any $p \in \mathbf{N}$. Then $\natural(\mathbf{U}(p)) \leq (\delta^{t+p})^{-s}$. Let diam $[\underline{\gamma}^j] = \delta^{t+p_j} (j = 1, \ldots, n_0)$ and put $p_0 = \max\{p_j \mid j = 1, \ldots, n_0\}$. By Lemma 2.4, there exists $c > 0$ such that

$$\natural(A(F;t+N,\underline{\gamma})) \leq c\lambda_0^{N-p},$$

for $\underline{\gamma} \in \mathbf{U} \cap E_k^{(t+p)}$ and $N > p_0$. Hence

$$\natural(\{\underline{y} \in A(F;t+N) \mid \underline{y} \succ \underline{\gamma} \text{ for some } \underline{\gamma} \in \mathbf{U}\})$$

$$= \sum_{j=1}^{n_0} \natural(A(F;t+N,\underline{\gamma}^j))$$

$$= \sum_{p=1}^{p_0} \sum_{\underline{\gamma} \in \mathbf{U}(p)} \natural(A(F;t+N,\underline{\gamma})) \leq c\lambda_0^{N+t} \sum_{p=1}^{p_0} (\lambda_0^{-1}\delta^{-s})^{p+t}.$$

Since $0 < \lambda_0^{-1}\delta^{-s} < 1$, we have $\sum_{n \geq t}(\lambda_0^{-1}\delta^{-s})^n < \frac{1}{2c}\lambda_0^{-r}$ by taking t large enough and we have

$$\natural(\{\underline{y} \in A(F;t+N) \mid \underline{y} \succ \underline{\gamma} \text{ for some } \underline{\gamma} \in \mathbf{U}\}) \leq \frac{1}{2}\lambda_0^{N+t-r}.$$

Lemma 2.4 implies $\natural(A(F;t+N)) \geq \lambda_0^{t+N-r}$ and so $A(F;t+N) \supsetneq \{\underline{y} \in A(F;t+N) \mid \underline{y} \succ \underline{\gamma} \text{ for some } \underline{\gamma} \in \mathbf{U}\}$, which implies $\mathcal{A}(F) \not\subset \bigcup_{j=1}^{n_0} [\underline{\gamma}^j]$. Since $\mathcal{A}(F)$ is a compact subset of $(E_k^{(\omega)}, \rho_\delta)$, dim $\mathcal{A}(F) \geq s$ follows from Lemma 2.3 and so dim $\mathcal{A}(F) \geq \log_{\frac{1}{\delta}} \lambda_0$. Therefore dim $\mathcal{A}(F) = \log_{\frac{1}{\delta}} \lambda_0$. $\square$

Even if M is reducible, by using Lemma 2.4(2) we can arrive at the same conclusion as Proposition 2.5. The following is the main theorem of this section.

THEOREM 2.6. *Let F be a finite subset of $E_k^{(r+1)}$ with $F' = P_r(F) = \{\underline{\beta}^1, \ldots, \underline{\beta}^m\}$ such that $\mathcal{A}(F) = \{\underline{x} = (x_j)_{j=1}^\infty \in E_k^{(\omega)} \mid (x_n x_{n+1} \ldots x_{n+r}) \in F$ for any*

$n \in \mathbf{N}\}$ *is non-empty. Let M be the induced $m \times m$ matrix from F and let λ_0 be the spectral radius of M.*

Then the Hausdorff dimension $\dim \mathcal{A}(F)$ *in the metric space* $(E_k^{(\omega)}, \rho_\delta)$ *with $0 < \delta < 1$ is computed by the following formula*

$$\dim \mathcal{A}(F) = \log_{\frac{1}{\delta}} \lambda_0.$$

PROOF. At first we shall show that

(*) $\dim \mathcal{A}(F) \geq \log_{\frac{1}{\delta}} \lambda_0$

holds by induction. When $m = 1$, the matrix M is irreducible and we have $\dim \mathcal{A}(F) = \log_{\frac{1}{\delta}} \lambda_0$ by Proposition 2.5 and (*) holds.

Suppose (*) holds for all m satisfying $m \leq n$. Let $m = n + 1$. When M is irreducible, (*) holds by Proposition 2.5. When M is reducible, there exists an $m \times m$ permutation matrix Q for which QMQ^T is $\begin{pmatrix} M_{11} & 0 \\ M_{21} & M_{22} \end{pmatrix}$, where M_{11} is a $m_0 \times m_0$ matrix and the spectral radius of either M_{11} or M_{22} is λ_0. If $r(M_{11}) = \lambda_0$, let F_1 be the subset of F corresponding to M_{11}. If $r(M_{11}) \leq \lambda_0$, then $r(M_{22}) = \lambda_0$ and let F_1 be the subset of F corresponding to M_{22}. Then $\mathcal{A}(F_1)$ is non-empty by Lemma 2.1. Since both m_0 and $m - m_0$ are less that n, we have $\dim \mathcal{A}(F_1) \geq \log_{\frac{1}{\delta}} \lambda_0$ by induction. So (*) holds for any $m \in \mathbf{N}$.

To prove the reversed inequality, suppose $s > \log_{\frac{1}{\delta}} \lambda_0$. Then $\delta^s \lambda_0 < 1$. For any $\varepsilon > 0$, find $p \in \mathbf{N}$ such that $\delta^p < \varepsilon$. Since $\mathcal{A}(F) \subset \mathcal{A}_p(F) \subset \bigcup_{\gamma \in A(F;p)} [\gamma]$ holds, $\{[\gamma]; \gamma \in A(F;p)\}$ is an ε-covering of $\mathcal{A}(F)$. So by Lemma 2.4, there exists $c_0 > 0$ satisfying

$$H_\varepsilon^s(\mathcal{A}(F)) \leq \natural(A(F;p))(\delta^p)^s \leq c_0 (p - r)^m \lambda_0^{-r} (\delta^s \lambda_0)^p.$$

Since $\delta^s \lambda_0 < 1$ implies $\lim_{p \to \infty} p^m (\delta^s \lambda_0)^p = 0$, we have $H_\varepsilon^s(\mathcal{A}(F)) < 1$ for any $\varepsilon > 0$, by taking p so large. So $\dim \mathcal{A}(F) \leq s$, which implies

$$\dim \mathcal{A}(F) \leq \log_{\frac{1}{\delta}} \lambda_0.$$

Hence $\dim \mathcal{A}(F) = \log_{\frac{1}{\delta}} \lambda_0$ is obtained. $\square$

3. HAUSDORFF DIMENSION OF FRACTALS IN $\mathbf{R}^d$

Throughout this section suppose that $(f_1, \ldots, f_k)$ is an iterated function system of similitudes with ratio δ in $\mathbf{R}^d$, satisfying $\|f_j(u) - f_j(v)\| = \delta \|u - v\|$ for

$u, v \in \mathbf{R}^d$, with $0 < \delta < 1$. It is known $[3, \S 3.1]$ that there is a unique compact invariant set K with respect to $(f_1, \ldots, f_k)$, that is, $K = \bigcup_{j=1}^{k} f_j(K)$. Some typical fractals such as the Cantor set, the Koch curve and the Sierpinski gasket are the compact invariant set K with respect to a certain iterated function system of similitudes in $\mathbf{R}^d$.

In this section, we shall compute the Hausdorff dimension of subsets of K of certain type. Such a subset may or may not be a fractal.

Usefulness of the theory developed in the previous section is seen in the following fact (see $[3, \S 3]$): there is a natural embedding φ from $E_k^{(\omega)}$ onto K such that $\varphi(\underline{x}) = \bigcap_{n=1}^{\infty} f_{x_1} f_{x_2} \ldots f_{x_n}(K)$ for $\underline{x} = (x_n)_{n=1}^{\infty} \in E_k^{(\omega)}$.

At first we shall establish a sufficient condition for dim A of a subset A of $(E_k^{(\omega)}, \rho_\delta)$ to coincide with $\dim\varphi(A)$ in $\mathbf{R}^d$. For $\underline{\gamma} = (\gamma_j)_{j=1}^{n} \in E_k^{(n)}$ and $B \subset \mathbf{R}^d$, we shall write $f_{\underline{\gamma}}(B)$ for $f_{\gamma_1} f_{\gamma_2} \ldots f_{\gamma_n}(B)$ and $\overline{B}$ for the closure of B.

THEOREM 3.1. *Let* $(f_1, \ldots, f_k)$ *be an iterated function system of similitudes with ratio* δ *(*$0 < \delta < 1$*) in* $\mathbf{R}^d$, *K the compact invariant set with respect to* $(f_1, \ldots, f_k)$ *and* $\varphi : E_k^{(\omega)} \to K$ *the natural embedding.*

Let F be a finite subset of $E_k^{(r+1)}$ *(*$r \geq 1$*) with* $F' = P_r(F) = \{\underline{\beta}^1, \ldots, \underline{\beta}^m\}$ *such that* $\mathcal{A}(F)$ *is non-empty.*

If there exists a non-empty bounded open subset U of $\mathbf{R}^d$ *such that*

$$(3.1) \qquad f_{\underline{\beta}^i}(U) \cap f_{\underline{\beta}^j}(U) = \emptyset \quad \text{for } \underline{\beta}^i, \underline{\beta}^j \in F' \text{ with } i \neq j$$

and

$$(3.2) \qquad \varphi(\mathcal{A}(F)) \subset \overline{U},$$

then for any subset A of $\mathcal{A}(F)$ *in the metric space* $(E_k^{(\omega)}, \rho_\delta)$,

$$\dim A = \dim \varphi(A).$$

PROOF. For $\underline{x}, \underline{y} \in A$ with $\underline{x} \neq \underline{y}$, take $n \in \mathbf{N} \cup \{0\}$ such that $P_n\underline{x} = P_n\underline{y}$ and $x_{n+1} \neq y_{n+1}$. Then $\rho_\delta(\underline{x}, \underline{y}) = \delta^n$. Since K is the set $\varphi(E_k^{(\omega)})$, we have $d(\varphi(\underline{x}), \varphi(\underline{y})) = \delta^n d(\varphi(\sigma^n\underline{x}), \varphi(\sigma^n\underline{y})) \leq (\text{diam } K)\delta^n = (\text{diam } K)\rho_\delta(\underline{x}, \underline{y})$. So φ has bounded increase, which implies $\dim \varphi(A) \leq \dim A$ $[2, \S 6.1]$.

When dim $A = 0$, dim $\varphi(A) = $dim A follows from the inequality $0 \leq \dim \varphi(A) \leq \dim A$. Suppose $0 <$dim A and take $0 < s < \dim A$. Take ε_1 such that

$\delta^r > \varepsilon_1 > 0$ and $H^s_{\varepsilon_1}(A) > 1$. Let $0 < \varepsilon < \varepsilon_1 \operatorname{diam} U$ and let $\{K_j\}$ be a countable family of subsets of $\mathbf{R}^d$ such that

$$\varphi(A) \subset \bigcup_{j \in \mathbf{N}} K_j \quad \text{and} \quad 0 \leq \operatorname{diam} K_j \leq \varepsilon \text{ for any } j \in \mathbf{N}.$$

For each $j \in \mathbf{N}$ there is $n_j \in \mathbf{N}$ satisfying

$$(3.3) \qquad\qquad \delta^{n_j} \operatorname{diam} U < \operatorname{diam} K_j \leq \delta^{n_j - 1} \operatorname{diam} U.$$

Then $\delta^r > \varepsilon_1 > \delta^{n_j}$ implies $n_j > r$. Let

$$T_j = \{\gamma \in A(F; n_j) \mid f_\gamma(\overline{U}) \cap K_j \neq \emptyset\} \quad \text{for each } j \in \mathbf{N}.$$

We show that there exists $b > 0$ such that $b \geq \sharp(T_j)$ for any $j \in \mathbf{N}$ and b is independent of ε-cover $\{K_j\}$ of $\varphi(A)$. Let $u_0 \in K_j$ be fixed and let r_j be $\operatorname{diam} K_j$. Then for $\gamma \in T_j$, $\|f_j(u) - f_j(v)\| = \delta\|u - v\|$ $(u, v \in \mathbf{R}^d)$ implies $\operatorname{diam} f_\gamma(U) = \delta^{n_j} \operatorname{diam} U$ and $f_\gamma(U)$ is contained in the ball $B(u_0, 2r_j)$ of radius $2r_j$ with center u_0. For $\gamma^1, \gamma^2 \in T_j$ $(\gamma^1 \neq \gamma^2)$, there exists $n \in \mathbf{N} \cup \{0\}$ such that $n \leq n_j - r$, $P_n \gamma^1 = P_n \gamma^2$ and $P_r \sigma^n \gamma^1 \neq P_r \sigma^n \gamma^2$. By assumption, $f_{P_r \sigma^n \gamma^1}(U) \cap f_{P_r \sigma^n \gamma^2}(U) = \emptyset$ holds and we have $f_{\gamma^1}(U) \cap f_{\gamma^2}(U) = \emptyset$ since f_j are similitudes $(j = 1, \ldots, k)$. Let v and w be the volumes of U and of $B(0, 1)$ respectively and let r_0 be $\operatorname{diam} U$. Since all disjoint sets $\{f_\gamma(U) \mid \gamma \in T_j\}$ are contained in $B(u_0, 2r_j)$, we have

$$\sharp(T_j)(\delta^{n_j})^d v \leq (2r_j)^d w.$$

From (3.3), it follows that $r_j = \operatorname{diam} K_j \leq \delta^{n_j - 1} r_0$. By putting $b = (\frac{2}{\delta} r_0)^d \frac{w}{v}$, we have

$$(3.4) \qquad\qquad b \geq \sharp(T_j) \quad \text{for any } j \in \mathbf{N}.$$

For $\underline{x} \in A$, we have $\underline{x} \in \mathcal{A}(F)$ and $\varphi(\underline{x}) \in K_j$ for some $j \in \mathbf{N}$. By putting $P_{n_j} \underline{x} = \gamma$, we have $\gamma \in A(F; n_j)$ and $\sigma^{n_j} \underline{x} \in \mathcal{A}(F)$, which implies $\varphi(\underline{x}) = f_\gamma(\varphi(\sigma^{n_j} \underline{x})) \in f_\gamma(\overline{U})$ and $\gamma \in T_j$. Therefore

$$A \subset \bigcup_{j \in \mathbf{N}} \bigcup_{\gamma \in T_j} [\gamma].$$

Since $\delta^{n_j} < \varepsilon_1$ for any $j \in \mathbf{N}$ and $H^s_{\varepsilon_1}(A) > 1$, we have

$$\sum_{j \in \mathbf{N}} \sum_{\gamma \in T_j} (\operatorname{diam} [\gamma])^s > 1$$

and so $b \sum_{j \in \mathbf{N}} (\frac{\operatorname{diam} K_j}{\operatorname{diam} U})^s \geq 1$ follows from (3.3), (3.4) and the relation $\operatorname{diam}[\gamma] = \delta^{n_j}$ for $\gamma \in T_j \subset E_k^{(n_j)}$. Hence $\sum_{j \in \mathbf{N}} (\operatorname{diam} K_j)^s \geq \frac{(\operatorname{diam} U)^s}{b} > 0$. Since the above inequality holds for any ε-cover $\{K_j\}_{j \in \mathbf{N}}$ of $\varphi(A)$, we have $H^s(\varphi(A)) > 0$, which implies $s \leq \dim \varphi(A)$. Therefore $\dim A = \dim \varphi(A)$. $\square$

To guarantee the condition in Theorem 3.1, let us introduce a notion. The iterated function system $(f_1, \ldots, f_k)$ is said to satisfy *Moran's open set condition* if there exsits a non-empty bounded open set U for which we have $f_i(U) \cap f_j(U) = \emptyset$ for $i \neq j$ and $U \supset f_i(U)$ for all i.

COROLLARY 1. *If $(f_1, \ldots, f_k)$ satisfies Moran's open set condition, then*

$$\dim A = \dim \varphi(A)$$

for any subset A of $(E_k^{(\omega)}, \rho_\delta)$.

PROOF. The condition $U \supset f_i(U)$ $(i = 1, \ldots, k)$ implies $\overline{U} \supset K = \varphi(E_k^{(\omega)})$. By putting $F = \{(11), (12), \ldots, (kk)\}$ with $\natural(F) = k^2$, we have $\mathcal{A}(F) = E_k^{(\omega)}$. Then by applying Theorem 3.1, we obtain the result. $\square$

This corollary, combined with Theorem 2.6, yields a well-known result (see [5]).

COROLLARY 2. *If $(f_1, \ldots, f_k)$ satisfies Moran's open set condition, then*

$$\dim K = \log_{\frac{1}{\delta}} k$$

for the compact invariant set K.

PROOF. With F in the proof of Corollary 1, its induced matrix is the $k \times k$ matrix with all entries equal to one. Hence it has k as its spectral radius. Now since $\varphi(\mathcal{A}(F)) = K$, the assertion follows from Corollary 1 and Theorem 3.6. $\square$

In some cases shown in Examples (1)$\sim$(2) at Section 4, Corollary 1 is applicable. But in a complicated case such as the dragon boundary (Example (3) at Section 4), the corollary does not work but Theorem 3.1 is applicable. In Example (4) at Section 4, we can not find a set U satisfying (3.1) and (3.2) in Theorem 3.1. If we assume (3.1) holds for some members $\underline{\beta}^i, \underline{\beta}^j$ of F' but not all of F', it is easier to find the set U. The following theorem is more applicable than Theorem 3.1.

THEOREM 3.2. *Let $(f_1, \ldots, f_k)$ be an iterated function system of similitudes with ratio δ $(0 < \delta < 1)$ in $\mathbf{R}^d$, K the compact invariant set with respect to*

$(f_1, \ldots, f_k)$ and $\varphi : E_k^{(\omega)} \to K$ the natural embedding.

Let F be a subset of $E_k^{(r+1)}$ such that $\mathcal{A}(F)$ is non-empty and let $\underline{\beta}^1, \ldots, \underline{\beta}^m$ be the different elements of the set $F' = P_r(F)$.

Let M be the induced $m \times m$ matrix from F with spectral radius λ_0 and let $v = (v_j)$ be a non-negative eigenvector corresponding to λ_0.

Put $J_0 = \{j \in \{1, \ldots, m\} \mid v_j > 0\}$ and $F_0 = \{\underline{\alpha} \in F \mid$ there exist $i, j \in J_0$ such that $\underline{\alpha} \succ \underline{\beta}^i, \sigma\underline{\alpha} \succ \underline{\beta}^j\}$. Suppose that there exists a non-empty bounded open subset U of $\mathbf{R}^d$ such that

$$f_{\underline{\beta}^i}(U) \cap f_{\underline{\beta}^j}(U) = \emptyset \ \text{ for } i, j \in J_0 \ \text{ with } \ i \neq j$$
$$\text{and} \quad \varphi(\mathcal{A}(F_0)) \subset \overline{U}.$$

Then we have

$$(3.5) \qquad\qquad \dim \varphi(\mathcal{A}(F)) = \dim \mathcal{A}(F),$$

where $\mathcal{A}(F)$ is considered as a subset of the metric space $(E_k^{(\omega)}, \rho_\delta)$.

PROOF. In the same way as Theorem 3.1, we have

$$(3.6) \qquad\qquad \dim \varphi(\mathcal{A}(F)) \leq \dim \mathcal{A}(F).$$

We shall show that (3.5) holds by induction. If $m = 1$, then $\lambda_0 = 1$ and $\dim \mathcal{A}(F) = 0$ follows from Theorem 2.6. By using (3.6), we have $\dim \varphi(\mathcal{A}(F)) = \dim \mathcal{A}(F) \ (=0)$.

Suppose that (3.5) holds for all m satisfying $m \leq n$. Let $m = n + 1$. If M is irreducible, $v > 0$ follows from Theorem A (2) and $J_0 = \{1, \ldots, m\}$. Hence the assumption of Theorem 3.1 is satisfied and so we have $\dim \varphi(\mathcal{A}(F)) = \dim \mathcal{A}(F)$. When M is reducible, there exists a permutation matrix Q such that $QMQ^T = \begin{pmatrix} M_{11} & 0 \\ M_{21} & M_{22} \end{pmatrix}$, where M_{11} is an irreducible $m_0 \times m_0$-matrix $(1 \leq m_0 \leq m-1)$. Put $(u_1, \ldots, u_m) = v^T Q$, $(u^1)^T = (u_1, \ldots, u_{m_0})$ and $(u^2)^T = (u_{m_0+1}, \ldots, u_m)$. Then $M_{11} u^1 = \lambda_0 u^1$.

If $u^1 \neq 0$, then $r(M_{11}) = \lambda_0$. Let F_1 and J_1 be subsets of F and of $\{1, \ldots, m\}$ respectively, corresponding to M_{11}. If $m_0 = 1$, $u^1 \neq 0$ implies $u^1 > 0$. If $m_0 \geq 2$, then $u^1 > 0$ by Theorem A (2). Hence it follows that $J_1 \subset J_0$, which implies $\mathcal{A}(F_1) \subset \mathcal{A}(F_0)$. $r(M_{11}) = \lambda_0$ implies that $\mathcal{A}(F_1)$ is non-empty by Lemma 2.1. So by applying Theorem 3.1 to $\mathcal{A}(F_1)$, we have

$$\dim \mathcal{A}(F_1) = \dim \varphi(\mathcal{A}(F_1)).$$

By Theorem 2.6, we have

$$\dim \mathcal{A}(F) = \log_{\frac{1}{4}} \lambda_0 \quad \text{and} \quad \dim \mathcal{A}(F_1) = \log_{\frac{1}{8}} \lambda_0.$$

On the other hand the relation $\varphi(\mathcal{A}(F_1)) \subset \varphi(\mathcal{A}(F))$ implies that

$$\dim \varphi(\mathcal{A}(F_1)) \leq \dim \varphi(\mathcal{A}(F)).$$

Therefore

$$\dim \varphi(\mathcal{A}(F)) = \dim \mathcal{A}(F).$$

If $u^1 = 0$, then $r(M_{22}) = \lambda_0$. Let F_2 and J_2 be subsets of F and of $\{1, \ldots, m\}$ respectively, corresponding to M_{22}. $r(M_{22}) = \lambda_0$ implies that $\mathcal{A}(F_2)$ is non-empty by Lemma 2.1. $u^1 = 0$ implies $J_0 \subset J_2$, which implies $F_0 \subset F_2$. Since $m - m_0 \leq n$ holds, by induction (3.5) holds for F_2, that is, $\dim \varphi(\mathcal{A}(F_2)) = \dim \mathcal{A}(F_2)$. Hence by using the relation $F_2 \subset F$ and Theorem 2.6, we have $\dim \varphi(\mathcal{A}(F)) = \dim \mathcal{A}(F)$ in the same way as above. $\square$

4. EXAMPLES

We shall compute the Hausdorff dimension of some fractals by using Theorems 3.1 and 3.2.

(1) The Cantor set and the Koch curve.

We compute Hausdorff dimensions of the Cantor set and the Koch curve by using Theorem 3.1, though they have been obtained from the similarity dimension [2,§6.3].

(i) Let $f_1(u) = \frac{u}{3}$ and $f_2(u) = \frac{u}{3} + \frac{2}{3}$ for $u \in \mathbf{R}$. Then the Cantor set $C = \{u = \sum_{n=1}^{\infty} \frac{a_n}{3^n} \mid a_n \in \{0,2\}$ for $n \in \mathbf{N}\}$ is the image of the natural embedding $\varphi : E_2^{(\omega)} \to \mathbf{R}$ described in Section 3. Let F be the set $\{(11),(12),(21),(22)\}$. Then the induced matrix $M(F)$ is $\begin{pmatrix} 1 & 1 \\ 1 & 1 \end{pmatrix}$ and its spectral radius is $\lambda_0 = 2$. The Hausdorff dimension of the set $\mathcal{A}(F)$ in the metric space $(E_2^{(\omega)}, \rho_{\frac{1}{3}})$ is $\log_3 2$ by Theorem 2.6. Since $U = (0,1)$ satisfies the assumption of Corollary to Theorem 3.1, we have $\dim C = \log_3 2$.

(ii) Let $f_1(z) = \mu \bar{z}$ and $f_2(z) = \overline{\mu z} + \mu$ for $z \in \mathbf{C}$, where $\bar{z}$ is the complex conjugate of z and $\mu = \frac{1}{2} + \frac{1}{2\sqrt{3}} i$. Then the Koch curve K shown in Fig.1 is the image of the natural embedding $\varphi : E_2^{(\omega)} \to \mathbf{C}$.

Let F be the set $\{(11),(12),(21),(22)\}$. Then the induced matrix $M(F)$ is a 2×2 matrix with all entries equal to one and $r(M)$ is $\lambda_0 = 2$. The Hausdorff dimension of the set $\mathcal{A}(F)$ in the metric space $(E_2^{(\omega)}, \rho_{\frac{1}{\sqrt{3}}})$ is $\log_{\sqrt{3}} 2$ by Theorem 2.6. Since $U = \{z = \xi + i\eta \mid 0 < \eta < \frac{1}{2\sqrt{3}}, \sqrt{3}\eta < \xi < 1 - \sqrt{3}\eta\}$ satisfies the assumption of Corollary to Theorem 3.1, we have dim $K = \log_3 4$.

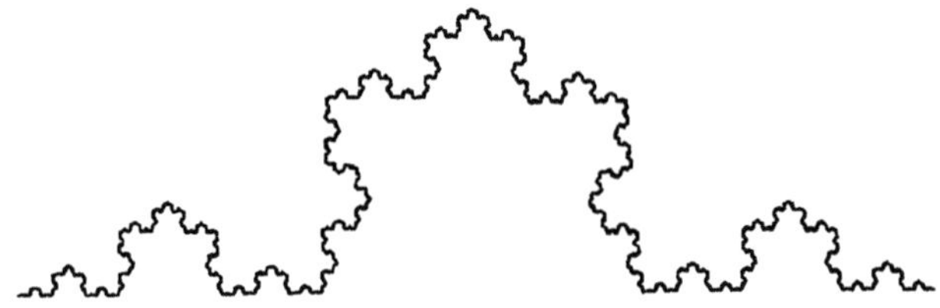

Fig.1

(2) The subset $T_2(1,2)$ in $(0,1)$.

Consider the set

$$T_2(1,2) = \{u \in (0,1) \mid u_n + u_{n+1} \geq 1, \text{for } n = 1, 2, \dots\},$$

where $u = \sum_{n=1}^{\infty} u_n 2^{-n}$ with $u_n \in \{0,1\}$. The Hausdorff dimension of the set $T_2(1,2)$ in $(0,1)$ is calculated in [1], where they use 3×3 matrix. We compute its Hausdorff dimension by applying Theorem 3.1 where we use 2×2 matrix. Let $f_j(u) = \frac{u+j-1}{2}$ $(j = 1, 2)$ for $u \in \mathbf{R}$ and let F be the set $\{(12), (21), (22)\}$. Then the set $T_2(1,2)$ is the image of the natural embedding $\varphi : \mathcal{A}(F) \to \mathbf{R}$. We have $F' = \{(1), (2)\}$, the induced matrix $M(F) = \begin{pmatrix} 0 & 1 \\ 1 & 1 \end{pmatrix}$ and its spectral radius $\lambda_0 = \frac{1+\sqrt{5}}{2}$. The Hausdorff dimension of the set $\mathcal{A}(F)$ in the metric space $(E_2^{(\omega)}, \rho_{\frac{1}{2}})$ is $\log_2 \frac{1+\sqrt{5}}{2}$ by Theorem 2.6. Since $U = (0,1)$ satisfies the assumption of Corollary to Theorem 3.1, we have dim $T_2(1,2) = \log_2 \frac{1+\sqrt{5}}{2}$.

(3) The dragon boundary.

Let $f_1(z) = (\frac{1}{2} + \frac{1}{2}i)z$ and $f_2(z) = (-\frac{1}{2} + \frac{1}{2}i)z + \frac{3}{2} - \frac{1}{2}i$ for $z \in \mathbf{C}$. Then the invariant set with respect to (f_1, f_2) is a set (Fig.2(a)) which is so called a dragon. The Hausdorff dimension of this set is 2. But the Hausdorff dimension of its boundary (Fig.2(b)) is not an integer. Consider $f_3(z) = (\frac{1}{2} - \frac{1}{2}i)z$ and let $F = \{(111), (112), (122), (123), (211), (212), (221), (231), (311), (312)\}$. Then the image DB of the natural embedding from $\mathcal{A}(F)$ is the dragon boundary (Fig. 1(b)). It follows that $\delta = \frac{1}{\sqrt{2}}, F' =$

$$\{(11),(12),(21),(22),(23),(31)\} \text{ and } M = \begin{pmatrix} 1 & 1 & 0 & 0 & 0 & 0 \\ 0 & 0 & 0 & 1 & 1 & 0 \\ 1 & 1 & 0 & 0 & 0 & 0 \\ 0 & 0 & 1 & 0 & 0 & 0 \\ 0 & 0 & 0 & 0 & 0 & 1 \\ 1 & 1 & 0 & 0 & 0 & 0 \end{pmatrix}. \text{ The spectral radius}$$

λ_0 of M satisfies the equation

$$(\lambda_0^3 - \lambda_0^2 - 2)\lambda_0^3 = 0.$$

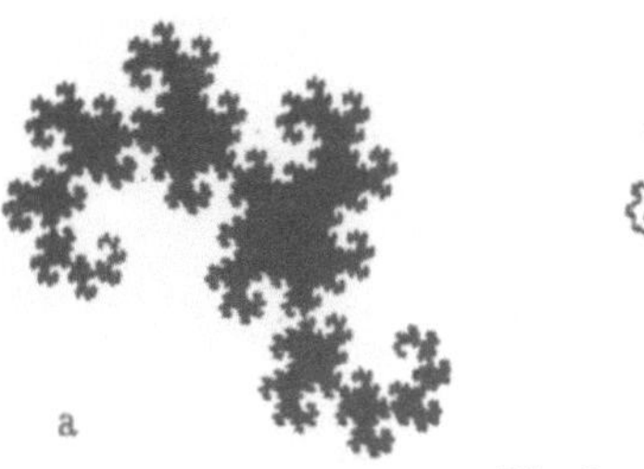

Fig.2

The approximate value $\lambda_0 \approx 1.6956$. We cannot easily find a non-empty bounded open subset U of $\mathbf{R}^d$ satisfying the condition of Corollary to Theorem 3.1. Let U be the interior set of the dragon. Then U satisfies the assumption of Theorem 3.1, i.e. $f_{\underline{\beta^i}}(U) \cap f_{\underline{\beta^j}}(U) = \emptyset$ for $\underline{\beta^i}, \underline{\beta^j} \in F'$ with $\underline{\beta^i} \neq \underline{\beta^j}$ and $\varphi(\mathcal{A}(F)) \subset \overline{U}$. So by using Theorem 3.1, we have $\dim DB = \log_{\sqrt{2}}\lambda_0 \approx 1.5236$.

(4) Fractals of combinations of Koch curve and others.

We calculate the Hausdorff dimension of some fractals which are images K of $\mathcal{A}(F)$ with respect to (f_1, f_2, f_3, f_4) such that $f_i(K)$ and $f_j(K)$ may overlap so much for some $i, j (i \neq j)$.

(i) Consider maps $f_1(z) = \delta z$, $f_2(z) = \delta z + 1 - \delta$, $f_3(z) = \mu \bar{z}$, $f_4(z) = \overline{\mu z} + \mu$ for $z \in \mathbf{C}$, where $\mu = \frac{1}{2} + \frac{1}{2\sqrt{3}}i$ and $\delta = \frac{1}{\sqrt{3}}$. Let $F = \{(11),(12),(21),(22),(31),(32),$ $(33),(34),(41),(42),(43),(44)\}$. Then $F' = \{(1),(2),(3),(4)\}$, $M = \begin{pmatrix} 1 & 1 & 0 & 0 \\ 1 & 1 & 0 & 0 \\ 1 & 1 & 1 & 1 \\ 1 & 1 & 1 & 1 \end{pmatrix}$ and $r(M) = 2$. Fig.3(a) is an image K of the natural embedding of $E_4^{(\omega)}$ with respect to (f_1, f_2, f_3, f_4) and Fig.3(b) is an image of the natural embedding of $\mathcal{A}(F)$. The Hausdorff dimensions of $(E_4^{(\omega)}, \rho_\delta)$ and $\mathcal{A}(F)$ are $\log_3 16$ and $\log_3 4$ respectively by Theorem 2.6. Since both $f_3 f_3$ and $f_1 f_1$ are the same, we cannot find the set U satisfying (3.1) and (3.2) in Theorem 3.1, but as for Fig.3(b), Theorem 3.2 may be applicable. Let $v^T = (0011)$, $J_0 = \{3,4\}$ and $F_0 = \{(33),(34),(43),(44)\}$. Then $Mv = 2v$ and $U = \{z = \xi + i\eta \mid 0 < \eta < \frac{1}{2\sqrt{3}}, \sqrt{3}\eta < \xi < 1 - \sqrt{3}\eta\}$ satisfies the assumption of Theorem 3.2. So by Theorem 3.2 we have dim $\varphi(\mathcal{A}(F)) = \log_3 4$, which is the same as that of Fig.1.

a   b

Fig.3

       (ii) We shall show some examples, which are the images of the same $\mathcal{A}(F)$ in $(E_4^{(\omega)}, \rho_\delta)$ as that of Fig.3(b) and whose Hausdorff dimensions are $\log_3 4$. The corresponding similitudes are as follows:

$$4(a): f_1(z) = \overline{\mu z}, \quad f_2(z) = \mu\bar{z} + 1 - \mu, \quad f_3(z) = \mu\bar{z} \quad \text{and} \quad f_4(z) = \overline{\mu z} + 1 - \bar{\mu}$$

$$4(b): f_1(z) = \overline{\mu}z, \quad f_2(z) = \mu z + 1 - \mu, \quad f_3(z) = \mu\bar{z} \quad \text{and} \quad f_4(z) = \overline{\mu z} + 1 - \bar{\mu}$$

$$4(c): f_1(z) = \delta\bar{z}, \quad f_2(z) = \mu z + 1 - \mu, \quad f_3(z) = \mu\bar{z} \quad \text{and} \quad f_4(z) = \overline{\mu z} + 1 - \bar{\mu}$$

$$4(d): f_1(z) = \delta z, \quad f_2(z) = \alpha z + 1 - \alpha, \quad f_3(z) = \mu\bar{z} \quad \text{and} \quad f_4(z) = \overline{\mu z} + 1 - \bar{\mu},$$

for $z \in \mathbb{C}$, where $\mu = \frac{1}{2} + \frac{1}{2\sqrt{3}}i$, $\delta = \frac{1}{\sqrt{3}}$ and $\alpha = 0.57 + \sqrt{\frac{1}{3} - 0.57^2}\,i$.

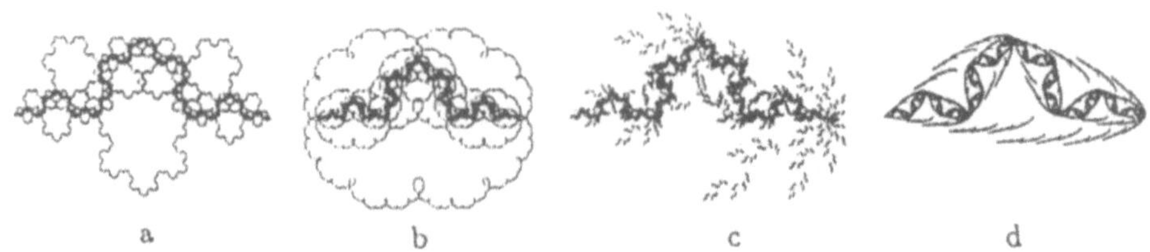

a   b   c   d

Fig.4

In these cases, both $f_3 f_3$ and $f_1 f_1$ are the same and Theorem 3.1 cannot be applied, but Theorem 3.2 can be applicable.

REFERENCES

[1]   V. Drobot and J. Turner, *Hausdorff Dimension and Perron-Frobenius Theory*, Illinois J. of Math. **33** (1989), 1-9.

[2]   G. A. Edgar, *Measure, Topology, and Fractal Geometry*, Springer-Verlag, 1990.

[3]   J. E. Huchinson, *Fractals and self-similarity*, Indiana Univ. Math. J. **30** (1981), 713-747.

[4]   Benoit B. Mandelbrot, *The Fractal Geometry of Nature*, W. H. Freeman and Company, 1982.

[5]   P. A. Moran, *Additive functions of intervals and Hausdorff measure*, Proc. Camb. Phil. Soc. **42** (1946), 15-23.

Department of Information Sciences, Faculty of Science
Ochanomizu University
2-1-1, Otsuka, Bunkyo-ku, Tokyo 112
Japan.

AMS Subject Classification: Primary 47N99, Secondary 15A48, 54F45

Operator Theory:
Advances and Applications, Vol. 62
© 1993 Birkhäuser Verlag Basel

OPERATORS WHICH HAVE COMMUTATIVE POLAR DECOMPOSITIONS

Mitsuru Uchiyama

Dedicated to Professor Tsuyoshi Ando on his sixtieth birthday

Let T be an operator on a separable Hilbert space. We show that if T is a compact operator and satisfies $|T^n| = |T|^n$ for some n, then T is normal, and that if T is a bounded operator and satisfies $|T^n| = |T|^n$ for $n = i, i+1, k, k+1$ ($1 \leq i < k$), then the polar decomposition of T is commutative. For a closed operator we obtain the analogous results.

Introduction

An operator which has commutative polar decomposition is called a quasi-normal operator. In this note we consider conditions on powers of an operator on a Hilbert space which imply that the operator is quasi-normal. In Section 1 we deal with a finite matrix or a compact operator. In Section 2 we deal with a bounded operator. Section 3, the main part of this note, extends the results in section 2 to the case of an unbounded closed operator. Specifically we define a notion of closed quasi-normal operator and give a model for such operators of the same form as the model for a bounded quasi-normal operator given by A.Brown.

Parts of this work were carried out while the author stayed at the University of California, San Diego as a visiting scholar. He would like to express his gratitude to Professor J.W. Helton. He is also grateful to Professors J. Agler, J. Ball and O. Merino.

§1.

First of all we will give a simple proposition about a finite matrix or a compact operator. It may be known to other mathematicians, but I could not find it in any books or papers. We recall that a quasi-normal compact operator is normal.

Proposition. *Let T be a finite matrix or a compact operator on a separable Hilbert space. If $|T^\kappa| = |T|^\kappa$ for some integer $\kappa \geq 2$, then T is normal.*

Proof. We may assume that T is represented by Schatten's formula as

$$T = \sum_{n=1}^{\infty} a_n e_n \otimes d_n,$$

where $\{e_n\}$ and $\{d_n\}$ are orthonormal families and $|a_n| \downarrow 0$ $(n \to \infty)$. Then it follows that

$$|T| = \sum |a_n| d_n \otimes d_n, \quad |T^*| = \sum |a_n| e_n \otimes e_n.$$

Since the eigenspace corresponding to $|a_1|$ is finite dimensional, there is an $i \in \mathbb{N}$ such that $|a_1| = \cdots = |a_i| > |a_{i+1}|$. Then we have

$$|a_1|^{2k} = ((T^*T)^k d_1, d_1) = (T^{*k} T^k d_1, d_1) = |a_1|^2 (T^{*k-2} T^{k-2} T e_1, T e_1)$$
$$\leq |a_1|^{2k-2} (T e_1, T e_1) \leq |a_1|^{2k},$$

from which it follows that

$$(T^*T e_1, e_1) = |a_1|^2 \quad \text{and hence} \quad (|a_1|^2 - T^*T) e_1 = 0,$$

because $|a_1|^2 - T^*T \geq 0$. In the same way as above we get

$$\{e_1, \cdots, e_i\} \subset \mathfrak{N}(T^*T - |a_1|^2),$$

where $\mathfrak{N}(X)$ denotes the null space of X, and hence

$$\mathfrak{N}(TT^* - |a_1|^2) = \mathfrak{N}(T^*T - |a_1|^2),$$

which reduces T to the normal operator, that is $T^*T e_n = TT^* e_n$ for $1 \leq n \leq i$. Repeating this procedure in the same way to the other restrictions of T, we derive $T^*T e_n = TT^* e_n$ for every n. $\{e_n\}$ and $\{d_n\}$ span the same space. Thus the proof is complete.

§2.

Secondly we consider a bounded operator T on a separable Hilbert space. We start with some definitions. T is called quasi-normal if T satisfies $TT^*T = T^*TT$ or equivalently $V|T| = |T|V$, where $T = V|T|$ is the polar decomposition of T. T is called subnormal if T has a normal extension. A quasi-normal operator is subnormal (c.f. [3]).

In [2] Brown showed that every quasi-normal operator is unitarily equivalent to the following:

$$\text{normal } \oplus (A \otimes S),$$

where A is a bounded positive operator with $\mathfrak{N}(A) = \{0\}$, and S the unilateral shift. $A \otimes S$ may be interpreted as the subdiagonal matrix, whose $(i+1, i)$ element is A.

Now notice that the operator

$$T = \begin{pmatrix} 0 & 0 & 0 & 0 \\ 1 & 1 & 0 & \ddots \\ 0 & 1 & 1 & \ddots \\ 0 & \ddots & \ddots & \ddots \end{pmatrix}$$

satisfies $T^{*2}T^2 = (T^*T)^2$, but T is not quasi-normal.

Theorem 2.1. *Let T be a bounded operator on a separable Hilbert space. Then the following are equivalent:*

(i) *T is quasi-normal.*

(ii) *$|T^n| = |T|^n$ for every $n \in \mathbb{N}$.*

(iii) *there are integers i and k such that $|T^n| = |T|^n$ for $n = i, i+1, k$ and $k+1$, where $1 \le i < k$.*

To prove this theorem we need two lemmas.

Lemma 2.1. *Let A be an invertible positive operator and P a projection. Then we have $PA^{-1}|_{\mathcal{L}} \ge (PA|_{\mathcal{L}})^{-1}$, where $\mathcal{L}$ is the range of P. Moreover the equality holds if and only if $PA = AP$.*

Proof. Let B be the inverse of A. Then $PA|_{\mathcal{L}}$ and $(I-P)B|_{\mathcal{L}^\perp}$ are invertible, and we have

$$(PA|_{\mathcal{L}})^{-1} = PB|_{\mathcal{L}} - PB(I-P)((I-P)B|_{\mathcal{L}^\perp})^{-1}(I-P)B|_{\mathcal{L}}$$

from which we easily obtain the inequality in the lemma.

Suppose $(PA|_{\mathcal{L}})^{-1} = PB|_{\mathcal{L}}$. Then we have $(I-P)B|_{\mathcal{L}} = 0$, which implies $PB = BP$, and hence $PA = AP$. The converse assertion is obvious.

A continuous function f on $[0, \infty)$ is called an operator monotone function if $0 \le A \le B$ implies $f(A) \le f(B)$. The first half of the next lemma was shown in [5].

Lemma 2.2.   *Let f be an operator monotone function on $[0, \infty)$ with $f(0) \ge 0$. Suppose that A is a bounded positive operator and P a non trivial projection. Then we have $Pf(A)P \le f(PAP)$. Moreover the equality holds only in the case of $PA = AP$ and $f(0) = 0$, if f is not a linear function.*

Proof. By the Löwner's theory, f can be represented as

$$f(x) = a + bx + \int_{+0}^{\infty} (\frac{1}{t} - \frac{1}{t+x})d\mu(t)$$

where $a = f(0)$, $b \geq 0$ and μ is a positive Borel measure such that

$$\int_{+0}^{\infty} \frac{1}{1+t^2} d\mu(t) < \infty.$$

Thus we have

$$(Pf(A)Ph, h) = ((a + bA)Ph, Ph) + \int_{+0}^{\infty} ((\frac{1}{t}I - (tI + A)^{-1})Ph, Ph)d\mu(t)$$

and

$$(f(PAP)h, h) = ((a + bPAP)h, h) + \int_{+0}^{\infty} ((\frac{1}{t}I - (tI + PAP)^{-1})h, h)d\mu(t)$$
$$= ((a + bPAP)h, h) + \int_{+0}^{\infty} ((\frac{1}{t}P - (P(tI + A)|_{\mathfrak{L}})^{-1})Ph, Ph)d\mu(t),$$

where $\mathfrak{L}$ is the range of P. By Lemma 1 $\frac{1}{t}P - (P(tI + A)|_{\mathfrak{L}})^{-1} \geq \frac{1}{t}P - P(tI + A)^{-1}P$ for $t > 0$ implies that $f(PAP) \geq Pf(A)P$. If $f(PAP) = Pf(A)P$, then it follows that, for every h, $(ah, h) = (aPh, Ph)$ and $((P(tI + A)|_{\mathfrak{L}})^{-1}Ph, Ph) = ((tI + A)^{-1}Ph, Ph)$ for almost every $t > 0$ $w.r.t.\mu$. Since the whole space is separable, we obtain

$$(P(tI + A)|_{\mathfrak{L}})^{-1} = P(tI + A)^{-1}P \quad \text{for almost every } t > 0.$$

Lemma 2.1 implies $PA = AP$. Clearly we have $f(0) = a = 0$.

Proof of Theorem 2.1. (i)$\Rightarrow$(ii)$\Rightarrow$(iii) are trivial, so we show only (iii) $\Rightarrow$ (i). From

$$T^*(T^*T)^k T = T^*(T^{*k}T^k)T = (T^*T)^{k+1} = T^*(TT^*)^k T$$

it follows that $P(T^*T)^k P = P(TT^*)^k P = (TT^*)^k$, where P is the projection on the closure of the range of T. Similarly we obtain $P(T^*T)^i P = (TT^*)^i$, and hence

$$(P(T^*T)^k P)^{i/k} = P(T^*T)^i P = P((T^*T)^k)^{i/k} P.$$

Since $f(x) = x^{i/k}$ is an operator monotone function, by Lemma 2.2 P commutes to $(T^*T)^k$ and hence to T^*T. Consequently we obtain $(PT^*TP)^k = (TT^*)^k$ and hence $PT^*TP = TT^*$. Therefore

$$T^*TT = T^*TPT = PT^*TPT = TT^*T.$$

This concludes the proof.

Remark. In [4], Embry showed that (i) and (ii) are equivalent using her theorem about subnormal operators. From lemma 2.2, it follows that

$$PAP \le (PA^2P)^{1/2} \le \cdots \le (PA^nP)^{1/n} \le \cdots, \tag{2.1}$$

which was shown in [1].

Proposition 2.2.   *If a subnormal operator T satisfies $|T^n| = |T|^n$ for some $n \ge 2$, then T is quasi-normal.*
Proof. First we show that a subnormal operator T on $\mathfrak{H}$ satisfies

$$|T| \le |T^2|^{1/2} \le \cdots \le |T^n|^{1/n} \le \cdots \tag{2.2}$$

In fact, T has a normal extension N on $\mathfrak{K} \supset \mathfrak{H}$. Let Q be the projection from $\mathfrak{K}$ onto $\mathfrak{H}$. Then, we have

$$(Q|N^{2n}|Q)^{\frac{1}{2n}} = (QN^{*n}N^nQ)^{\frac{1}{2n}} = \begin{bmatrix} T^{*n}T^n, & 0 \\ 0 & 0 \end{bmatrix}^{\frac{1}{2n}} = \begin{bmatrix} |T^n|^{1/n}, & 0 \\ 0 & 0 \end{bmatrix}.$$

Thus by (2.1) we get (2.2). If $|T^n| = |T|^n$, then (2.2) implies that $|T|^2 = |T^2|$, which means that T is a quasi-normal (cf. [4]). The proof is complete.

Proposition 2.3 *A hyponormal T is quasi-normal if $|T^n| = |T|^n$, $|T^{n+1}| = |T|^{n+1}$ for some $n \ge 2$.*
Proof. From the assumption we have

$$P(T^*T)^n P = P(TT^*)^n P = (PTT^*P)^n,$$

where P is the projection onto $\overline{T\mathfrak{H}}$. From (2.1) we obtain

$$0 \le PT^*TP \le (P(T^*T)^n P)^{1/n} = PTT^*P,$$

from which $PT^*TP = PTT^*P$ follows, because T is hyponormal. Consequently we get $TT^*T = T^*TT$, in the same way as the proof of Theorem 2.1. The proof is complete.

§3.

Finally we will consider a closed operator. We begin by giving a couple of lemmas.

Lemma 3.1. *Let A be an unbounded selfadjoint operator with the dense domain $\mathfrak{D}(A)$ in $\mathfrak{H}$, P an orthogonal projection. Then PAP is self-adjoint if P and A satisfy the following properties:*

(i) $P\mathfrak{D}(A) \subset \mathfrak{D}(A)$.

(ii) *There is a dense subset $\mathfrak{L} \subset \mathfrak{H}$ such that $(1 - P)APx = 0$ for every $x \in \mathfrak{L}$.*

Proof. Since $(1 - P)AP$ has the dense domain, its adjoint operator is well defined and satisfies $\mathfrak{D}(((1-P)AP)^*) \supset \mathfrak{D}(PA(1-P)) = \mathfrak{D}(A(1-P)) \supset \mathfrak{D}(A)$. Therefore $(1 - P)AP$ has a closed extension (cf. [8]), which vanishes on closed subspace including $\mathfrak{L}$. Thus we have $(1 - P)AP = 0$, from which it follows that

$$0 = ((1 - P)AP)^* \supset PA(1 - P) \supset PA - PAP \quad \text{and hence} \quad AP = PAP \supset PA,$$

where, for operators X and Y, $X \supset Y$ means that X is an extension of Y. This completes the proof.

Lemma 3.2. *Let A be an unbounded symmetric operator, P an orthogonal projection. Suppose $\mathfrak{N}(PAP)$ includes a dense set. Then $PAP = 0$.*

Proof. PAP is symmetric and its adjoint operator is closed. So we have $PAP = 0$.

Lemma 3.3. *Let T be a closed operator with the dense domain. Then, for $k \in \mathbf{N}$, $\mathfrak{D}((T^*T)^k)$ is a core of T.*

Proof. The case of $k = 1$ was shown by von Neumann. In general, we have

$$n(n + (TT^*)^k)^{-1}T \subset nT(n + (T^*T)^k)^{-1} \quad \text{for every} \quad n \in \mathbf{N};$$

in fact, for any x in $\mathfrak{D}(T)$ set $x_n = n(n + (T^*T)^k)^{-1}x$; then $x_n \in \mathfrak{D}((T^*T)^k)$ and $Tx_n = n(n + (TT^*)^k)^{-1}Tx$. Let $\{E_t\}$ be the resolution of the identity corresponding to T^*T. Then we have

$$\|x_n - x\|^2 = \|(T^*T)^k(n + (T^*T)^k)^{-1}x\|^2 = \int_0^\infty |\frac{t^k}{n + t^k}|^2 d\|E_t x\|^2,$$

which, by the Lebesgue's theorem, implies that $x_n \to x$. Similarly we get $Tx_n \to Tx$ (cf. [6] page 282). This completes the proof.

Theorem 3.1. *Let T be a closed operator with the dense domain in $\mathfrak{H}$ and P be the projection onto $\overline{T\mathfrak{H}}$. Then the following are equivalent:*

(i) $T^{*n}T^n = (T^*T)^n, \mathfrak{D}((T^*T)^n) \subset \mathfrak{D}((TT^*)^n) = \mathfrak{D}((T^*T)^n P)$ *for every $n \in \mathbf{N}$.*

(ii) *there are integers i and k such that $T^{*n}T^n = (T^*T)^n$ for $n = i, i+1, k, k+1$,*

$$\mathfrak{D}((T^*T)^i) \subset \mathfrak{D}((TT^*)^i) = \mathfrak{D}((T^*T)^i P),$$
$$\mathfrak{D}((T^*T)^k) \subset \mathfrak{D}((TT^*)^k) = \mathfrak{D}((T^*T)^k P) \quad \text{and} \quad 2i < k.$$

(iii) $TT^*T = T^*TT$ *and* $\mathfrak{D}(T^*T) \subset \mathfrak{D}((TT^*)) = \mathfrak{D}(T^*TP)$.

(iv) $V|T| = |T|V$, *where* $T = V|T|$ *is the polar decomposition.*

Remark. We do not know if Lemma 2.2 is valid for unbounded A. The condition on i and k in Theorem 3.1 (ii) is different from that in Theorem 2.1.

Proof. (i) $\implies$ (ii). It is obvious.

(ii) $\implies$ (iii). Since $(TT^*)^n(1 - P) = 0$ on $\mathfrak{H}$, for every n, we have

$$(TT^*)^i = (TT^*)^i P = (PTT^*P)^i \quad \text{and} \quad (TT^*)^k = (TT^*)^k P = (PTT^*P)^k. \tag{3.1}$$

Notice that the above operators are selfadjoint. From (ii) it follows that

$$T^{*i+1}T^{i+1} - (T^*T)^{i+1} \subset T^*((T^*T)^i - (TT^*)^i)T,$$

and

$$T^{*k+1}T^{k+1} - (T^*T)^{k+1} \subset T^*((T^*T)^k - (TT^*)^k)T.$$

Thus we get

$$P((T^*T)^i - (TT^*)^i)P = 0 \quad \text{on} \quad T\mathfrak{D}((T^*T)^{i+1}) \oplus \mathfrak{N}(T^*).$$

and

$$P((T^*T)^k - (TT^*)^k)P = 0 \quad \text{on} \quad T\mathfrak{D}((T^*T)^{k+1}) \oplus \mathfrak{N}(T^*).$$

Since $(T^*T)^i - (TT^*)^i$, and $(T^*T)^k - (TT^*)^k$ are symmetric, Lemmas 3.2 and 3.3 imply

$$P((T^*T)^i - (TT^*)^i)P = 0, \quad \text{and} \quad P((T^*T)^k - (TT^*)^k)P = 0,$$

from which it follows that

$$P(T^*T)^i P = P(TT^*)^i P \quad \text{and} \quad P(T^*T)^k P = P(TT^*)^k P; \tag{3.2}$$

in fact, $\mathfrak{D}((T^*T)^k P) = \mathfrak{D}((TT^*)^k) = \mathfrak{D}((TT^*)^k P)$. Let $\{E_t\}$ be the resolution of the identity corresponding to T^*T. Then we get

$$P(T^*TE_t)^k P \leq P(T^*T)^k P.$$

Since both operators are selfadjoint, we can apply the Heinz's inequality (cf., [7]) as follows:

$$(P(T^*TE_t)^k P)^{2i/k} \leq (P(T^*T)^k P)^{2i/k} = (PTT^*P)^{2i},$$

where we used (3.2) and (3.1) for the last equality. Thus for an arbitrary x in $\mathfrak{D}(P(T^*T)^i P) \cap \mathfrak{D}((PTT^*P)^{2i})$ we have

$$\begin{aligned}
\|P(T^*T)^i Px\|^2 &= \lim_{t \to \infty} \|P(T^*TE_t)^i Px\|^2 \leq \lim \|(T^*TE_t)^i Px\|^2 \\
&= \lim(P(T^*TE_t)^{2i} Px, x) \leq \lim((P(T^*TE_t)^k P)^{2i/k} x, x) \\
&\leq ((PTT^*P)^{2i} x, x) = \|P(T^*T)^i Px\|^2,
\end{aligned}$$

where we used Lemma 2.2, and the last equality follows from (3.2). Consequently we obtain

$$\|P(T^*T)^i Px\|^2 = \lim_{t \to \infty} \|(T^*TE_t)^i Px\|^2 = \|(T^*T)^i Px\|^2,$$

which means $(1-P)(T^*T)^i Px = 0$. Since $\mathfrak{D}(P(T^*T)^i P) \cap \mathfrak{D}((PTT^*P)^{2i}) = \mathfrak{D}((PTT^*P)^{2i})$ is dense, where we used (3.2), Lemma 3.1 implies that $P(T^*T)^i \subset (T^*T)^i P$ and hence $PT^*T \subset T^*TP$. PT^*TP is a selfadjoint operator. By (3.1) and (3.2), we get

$$T^*TP = PT^*TP = PTT^*P = TT^*P,$$

from which it follows that

$$T^*TT = T^*TPT = TT^*PT = TT^*T.$$

Thus we have obtained (iii).

(iii) $\Longrightarrow$ (iv). For any $x \in T\mathfrak{D}(TT^*T) \oplus \mathfrak{N}(T^*)$ we have $TT^*Px = T^*TPx$, and hence $PT^*TPx = PTT^*Px = TT^*Px = T^*TPx$. Since $T\mathfrak{D}(TT^*T) \supset T\mathfrak{D}((T^*T)^2)$, Lemma 3.3 implies that $T\mathfrak{D}(TT^*T) \oplus \mathfrak{N}(T^*)$ is dense. Thus from Lemma 3.1 it follows that $PT^*T \subset T^*TP$ and that PT^*TP is selfadjoint. On the other hand, the above equalities and Lemma 3.2 give $P(T^*T - TT^*)P = 0$ and hence $PT^*TP \subset PTT^*P$; both of them are selfadjoint; thus they are coincident. Therefore we have

$$|T|P = P|T|P = (PT^*TP)^{1/2} = (PTT^*P)^{1/2} = |T^*|P,$$

and hence $|T|V = |T^*|V$. Since $V|T| = |T^*|V$ (cf. [7]), we have $V|T| = |T|V$. Thus we have obtained (iv).

(iv) $\Longrightarrow$ (i). It is easy to show that $TT^*T = T^*TT$, from which, by induction, it follows that $T^{*n}T^n = (T^*T)^n$. Let $\{E_t\}$ and $\{F_t\}$ be the resolutions of the identity corresponding to $|T|$ and $|T^*|$ respectively. Then it follows that $F_t = F_0 + VE_tV^* =$

$F_0 + E_t VV^* = (1 - P) + E_t P$ and $E_t = E_t(1 - P) + E_t P$, where the first equality was shown in [7]. Thus we have

$$\int_0^\infty t^{2n} d\|F_t x\|^2 = \int_0^\infty t^{2n} d\|E_t P x\|^2 \le \int_0^\infty t^{2n} d\|E_t x\|^2$$

for every x. This shows that

$$\mathfrak{D}(|T|^n) \subset \mathfrak{D}(|T^*|^n) = \mathfrak{D}(|T|^n P),$$

and for $x \in \mathfrak{D}(|T|^n)$

$$\||T^*|^n x\| = \||T|^n P x\| \le \||T|^n x\|. \tag{3.3}$$

This completes the proof.

Definition. Let T be a densely defined, closed operator. Then T will be called a quasi-normal operator if $V|T| = |T|V$, where $T = V|T|$ is the polar decomposition of T.

Corollary 3.1. *If T is a quasi-normal operator, then we have*

$$T^{*n}T^n = (T^*T)^n \ge (TT^*)^n \ge T^n T^{*n} \quad \text{for every} \quad n \in \mathsf{N}.$$

Proof. We obtain the last inequality by induction. (3.3) implies the second relation.

Corollary 3.2. *Let T be a quasi-normal operator. Then T is normal if* $\mathfrak{N}(T) = \mathfrak{N}(T^*)$.

Proof. For $\{E_t\}$ and $\{F_t\}$ given in the proof of Theorem 3.1 we have $1 - P = E_0 = F_0$, from which we get $F_t = F_0 + VE_t V^* = E_0 + E_t VV^* = E_0 + E_t P = E_t(1 - P) + E_t P = E_t$. Thus we have $|T| = |T^*|$, which means $T^*T = TT^*$.

Corollary 3.3. *Let T be a quasi-normal operator, and suppose* $\mathfrak{N}(T^*) \subset \mathfrak{D}(T^*T)$ *and* $\mathfrak{N}(T^*) = \mathfrak{N}(T^{*2})$. *Then T is normal.*

Proof. From Corollary 3.1 it follows that $\mathfrak{N}(T) \subset \mathfrak{N}(T^*)$. Since

$$|T|^2 V^* = (V|T|^2)^* = (|T|^2 V)^* \supset V^*|T|^2,$$

for any x in $\mathfrak{N}(T^*) \subset \mathfrak{D}(|T|^2)$ we have

$$0 = |T|^3 V^* x = |T|V^*|T|^2 x = T^*|T|^2 x = T^{*2}Tx,$$

from which it follows that $T^*Tx = 0$ and hence $Tx = 0$. Thus we have $\mathfrak{N}(T) = \mathfrak{N}(T^*)$. Corollary 3.2 gives the conclusion.

Corollary 3.4. *Let T be a quasi-normal operator, and suppose $\mathfrak{N}(T^*) \subset \mathfrak{D}(T^*T)$ and $T^kT^{*k} = (TT^*)^k$ for some $k \geq 2$. Then T is normal.*

Proof. We have $\mathfrak{N}(T^{*k}) = \mathfrak{N}(T^{*2}) = \mathfrak{N}(T^*)$, which completes the proof.

Definition. Let A be a closed operator with the dense domain in $\mathfrak{L}$. Then $A \otimes S$ is the closed operator in $l^2(\mathfrak{L})$ defined by $\mathfrak{D}(A \otimes S) = \{(x_0, x_1, \cdots) \in l^2(\mathfrak{L}) : x_n \in \mathfrak{D}(A), \sum_{n=1}^{\infty} ||Ax_n||^2 < \infty\}$, $A \otimes S(x_0, x_1, \cdots) = (0, Ax_0, Ax_1, \cdots)$.

Theorem 3.2. *Let T be a closed operator with the dense domain in $\mathfrak{H}$. Then T is quasi-normal if and only if T is unitarily equivalent to*

$$\text{normal} \oplus (A \otimes S),$$

where A is a positive operator with the dense domain in an appropriate Hilbert space $\mathfrak{L}$ and $\mathfrak{N}(A) = 0$.

Proof. Suppose that T is quasi-normal. Corollary 3.1 implies $\mathfrak{N}(T) \subset \mathfrak{N}(T^*)$. Thus $\mathfrak{N}(T)$ reduces T to 0. So we may assume $\mathfrak{N}(T) = 0$. In this case we have $\mathfrak{N}(V) = 0$, where $T = V|T|$ is the polar decomposition. Since V is an isometry, we can apply the Wold's decomposition to V;

$$V = V_0 \oplus V_1 \quad \text{on} \quad \mathfrak{H} = \mathfrak{H}_0 \oplus \mathfrak{H}_1,$$

where $\mathfrak{H}_0 = \bigcap_{n=0}^{\infty} V^n \mathfrak{H}$, V_0 is the unitary part of V, and V_1 the unilateral shift part.

We show $\mathfrak{H}_0$ reduces T too. Let Q be the projection from $\mathfrak{H}$ to $\mathfrak{H}_0$, and $\{E_t\}$ the resolution of the identity corresponding to $|T|$. From $V|T| = |T|V$ it follows that $VE_t = E_tV$ and hence $(1-Q)VE_tQ = (1-Q)E_tVQ$. Since Q commutes to V, we have $V_1X = XV_0$, where $X = (1-Q)E_t|\mathfrak{H}_0$. Making use of $V_1^{*n} \to 0$ and $V_0^{*n} \nrightarrow 0$ $(n \to \infty)$ gives $X = 0$, which means $(1-Q)E_tQ = 0$ and hence $QE_t = E_tQ$. Thus we have $Q|T| \subset |T|Q$, from which it follows that $QT \subset TQ$ and hence that $\mathfrak{H}_0$ reduces T. Let $T = T_0 \oplus T_1$ be the decomposition of T corresponding to $\mathfrak{H} = \mathfrak{H}_0 \oplus \mathfrak{H}_1$. Then we have $T_0 = V_0|T_0|$ and $T_1 = V_1|T_1|$, which are the polar decompositions. Since T_0 is quasi-normal and $\mathfrak{N}(T_0) = \mathfrak{N}(T_0^*) = 0$, Corollary 3.2 implies that T_0 is normal. We will show that T_1 is unitarily equivalent to $A \otimes S$. Set $\mathfrak{L} = \mathfrak{H}_1 \ominus V_1\mathfrak{H}_1$. Then we get

$$\mathfrak{H}_1 = \mathfrak{L} \oplus V_1\mathfrak{L} \oplus \cdots \oplus V_1^n\mathfrak{L} \oplus \cdots. \tag{3.4}$$

Each $V_1^n \mathcal{L}$ reduces $|T_1|$; in fact, $V_1^* |T_1| \subset |T_1| V_1^*$ implies $V_1 V_1^* |T_1| \subset V_1 |T_1| V_1^* = |T_1| V_1 V_1^*$; since $V_1 V_1^*$ is the projection onto $V_1 \mathfrak{H}_1$, $\mathcal{L}$ reduces $|T_1|$. Similarly we can show that $V_1^n \mathcal{L}$ reduces $|T_1|$. Set $A = |T_1| \|_{\mathcal{L}}$. Then it is easy to see that $\mathfrak{D}(A)$ is dense and $\mathfrak{N}(A) = 0$. We show

$$\mathfrak{D}(|T_1|) = \{\sum \oplus V_1^n x_n : x_n \in \mathfrak{D}(A), \sum_{n=0}^{\infty} \|A x_n\|^2 < \infty\}. \tag{3.5}$$

Take any x in $\mathfrak{D}(|T_1|)$. Then (3.4) gives

$$x = x_0 \oplus V_1 x_1 \oplus \cdots \oplus V_1^n x_n \oplus \cdots .$$

Since each $V_1^n \mathcal{L}$ reduces $|T_1|$, $V_1^n x_n \in \mathfrak{D}(|T_1|)$; this implies $x_n \in \mathfrak{D}(A)$, because $|T_1| V_1 = V_1 |T_1|$. $\sum_{n=0}^{\infty} \|A x_n\|^2 = \| |T_1| x\|^2 < \infty$. Thus we have shown that x belongs to the right side of (3.5). Since $|T_1|$ is closed, we obtain the converse inclusion as well. From the definition and (3.5) it is easy to see that the unitary U defined by

$$U : \ell^2(\mathcal{L}) \ni (x_0, x_1, \cdots) \to \sum \oplus V_1^n x_n \in \mathfrak{H}_1$$

satisfies $U(A \otimes S) = T_1 U$. Thus T is unitarily equivalent to $0 \oplus T_0 \oplus (A \otimes S)$.

To prove the converse assertion we have only to show that $A \otimes S$ is quasi-normal.

$$\mathfrak{D}((A \otimes S)^*) = \{(x_0, x_1, \cdots) \in \ell^2(\mathcal{L}) : x_n \in \mathfrak{D}(A) \quad \text{for} \quad n \geq 1, \sum_{n=1}^{\infty} \|A x_n\|^2 < \infty\}$$

and

$$(A \otimes S)^*(x_0, x_1, \cdots) = (A x_1, A x_2, \cdots)$$

implies that

$$\mathfrak{D}(|A \otimes S|) = \{(x_0, x_1, \cdots) \in \ell^2(\mathcal{L}) : x_n \in \mathfrak{D}(A) \quad \text{for} \quad n \geq 0, \sum_{n=0}^{\infty} \|A x_n\|^2 < \infty\}$$

and

$$|A \otimes S|(x_0, x_1, \cdots) = (A x_0, A x_1, \cdots).$$

It is easy to see that $(I \otimes S)|A \otimes S|$ is the polar decomposition of $A \otimes S$ and that $(I \otimes S)|A \otimes S| = |A \otimes S|(I \otimes S)$. This completes the proof.

Theorem 3.3. *A closed quasi-normal operator T with the dense domain in $\mathfrak{H}$ has a normal extension, that is, there is a normal operator N on $\mathfrak{K} \supset \mathfrak{H}$ such that*

$$N \mathfrak{H} \subset \mathfrak{H} \quad \text{and} \quad N|\mathfrak{H} = T.$$

Proof. By Theorem 3.2 we need only to show that $A \otimes S$ has a normal extension. But it is easy to see that $A\otimes$ (the bilateral shift) is the normal extension. Thus the proof is complete.

Corollary. *If a closed symmetric operator T is quasi-normal, then T is selfadjoint.*

Proof. Since $A\otimes S$ is not symmetric, T is normal. This means $\mathfrak{D}(T) = \mathfrak{D}(T^*)$. Thus T is selfadjoint.

Acknowledgement. The author would like to thank the referee.

References

1.   T, Ando, *Topics on operator inequalities* , Lecture notes. Hokkaido Univ. Sapporo 1978

2.   A. Brown, *On a class of operators*, P.A.M.S.4(1953),723-728

3.   J. B. Conway, *Subnormal Operators*, (1981), Pitman Research Notes in Math.

4.   M. R. Embry, *A generalization of the Halmos-Bram criterion for subnormality*, Acta Sci.Math.(Szeged)35(1973), 61-64

5.   F. Hansen , *An operator inequality*, Math.Ann.246(1980), 249-250

6.   T. Kato, *Perturbation Theory for Linear Operators*, Springer Verlag,(1966)

7.   T. Kato, *Notes on some inequalities for linear operators*, Math. Ann.125 (1952), 208-212

8.   F. Riesz, and Sz. Nagy, *Functional Analysis*, (1955), Frederick Unger

Department of Mathematics
Fukuoka University of Education
Akama, Munakata
Fukuoka 811-41
Japan

AMS Subject Classification: Primary 47B20, Secondary 47A65

Operator Theory:
Advances and Applications, Vol. 62
© 1993 Birkhäuser Verlag Basel

TRACE FORMULA FOR THE PERTURBATION
OF PARTIAL DIFFERENTIAL OPERATOR
AND CYCLIC COCYCLE ON A
GENERALIZED HEISENBERG GROUP

Dedicated to the 60th birthday of Professor T. Ando

Daoxing Xia

A trace formula related to the operator $P\left(i\frac{\partial}{\partial x_1},\cdots,i\frac{\partial}{\partial x_n}\right)$ $+Q$ is established, where $P(\cdot)$ is a homogeneous real polynomial of degree m and Q is a bounded self-adjoint operator satisfying $e^{itx}Qe^{-itx}-Q\in\mathcal{L}^1$ and $e^{iR}Qe^{-iR}-Q\in\mathcal{L}^1$ for every $R\in D_m$ where D_m is the set of all partial differential operator R of order $\leq m$ with real constant coefficients. The form of the related cyclic one-cocycle on the group of all the elements of $e^{iR}e^{ixt}e^{i\theta}$ for $R\in D_m, t\in\mathbf{R}^n, \theta\in\mathbf{R}^1$ is determined.

This paper is a continuation of previous works [4], [5] and [6] on almost unperturbed Schrödinger pairs of operators. In this paper , the trace formulas of the perturbation of partial differential operator $P\left(i\frac{\partial}{\partial x_1},\cdots,i\frac{\partial}{\partial x_n}\right)$ is investigated, where P is a homogeneous real polynomail.

These trace formulas are naturally related to the cyclic one-cocycles on a nilpotent analytic group, a "generalized" Heisenberg group $G_{m,n}$ of elements $e^{iR}e^{ixt}e^{i\theta}$ for $t\in\mathbf{R}^n$ and $R\in D_m$ which is the set of all partial differential operators of order m with constant real coefficients. The abstract definition and the structure of $G_{m,n}$ is described in §2. In §3-4, it is proved that on this analytic group every cyclic homogeneous Baire one-cocycle equals to the coboundary of a homogeneous Baire zero-cocycle almost everywhere.

In §5 as in [3], the almost unperturbed Lie group of operators $\{W(g) : g \in G_{m,n}\}$ is introduced and a trace formula is given by means of the cyclic cocycle

$$\text{tr}\left([W(g_1),W(g_2)] - W(g_1g_2) + W(g_2g_1)\right),$$

for $g_1, g_2 \in G$.

In §6, the explicit form of the operator $W(g)$ is given which is expressed by some singular integral operator.

Finally in §7, it is proved that if P is a homogeneous real polynomial of n variables of degree m and Q is a bounded self-adjoint operator on $L^2(\mathbf{R})$ which is almost commuting [1] with $e^{iR}e^{ixt}$ for $t \in \mathbf{R}^n$ and $R \in D_m$. Then

$$e^{i\rho(P(i\frac{d}{dx})+Q)} = \mathcal{F}^{-1}K(\rho)\mathcal{F} + c(\rho)e^{i\rho P(i\frac{d}{dx})}, \quad \rho \in \mathbf{R}^1$$

where $i\frac{d}{dx}$ stands for $(i\frac{\partial}{\partial x}, \cdots, i\frac{\partial}{\partial x_n})$, $\mathcal{F}$ is the Fourer transform on $L^2(\mathbf{R}^n)$, $c(\rho)$ is a complex function and $K(\rho)$ is a singular integral operator on $L^2(\mathbf{R})$,

$$(K(\rho)f)(t) = \int f(x)k(x,t;\rho)dx.$$

The kernel $k(\cdot,\cdot,\cdot)$ is determined by the trace formula

$$\mathrm{tr}\,\left(e^{ixt}[e^{i\rho(p(i\frac{d}{dx})+Q)}, e^{-y\frac{d}{dx}}]\right) = (1 - e^{iyt})\int e^{iyx}k(x,t+x;\rho)dx$$

for $y, t \in \mathbf{R}^n$ and $\rho \in \mathbf{R}^1$. Besides, the funciton $c(\cdot)$ is determined by the kernel $k(\cdot,\cdot,\cdot)$, if Q is not a multiple of identity.

2. Let $S_{k,n}$ be the linear space of all symmetric k-fold tensors in $\mathbf{R}^n \otimes \cdots \otimes \mathbf{R}^n$. Then $S_{1,n} = \mathbf{R}^n$. Denote $\mathbf{R}^1$ by $S_{o,n}$. For $s \in S_{k,n}$, let $s^{i_1 \cdots i_k}$, $i_j = 1, 2, \cdots, n$, $j = 1, 2, \cdots, k$ be the components of s. Then

$$s^{i_1 \cdots i_k} = s^{i'_1 \cdots i'_k}$$

where $(i'_1, \cdots i'_k)$ is any permutation of $(i_1, \cdots, i_k)$. For $s \in S_{k,n}$ and $s' \in S_{k',n}$ where $k \geq k'$, define the product $s's = ss'$ as a tensor in $S_{k-k',n}$ by

$$(ss')^{i_1 \cdots i_{k-k'}} = \sum_{j_1, \cdots, j_{k'}} s^{j_1 \cdots j_{k'}} s'^{i_1 \cdots i_{k-k'} j_1 \cdots j_{k'}}$$

if $k > k'$, and

$$ss' = \sum_{j_1, \cdots, j_k} s^{j_1 \cdots j_k} s'^{j_1 \cdots j_k},$$

if $k' = k$. For $t = (t_1, \cdots, t_n) \in \mathbf{R}^n$, let t^k be the tensor in $\mathbf{S}_{k,n}$ defined by $(t^k)^{i_1 \cdots i_k} = t_{i_1} \cdots t_{i_k}$. Let $\mathbf{S}^{m,n}$ be the linear space of all vectors $(s_m, \cdots, s_1)$ for $s_k \in \mathbf{S}_{k,n}$.

Let $G_{m,n}$ be the Lie group of all elements $g(\varsigma, t, z)$ for $\varsigma \in \mathbf{S}^{m,n}$, $t \in \mathbf{R}^n$, $z \in \mathbf{T} = \{z \in \mathbf{C} : |z| = 1\}$ endowed with the product

$$g(\varsigma_0, t_0, z_0)g(\varsigma_1, t_1, z_1) = g(\varsigma_0 + \varsigma_1 + \mu(\varsigma_1, t_0), t_0 + t_1, z_0 z_1 \nu(\varsigma_1, t_0))$$

where

$$\mu(s_m, \cdots, s_1, t) = (0, s_m t, s_{m-1} t + s_m t^2/2!, \cdots, s_2 t + s_3 t^2/2! + \cdots + s_m t^{m-1}/(m-1)!)$$

and

$$\nu(s_m, \cdots, s_1, t) = \exp\left(i \sum_{j=1}^{m} s_j t^j / j!\right)$$

for $s_k \in \mathbf{S}_{k,n}$. The group $G_{m,n}$ is a nilpotent analytic group and $G_{1,n}$ is a representation of the Heisenberg group. So $G_{m,n}$ is a sort of the generalization of the Heisenberg group.

There is a natural unitary representation of $G_{m,n}$ on $L^2(\mathbf{R}^n)$. Let u be the n-tuple of self-adjoint partial differential operator on $L^2(\mathbf{R})$

$$u = \left(\frac{1}{i}\frac{\partial}{\partial x_1}, \cdots, \frac{1}{i}\frac{\partial}{\partial x_n}\right)$$

such that for $t = (t_1, \cdots, t_n) \in \mathbf{R}^n$,

$$(e^{iut}f)(x) = f(x+t), \qquad f \in L^2(\mathbf{R}),$$

where $ut = \sum_j t_j \frac{1}{i}\frac{\partial}{\partial x_j}$. Let $v_m^{i_1 \cdots i_m}$ be the self-adjoint operator

$$(v_m^{i_1 \cdots i_m} f)(x) = \frac{1}{m!} x_{i_1} \cdots x_{i_m} f(x), \qquad x = (x_1, \cdots, x_n).$$

Therefore v_m is a symmetric m-fold tensor of self-adjoint operators. For $s_m \in \mathbf{S}_{m,n}$, let $s_m v_m = \sum s_m^{i_1 \cdots i_m} v_m^{i_1 \cdots i_m}$. For $\varsigma = (s_m, \cdots, s_1) \in \mathbf{S}^{m,n}$, $t \in \mathbf{R}^n$ and $z \in \mathbf{P}$, let

$$g_0(\varsigma, t, z) = z(\prod_{j=1}^{m} e^{is_m v_m})e^{iut}$$

Then $g(\varsigma, t, z) \mapsto g_0(\varsigma, t, z)$ is a faithful unitary representation of $G_{m,n}$ on $L^2(\mathbf{R}^n)$.

Let $\mathcal{F}$ be the Fourier transform from $L^2(\mathbf{R}^n)$ onto $L^2(\mathbf{R}^n)$ satisfying

$$\tilde{f}(x) = (\mathcal{F}f)(x) = \frac{1}{(2\pi)^{n/2}} \int e^{ixy} f(y) dy$$

for $f \in L^1(\mathbf{R}^n) \cap L^2(\mathbf{R}^n)$. Denote $\tilde{u} = \mathcal{F}^{-1} u \mathcal{F}$ and $\tilde{v}_m = \mathcal{F}^{-1} v_m \mathcal{F}, \tilde{g} = \mathcal{F}^{-1} g_0 \mathcal{F}$ Then

$$(\tilde{u}f)(x) = xf(x)$$

and

$$(\tilde{v}_m^{i_1 \cdots i_m} f)(x) = \frac{1}{m!} i^m \frac{\partial^m f}{\partial x_{i_1} \cdots \partial x_{i_m}},$$

Thus $\tilde{U} : g \mapsto \mathcal{F}^{-1} g_0 \mathcal{F}$ is another faithful unitary representation of $G_{m,n}$ on $L^2(\mathbf{R}^n)$ and the form of the elements in $\tilde{U} G_{m,n}$ is the operator

$$f \mapsto z \exp(iP(i\frac{\partial}{\partial x_1}, \cdots, i\frac{\partial}{\partial x_n})) e^{i(\cdot)t} f(\cdot), \quad f \in L^2(\mathbf{R}^n)$$

where $z \in \mathbf{T}$, $t \in \mathbf{R}^n$ and P is a real polynomial of degree $\leq m$.

Let $Z_{m,n} = \{g(\varsigma, 0, z) : \varsigma \in \mathbf{S}^{m,n}, z \in \mathbf{T}\}$ then $Z_{m,n}$ is a normal subgroup of $G_{m,n}$ with dimension less than $G_{m,n}$'s.

3. Let G be any group. A function $\phi(\cdot, \cdot)$ on $G \times G$ is said to be a cyclic one-cocycle ([2], [3]), if

$$\phi(g_0, g_1) = -\phi(g_1, g_0), \qquad g_j \in G, \tag{5}$$

and

$$\phi(g_0 g_1, g_2) + \phi(g_1 g_2, g_0) + \phi(g_2 g_0, g_1) = 0, \quad g_j \in G. \tag{6}$$

In the terminology of cyclic cohomology, a function on G is said to be a cyclic zero-cocycle. A cyclic one-cocycle ϕ is said to be the coboundary of a zero-cocycle σ on G and is denoted by $\phi = \partial\sigma$ if

$$\phi(g_0, g_1) = \sigma(g_0 g_1) - \sigma(g_1 g_0), \qquad g_j \in G.$$

A cyclic one-cocycle ϕ on $G_{m,n}$ is said to be homogeneous if

$$\phi(g(\varsigma_0,t_0,z_0),g(\varsigma_1,t_1,z_1)) = z_0 z_1 \phi(g(\varsigma_0,t_0,1),g(\varsigma_1,t_1,1))$$

for $z_j \in \mathsf{T}$. A zero-cocycle σ on $G_{m,n}$ is said to be homogeneous, if

$$\sigma(g(\varsigma,t,z)) = z\sigma(g(\varsigma,t,1)), \qquad z \in \mathsf{T}.$$

A function f is said to be a Baire function if $\{x : f(x) \in O\}$ is a Borel set for every open set O.

THEOREM 1. *For every homogeneous Baire cyclic one-cocycle ϕ on $G_{m,n}$ there is a Baire homogeneous zero-cocycle σ on $G_{m,n}$ such that*

$$\phi(g_1,g_2) = (\partial\sigma)(g_1,g_2) \qquad for \quad g_1 g_2 \in Z_{m,n}. \tag{7}$$

PROOF. (1) For a homogeneous cyclic one-cocycle ϕ on $G_{m,n}$, define a function (using the same letter ϕ)

$$\phi(\varsigma_0,t_0;\varsigma_1,t_1) = \phi(g(\varsigma_0,t_0,1),g(\varsigma_1,t_1,1)) \tag{8}$$

where $\varsigma_j \in \mathsf{S}^{m,n}$ and $t_j \in \mathsf{R}^n$ for $j = 0,1$. From (5), (6) and (8), we have

$$\phi(\varsigma_0,t_0;\varsigma_1,t_1) + \phi(\varsigma_1,t_1;\varsigma_0,t_0) = 0 \tag{9}$$

and

$$\sum_{j=0}^{2} \phi(\varsigma_j + \varsigma_{j+1} + \mu(\varsigma_{j+1},t_j),t_j + t_{j+1};\varsigma_{j+2},t_{j+2})\nu(\varsigma_{j+1},t_j) = 0, \tag{10}$$

where $\varsigma_{j+3} = \varsigma_j$ for $j = 0,1,2$. Define

$$\psi(\varsigma_1,t_1;\varsigma,t) = \phi(\varsigma - \varsigma_1, t - t_1;\varsigma_1,t_1) \tag{11}$$

Then (9) reduces to

$$\psi(\varsigma_0,t_0;\varsigma,t) + \psi(\varsigma_1,t_1;\varsigma,t) = 0, \tag{12}$$

where $\varsigma_1 = \varsigma - \varsigma_0$, and $t_1 = t - t_0$, and (10) reduces to

$$\psi(\varsigma_1 + \varsigma_2 + \mu(\varsigma_2, t_1), t_1 + t_2; \varsigma + \mu(\varsigma_2, t_1), t)$$
$$= \overline{\nu}(\varsigma_2, t_1)(\nu(\varsigma_0, t_2)\psi(\varsigma_1, t_1; \varsigma + \mu(\varsigma_0, t_2), t)$$
$$+ \nu(\varsigma_1, t_0)\psi(\varsigma_2, t_2; \varsigma + \mu(\varsigma_1, t_0), t)), \tag{13}$$

where $\varsigma_0 = \varsigma - \varsigma_1 - \varsigma_2$ and $t_0 = t - t_1 - t_2$. We try to solve the functional equations (12) and (13).

(2) For $\eta \in S_{m,n}$, let

$$\tau(\eta, t) = \mu((\eta, 0, \cdots, 0), t).$$

LEMMA 1. *Let* $\psi_0(r, \varsigma, t), r \in S_{m,n}, \varsigma \in S^{m,n}, t \in \mathbf{R}^n$ *be a Baire function satisfying the condition that*

$$\psi_0(r_1 + r_2, \varsigma, t) = e^{ir_1 t^m/m!}\psi_0(r_2, \varsigma + \tau(r_1, t), t) + \psi_0(r_1, \varsigma, t). \tag{14}$$

Then there is a Baire function F and $S_{m,n}$-valued Baire function $Z(\varsigma, t)$ such that

$$\psi_0(r, \varsigma, t) = F(\varsigma, t) - e^{irt^m/m!}F(\varsigma + \tau(r, t), t) + Z(\varsigma, t)r. \tag{15}$$

PROOF. Let $M_t = \{r \in S_{m,n} : rt = 0\}$ for $t \in \mathbf{R}^n$. Then $\tau(r, t) = 0$ for $r \in M_t$. Therefore

$$\psi_0(r_1 + r_2, \varsigma, t) = \psi_0(r_1, \varsigma, t) + \psi_0(r_2, \varsigma, t), \tag{16}$$

if r_1 or r_2 is in M_t. Hence, there is a M_t-valued Baire function $y(\cdot, \cdot)$ such that

$$\psi_0(r, \varsigma, t) = y(\varsigma, t)r, \qquad \text{for } r \in M_t. \tag{17}$$

By (16) again, if $r_1 - r_2 \in M_t$, then

$$\psi_0(r_1, \varsigma, t) = \psi_0(r_2, \varsigma, t) + \psi_0(r_1 - r_2, \varsigma, t)$$
$$= \psi_0(r_2, \varsigma, t) + y(\varsigma, t)(r_1 - r_2).$$

Thus the function $\psi_0(r,\varsigma,t) - y(\varsigma,t)r$ depends only on $r \cdot t, \varsigma$ and t. Define

$$F_0(rt,\varsigma,t) = \psi_0(r,\varsigma,t) - y(\varsigma,t)r$$

Then

$$\psi_0(r,\varsigma,t) = F_0(rt,\varsigma,t) + y(t,\varsigma)r. \tag{18}$$

For $\varsigma = (s_m, s_{m-1}, \cdots, s_1) \in S^{m,n}$ and $t \in \mathbf{R}^n$, $t \neq 0$, there is a $r_1 \in S_{m,n}$ such that $s_{m-1} = r_1 t$. In (18), let ς be replaced by $\varsigma - \tau(r_1,t)$. By (14) and (18), we have

$$\begin{aligned}
\psi_0(r_2,\varsigma,t) = e^{-is_{m-1}t^{m-1}/m!} \{ &F_0(r_1 t + r_2 t, \varsigma - \tau(r_1,t), t) - y(r_1 + r_2) \\
&- F_0(r_1 t, \varsigma - \tau(r_1,t), t) + yr_1 \},
\end{aligned} \tag{19}$$

where $y = y(\varsigma - \tau(r_1,t), t)$. Let $\sigma(s_{m-1},t) = (0, s_{m-1}, s_{m-1}t, \cdots, s_{m-1}t^{m-2}/(m-1)!)$. Then $\tau(r_1,t) = \sigma(s_{m-1},t)$. Define

$$F(\varsigma,t) = -e^{-is_{m-1}t^{m-1}/m!} F_0(s_{m-1}, \varsigma - \sigma(s_{m-1},t), t),$$

for $\varsigma = (\cdot, s_{m-1}, \cdots)$. Then (19) implies (15) where

$$Z(\varsigma,t) = e^{-is_{m-1}t^{m-1}/m!} y(\varsigma - \sigma(s_{m-1},t), t).$$

If $t = 0$, then $M_t = S_{m,n}$. Define $F \equiv 0$ and $Z(\varsigma,t) = y(\varsigma,t)$, then (17) implies (15).

4. Step (3). We have to prove another lemma.

LEMMA 2. *Let $\psi_m(\varsigma_1,\varsigma,t)$ be a Baire function satisfying the condition that*

$$\psi_m(\varsigma_1 + \varsigma_2, \varsigma, t) = \nu(\varsigma_1,t)\psi_m(\varsigma_2, \varsigma + \mu(\varsigma_1,t), t) + \psi_m(\varsigma_1, \varsigma, t) \tag{20$_m$}$$

for $\varsigma, \varsigma_j \in S^{m,n}$ and $t \in \mathbf{R}^n$. Then there is a Baire function F_m such that

$$\psi_m(\varsigma_1, \varsigma, t) = F_m(\varsigma, t) - \nu(\varsigma_1,t)F_m(\varsigma + \mu(\varsigma_1,t), t) \tag{21$_m$}$$

for $t \neq 0$.

PROOF. We prove this lemma by mathematical induction.

(i) Case $m = 1$. From $(20)_1$, we have

$$\psi_1(\varsigma_1 + \varsigma_2, \varsigma, t) = e^{i\varsigma_1 t}\psi_1(\varsigma_2, \varsigma, t) + \psi_1(\varsigma_1, \varsigma, t)$$
$$= e^{i\varsigma_2 t}\psi_1(\varsigma_1, \varsigma, t) + \psi_1(\varsigma_2, \varsigma, t),$$

where $\varsigma_j \in \mathbf{R}^n, j = 1, 2$ and $t \in \mathbf{R}^n$. Therefore

$$(1 - e^{i\varsigma_1 t})\psi_1(\varsigma_2, \varsigma, t) = (1 - e^{i\varsigma_2 t})\psi_1(\varsigma_1, \varsigma, t).$$

Thus there is a Baire function $F(\varsigma, t)$ such that

$$\psi_1(\varsigma_1, \varsigma, t) = (1 - e^{i\varsigma_1 t})F(\varsigma, t),$$

which proves $(21)_1$.

(ii) Assume Lemma 2 is true for $m - 1$ $(m \geq 2)$. Let ψ_m satisfy $(20)_m$. For convenience, we identify $(s_m, s_{m-1}, \cdots, s_1)$ with $(s_m, (s_{m-1}, \cdots, s_1))$. For fixed $r \in S_{m,n}$ define function

$$\psi_{m-1}(\xi_1, \xi, t; r) = \psi_m((0, \xi_1), (r, \xi), t)$$

for $\xi_1, \xi \in S^{m-1,n}, t \in \mathbf{R}^n$. Then $\psi_{m-1}(\cdot, \cdot, \cdot; r)$ satisfies $(20)_{m-1}$. Therefore there is a Baire function $F_{m-1}(\cdot, \cdot, \cdot; r)$ such that $(21)_{m-1}$ holds by the hypothesis of the mathematical induction. Hence

$$\psi_m((0, \xi_1), \varsigma, t) = F_m(\varsigma, t) - \nu(\xi_1, t)F_m(\varsigma + (0, \mu(\xi_1, t)), t). \tag{22}$$

where $F_m((r, \xi), t) = F_{m-1}(\xi, t; r)$ for $r \in S_{m,n}, \xi \in S^{m-1,n}$ and $t \in \mathbf{R}^n$. Define

$$\psi_0(r, \varsigma, t) = \psi_m((r, 0), \varsigma, t).$$

Then ψ_0 satisfies (14), since ψ_m satisfies $(20)_m$. By Lemma 1, there are Baire function F and vector-valued Baire function Z such that (15) is satisfied. By (15), $(20)_m$ and (22), we have

$$\psi_m(\varsigma_1, \varsigma, t) = e^{irt^m/m!}\psi_m((0, \xi), \varsigma + \tau(r, t), t) + \psi_0(r, \varsigma, t)$$
$$= e^{irt^m/m!}F_m(\varsigma + \tau(r, t), t) - \nu(\varsigma, t)F_m(\varsigma + \mu(\varsigma_1, t), t) \tag{23}$$
$$+ F(\varsigma, t) - e^{irt^m/m!}F(\varsigma + \tau(r, t), t) + Z(\varsigma, t)r,$$

where $\varsigma_1 = (r, \xi)$, since $e^{irt^m/m!}\nu(\xi,t) = \nu(\varsigma_1,t)$ and

$$(0, \mu(\xi,t)) + \tau(r,t) = \mu(\varsigma_1,t).$$

Define $G(\varsigma,t) = F_m(\varsigma,t) - F(\varsigma,t)$. Then (23) reduces to

$$\psi_m(\varsigma_1,\varsigma,t) = e^{irt^m/m!}G(\varsigma + \tau(r,t),t) - \nu(\varsigma_1,t)G(\varsigma + \mu(\varsigma_1,t),t) + F(\varsigma,t)$$
$$- \nu(\varsigma_1,t)F(\varsigma + \mu(\varsigma_1,t),t) + Z(\varsigma,t)r. \tag{24}$$

Substituting (24) for ψ_m into (20), we get

$$e^{i(r_1+r_2)t^m/m!}G(\varsigma + \tau(r_1 + r_2,t),t) - \nu(\varsigma_1,t)e^{ir_2^m/m!}G(\varsigma + \mu(r_1,t) + \tau(r_2,t),t)$$
$$+ \nu(\varsigma_1,t)G(\varsigma + \mu(r_1,t),t) - e^{ir_1 t^m/m!}G(\varsigma + \tau(r_1,t),t)+$$
$$(Z(\varsigma,t) - \nu(\varsigma_1,t)Z(\varsigma + \mu(\varsigma_1,t),t))r_2 = 0. \tag{25}$$

The equality (25) may be written as $Ae^{ist} + B = 0$ where $\varsigma_1 = (\cdots, s), s \in \mathbf{R}^n$, and A and B are independent of s. Therefore, $A = B = 0$, if $t \neq 0$. From $B = 0$, we get

$$e^{ir_2 t^m/m!}G(\varsigma + \tau(r_2,t),t) - e^{ir_1 t^m/m!}G(\varsigma + \tau(r_1,t),t) + Z(\varsigma,t)(r_2 - r_1) = 0.$$

Thus $e^{irt^m/m!}G(\varsigma + \tau(r,t),t) + Z(\varsigma,t)r$ is independent of r, i.e.

$$e^{irt^m/m!}G(\varsigma + \tau(r,t),t) + Z(\varsigma,t)r = G(\varsigma,t). \tag{26}$$

From (24) and (26), we get $(21)_m$.

(4) Now, let us prove this theorem. If ϕ is a Baire cyclic cocycle ϕ on $G_{m,n}$. Define $\phi(\cdot,\cdot;\cdot,\cdot)$ and $\psi(\cdot,\cdot;\cdot,\cdot)$ by (8) and (11). Then define

$$\psi_m(\varsigma_1;\varsigma,t) = \psi(\varsigma_1,0;\varsigma,t)$$

Therefore ψ_m satisfies $(20)_m$, since ψ satisfies (13). By Lemma 2, there is a Baire function F_m such that

$$\psi(\varsigma_1,0;\varsigma,t) = F_m(\varsigma,t) - \nu(\varsigma_1,t)F_m(\varsigma + \mu(\varsigma_1,t),t). \tag{27}$$

Define $f(t_1,\varsigma,t) = \psi(0,t_1;\varsigma,t)$ for $t_1 \in \mathbf{R}^n$. In (13), letting $\varsigma_2 = 0$ and $t_1 = 0$ and then changing t_2 to t_1, we have

$$\psi(\varsigma_1,t_1;\varsigma,t) = \nu(\varsigma_1,t_0)f(t_1;\varsigma+\mu(\varsigma_1,t_0),t) + \nu(\varsigma_0,t_1)\psi_m(\varsigma_1,0;\varsigma+\mu(\varsigma_0,t_1),t), \quad (28)$$

where $t_0 = t - t_1, \varsigma_0 = \varsigma - \varsigma_1$. From (12), (27) and (28), we have

$$\begin{aligned}
&\nu(\varsigma_1,t_0)f(t_1,\varsigma+\mu(\varsigma_1,t_0),t)+ \\
&+ \nu(\varsigma_0,t_1)F_m(\varsigma+\mu(\varsigma_0,t_1),t) - \nu(\varsigma_1,t)F_m(\varsigma+\mu(\varsigma_0,t_1)+\mu(\varsigma_1,t),t) \\
&+ \nu(\varsigma_0,t_1)f(t_0,\varsigma+\mu(\varsigma_0,t_1),t) \\
&+ \nu(\varsigma_1,t_0)F_m(\varsigma+\mu(\varsigma_1,t_0),t) - \nu(\varsigma_0,t)F_m(\varsigma+\mu(\varsigma_1,t_0)+\mu(\varsigma_0,t),t) = 0.
\end{aligned} \quad (29)$$

It is obvious that

$$f(t_1,\varsigma,t) = \nu(\varsigma,t_1)F_m(\varsigma+\mu(\varsigma,t_1),t) - F_m(\varsigma,t) \quad (30)$$

is a solution of (29), since $\mu(\varsigma,t_0+t_1) = \mu(\varsigma,t_0)+\mu(\varsigma,t_1)+\mu(\mu(\varsigma,t_0),t_1)$ and $\nu(\varsigma,t_0+t_1) = \nu(\varsigma,t_0)\nu(\varsigma,t_1)\nu(\mu(\varsigma,t_0),t_1)$. We have to prove that (30) is the unique solution for $f(t_1,\varsigma,t)$ in (29). Suppose f_0 is any other solution. Let $f_1 = f - f_0$. Then from (29), we have

$$\nu(\varsigma_1,t_0)f_1(t_1,\varsigma+\mu(\varsigma_1,t_0),t) + \nu(\varsigma_0,t_1)f_1(t_0,\varsigma+\mu(\varsigma_0,t_1),t) = 0. \quad (31)$$

Suppose $\varsigma_1 = (r_m,\cdots,r_1)$. For fixed $t_1,t_0,r_m,\cdots,r_2$ and ς, (31) may be rewritten as

$$e^{it_0 r_1}C + e^{it_1 r_1}D = 0,$$

where C and D are independent of r_1, since $\mu(\varsigma_1,t_0)$ and $\mu(\varsigma_0,t_1)$ are independent of r_1. Hence $C = 0$ which implies

$$f_1(t_0,\varsigma+\mu(\varsigma_0,t_1),t) = 0.$$

Thus (30) is the unique solution for $f(t_1,\varsigma,t)$ in (31), if $t \neq 0$. From (27), (28) and (30), it follows that if $t \neq 0$, then

$$\psi(\varsigma_1,t_1;\varsigma,t) = \nu(\varsigma_0,t_1)F_m(\varsigma+\mu(\varsigma_0,t_1),t) - \nu(\varsigma_1,t_0)F_m(\varsigma+\mu(\varsigma_1,t_0),t),$$

for $\varsigma_0 = \varsigma - \varsigma_1$ and $t_0 = t - t_1$, which proves (7), if we define $\sigma(g(\varsigma,t,z)) = -zF_m(\varsigma,t)$.

5. For the simplicity of notation, we identify $(s,0)$ with s, $(0,\xi)$ with ξ, and

$$(s,0,\cdots,0,t,z)$$

with $(s,0,t,z)$, etc. Therefore $G_{m-1,n}$ can be regarded as a subgroup of $G_{m,n}$. It is obvious that $G_n^m = \{g(s,0,1) : s \in \mathbf{S}_{m,n}\}$ is also a subgroup of $G_{m,n}$.

Suppose W is a continuous mapping from $G_{m,n}$ to the group of bounded linear operators on the Hilbert space $\mathcal{H}$. The mapping W is said to satisfy the condition (A), if it satisfies the following two conditions; (i) $W\big|_{G_{m-1,n}}$ is a group representation, and (ii)

$$W(g_1 g_2) - W(g_1)W(g_2) \in \mathcal{L}^1(\mathcal{H}) \tag{32}$$

for $g_1 \in G_n^m \cup G_{m-1,n}, g_2 \in G_n^m$, where $\mathcal{L}^1(\mathcal{H})$ is the trace ideal of the algebra of all bounded linear operators on $\mathcal{H}$. It is easy to see that under the conditions (A); (32) holds for any $g_1, g_2 \in G_{m,n}$. Thus $\{W(g) : g \in G_{m,n}\}$ is an almost unperturbed Lie group defined in [3].

For the mapping W satisfying the condition (A) define a function $\phi_W(\cdot\,,\cdot)$ on $G_{m,n} \times G_{m,n}$ as follows

$$\phi_W(g_1,g_2) = \mathrm{tr}\,([W(g_1),W(g_2)] - W(g_1 g_2) + W(g_2 g_1)),$$

where $[\cdot\,,\cdot]$ is the commutator. It is easy to see that ϕ_W is a Baire function and a homogeneous one-cocycle (cf. [3]) on G_{mn}. By theorem 1, there is a Baire function σ_W on G such that (7) holds for ϕ_W and σ_W. Define

$$F_W(\varsigma,t) = -\sigma_W(g(\varsigma,t,1)) \qquad \text{for} \quad \varsigma \in \mathbf{S}^{m,n},\ t \in \mathbf{R}^n.$$

Then F_W is a Baire function and (7) may be rewritten as

$$\phi_W(g(\varsigma_1,t_1,1),g(\varsigma_2,t_2,1))$$

$$= \nu(\varsigma_1,t_2)F_W(\varsigma + \mu(\varsigma_1,t_2),t) - \nu(\varsigma_2,t_1)F_W(\varsigma + \mu(\varsigma_2,t_1),t) \tag{33}$$

where $\varsigma_j \in \mathbf{S}^{m,n}, t_j \in \mathbf{R}^n, \varsigma = \varsigma_1 + \varsigma_2, t = t_1 + t_2$ and $t \neq 0$.

For this mapping W, define

$$q_W(s_0, t_0; s, s_1, t_1) = \tag{34}$$
$$\mathrm{tr}\ (W(g(0, s_0, t_0, 1))(W(g(0, s_1, t_1, 1))W(g(s, 0, 1)) - W(g(0, s_1, t_1, 1)g(s, 0, 1))))$$

for $s_j, t_j \in \mathbf{R}^n$ and $s \in \mathbf{S}_{m,n}$.

LEMMA 3. *For the mapping W satisfying the condition (A), the function q_W satisfies the identity*

$$q_W(s_0, t_0; \varsigma, s_1, t_1) = e^{is_1 t_0} F_W(\varsigma, 0, s_1, t) \tag{35}$$
$$- e^{i(s_0 t_1 + st_1^m/m!)} F_W((\varsigma, 0) + \tau(\varsigma, t_1) + (0, s), t),$$

where $\tau(\varsigma, t) = \mu((\varsigma, 0), t)$.

PROOF. For the simplicity of the notation, let $U(t) = W(g(0, t, 1)), V(s) = W(g(0, s, 0, 1))$ for $t, s \in \mathbf{R}^n, W(\varsigma) = W(g(\varsigma, 0, 1))$ for $\varsigma \in \mathbf{S}_{m,n}$, and $\nu(\varsigma, t) = \nu((\varsigma, 0), t)$. Then it is easy to verify that

$$q_W(s_0, t_0; \varsigma, s_1, t_1) = e^{is_1 t_0} \mathrm{tr}\ (V(s)(U(t)W(\varsigma) - W(g(\varsigma + \tau(\varsigma, t), t, \nu(\varsigma, t)))))$$
$$+ e^{is_1 t_0} \mathrm{tr}\ ([V(s), W(\varsigma)]W(g(\tau(\varsigma, t), 0, 1))U(t))\nu(\varsigma, t)$$
$$+ \mathrm{tr}\ (-V(s_0)U(t_0)W(\varsigma)W(g(\tau(\varsigma, t_1), 0, 1))V(s_1)U(t_1)\nu(\varsigma, t_1)$$
$$+ W(\varsigma)W(g(\tau(\varsigma, t), 0, 1))V(s)U(t)\nu(\varsigma, t)e^{is_1 t_0})$$
$$= e^{is_1 t_0}\phi(g(0, t_1, 1), g(\varsigma, 0, s, 0, 1)) + e^{i(s_1 t_0 + st^m/m!)}\phi(g(0, s, 0, 1), g(\varsigma + \tau(\varsigma, t), t, 1))$$
$$- e^{it_1^m/m!}\phi(g(0, s_0, t_0, 1), g(\varsigma + \tau(\varsigma, t_1) + s_1, t_1, 1)),$$

where $s = s_0 + s_1$ and $t = t_0 + t_1$. Therefore

$$q_W(s_0, t_0; \varsigma, s_1, t_1) = e^{is_1 t_0}(F_W(\varsigma, 0, s, t) - e^{i(st + st^m/m!)}F_W(\varsigma + \tau(\varsigma, t) + s, t))$$
$$+ e^{i(s_1 t_0 + st^m/m!)}(e^{ist} - 1)F_W(\varsigma + \tau(\varsigma, t) + s, t)$$
$$- e^{i\varsigma t_1^m/m!}e^{is_0 t_1}F_W(\varsigma + \tau(\varsigma, t) + s, t) + \nu(\varsigma, t)e^{is_1 t_0}F_W(\varsigma + \tau(\varsigma, t) + s, t))$$

which proves (35), where sometimes ς stands for $(\varsigma, 0)$ and s stands for $(0, s)$.

6. Now, let $W(g(0, \xi, t, z)) = g_0(0, \xi, t, z)$ in §2, for $\xi \in S^{m-1,n}$, $t \in \mathbf{R}^n$ and $z \in \mathbf{T}$. Therefore

$$(U(t)f)(x) = f(x + t), \qquad (V(s)f)(x) = e^{isx}f(x), \tag{36}$$

where $U(t) = W(g(0, t, 1))$ for $t \in \mathbf{R}^n$ and $V(s) = W(g(0, s, 0, 1))$ for $s \in \mathbf{R}^n$.

Using the same methods of proving Lemma 3 in [6], we may prove the following:

LEMMA 4. *Let $U(\cdot)$ and $V(\cdot)$ be defined by (36). Let $A \in \mathcal{L}^1(L^2(\mathbf{R}^n))$.*
Then

$$\frac{1}{(2\pi)^n} \int \left(\iint \mathrm{tr}\ (V(y)U(p - r)A)g(r)\overline{h(p)}e^{-iyr}\,dr\,dp \right)dy = (Ag, h) \tag{37}$$

for $g, h \in L^1(\mathbf{R}^n) \cap L^2(\mathbf{R}^n)$.

PROOF. The Lemma 3 in [6] may be generalized to the $\mathbf{R}^n$ in the following form. Let $\tilde{V}(y) = \mathcal{F}^{-1}V(y)\mathcal{F}$, $\tilde{U}(t) = \mathcal{F}^{-1}U(t)\mathcal{F}$, then

$$\frac{1}{(2\pi)^n} \int \left(\iint \mathrm{tr}\ (\tilde{U}(r - p)\tilde{V}(y)\mathcal{F}^{-1}A\mathcal{F})g(r)\overline{h(p)}e^{ipy}\,dr\,dp \right)dy = (\mathcal{F}^{-2}A\mathcal{F}^2g, h).$$

This formula is equivalent to (37), since $(\mathcal{F}^2g)(x) = g(-x)$ and

$$U(t)V(s) = e^{ist}V(s)U(t).$$

In (33), let $\varsigma_2 = (0, y)$ and ς_2 be denoted by y for $y \in \mathbf{R}^n$. Then

$$\phi_W(g(y, 0, 1), g(\varsigma, t, 1)) = (e^{iyt} - 1)F_W(\varsigma + y, t)$$

for $y, t \in \mathbf{R}^n$ and $\varsigma \in S^{m,n}$. On the other hand, it is easy to show that

$$\phi_W(g(\varsigma, t, 1), g(y, 0, 1)) = \mathrm{tr}\ (V(\xi)U(t)[W(s), V(y)]),$$

where $\varsigma = (s, \xi), s \in S_{m,n}, y \in \mathbf{R}^n$ and

$$V(\xi) = g_0(0, \xi, 0, 1), \qquad \xi \in S^{m-1,n}.$$

Therefore

$$\text{tr } (V(\xi)U(t)[V(y), W(s)]) = (e^{iyt} - 1)F_W(\varsigma + y, t), \tag{38}$$

for $y, t \in \mathbf{R}^n, \xi \in S^{m-1,n}$, and $s \in S_{m,n}$, where $\varsigma = (s, \xi)$,

LEMMA 5. *Let $A \in \mathcal{L}^1(L^2(\mathbf{R}^n))$, then there is a function $v(\cdot, \cdot) \in L^1(\mathbf{R}^n) \otimes L^\infty(\mathbf{R}^n)$ such that*

$$\text{tr } (V(\xi)U(t)A) = \int \nu(\xi, x)v(x, t)dx, \tag{39}$$

for $\xi \in S^{k,n}, k \geq 1$. Besides the function $(x, t) \mapsto v(x, t - x)$ belongs to $L^2(\mathbf{R}^{2n})$.

PROOF. Without loss of generality we may assume that there is an orthonormal basis $\{e_j\}$ for $L^2(\mathbf{R}^n)$ such that $Ae_j = \lambda_j e_j$, where λ_j satisfies $\sum |\lambda_j| < +\infty$. Let

$$v(x, t) = \sum \lambda_j e_j(x + t)\overline{e_j(x)}.$$

Then it is easy to calculate that (39) holds good.

It is easy to see that $v \in L^1(\mathbf{R}^n) \otimes L^\infty(\mathbf{R}^n)$. On the otherhand, by the Schwarz inequality, we have

$$|v(x, t - x)|^2 \leq \sum \lambda_j|e_j(t)|^2 \sum \lambda_j|e_j(x)|^2.$$

Hence $\int\int|v(x, t - x)|^2 dx dt \leq (\sum \lambda_j)^2$. Lemma 5 is proved.

Now, let $A = [V(y), W(s)], s \in S_{m,n}$ and $y \in \mathbf{R}^n$. Thus there is a function $v(\cdot, \cdot; y, s) \in L^1(\mathbf{R}^n) \otimes L^\infty(\mathbf{R}^n)$ such that

$$(e^{iyt} - 1)F_W(s, \xi + y, t) = \int \nu(\xi, x)v(x, t; y, s)dx. \tag{40}$$

Put $\xi = (0, s_0), s_0 \in \mathbf{R}^n$, then

$$(e^{is_1 t} - 1)F_W(s, 0, s_0 + s_1, t) = \int e^{is_0 x}v(x, t; s_1, s)dx.$$

Therefore

$$F_W(s, 0, y, t) = \int e^{iyx} v(x, t; s) dx,$$

where

$$v(x, t; s) = v(x, t; s_1, s) e^{-is_1 x} (e^{is_1 t} - 1)^{-1} \tag{41}$$

should be independent of s_1, provided that $e^{is_1 t} \neq 1$. Therefore (40) reduces to

$$F_W(s, \xi, t) = \int \nu(\xi, x) v(x, t; s) dx, \tag{42}$$

for $s \in S_{m,n}, \xi \in S^{m-1,n}$ and $t \in \mathbf{R}^n$, if $t \neq 0$.

From (35) and (42), we get

$$q_W(s_0, t_0; s, s_1, t_1) =$$
$$\int (e^{is_1 t_0} - e^{i(s_0 t_1 + s((t_1 + x)^m - x^m)/m!)}) e^{iyx} v(x, t; s) dx \tag{43}$$

for $s \in S_{m,n}, s_j, t_j \in \mathbf{R}^n$, where $y = s_0 + s_1$ and $t = t_0 + t_1$.

Let $L_C^\infty(\mathbf{R}^n)$ be the subspace of all functions in $L^\infty(\mathbf{R}^n)$ with compact supports. By Lemma 5 and (41), if $f \in L_C^\infty(\mathbf{R}^n)$, then the function

$$(x, t) \mapsto v(x, t - x; s) s_1(x - t) f(x)$$

belongs to $L^2(\mathbf{R}^{2n})$ for some $s_1 \in \mathbf{R}^n, (s_1 \neq 0)$. Besides, the Hilbert transform is a bounded operator. Therefore the operator

$$(K(s)f)(t) = \int f(x) v(x, t - x; s) dx \tag{44}$$

which is defined through the Hilbert transform is a linear bounded operator from $L_C^\infty(\mathbf{R}^n)$ to $L^2(\mathbf{R}^n)$.

Denote

$$W(s, y, t) = V(y) U(t) W(s) - W(g(0, y, t, 1) g(s, 0, 1)) \tag{45}$$

and

$$K(s, y, t) = V(y) U(t) K(s) - K(s) W(s)^{-1} W(g(0, y, t, 1) g(s, 0, 1))$$

for $s \in S_{m,n}$ and $y, t \in \mathbf{R}^n$.

LEMMA 6. *If f and h are smooth function with compact supports then*

$$(W(s,y,t)f,h) = (K(s,y,t)f,h) \tag{46}$$

for $s \in S_{m,n}, y, t \in \mathbf{R}^n$ and $t \neq 0$.

PROOF. Put $A = W(s,y,t)$ in (45). Then

$$
\begin{aligned}
(W(s,y,t)f,h) = & \\
\frac{1}{(2\pi)^n} \int (&\iint (\int (e^{i(y(p-r+x)+(z-r)z)} - e^{i(z(t-r+x)+yx+s((t+x)^m - x^m)/m!)} \\
& v(x,y+p-r,s)dx) f(r)\overline{h(p)}\, dr\, dp)\, dz.
\end{aligned} \tag{47}
$$

We have to prove that (47) equals to $I_1 - I_2$ where

$$
\begin{aligned}
I_1 &= \int (\int e^{iyp} f(r) v(r, t+p-r, s)\, dr\, \overline{h(p)}\, dp \\
&= (V(y)U(t)K(s)f,h)
\end{aligned}
$$

and

$$
\begin{aligned}
I_2 &= \int (\int f(x+t) e^{i(yx+s((t+x)^m - x^m)/m!} v(x, y+p-t-x, s)\overline{h(p)}\, dp \\
&= (K(s)W(s)^{-1}W(g(0,y,t,1)g(s,0,1))f,h).
\end{aligned}
$$

The basic point of this proof is to justify that $\int e^{izx}dz = (2\pi)^n \delta(x)$. As in the proof of Lemma 5, we may assume that there are an orthonormal basis $\{e_j\}$ of $L^2(\mathbf{R}^n)$ and a sequence $\{\lambda_n\}$ satisfying $\sum |\lambda_j| < +\infty$ which depend on s_1 and s such that

$$v(x,t,z) = \sum \lambda_j e_j(t+x)\overline{e_j(x)} e^{is_1 z}(e^{is_1 t} - 1)^{-1} \tag{48}$$

Choose $s_1 \neq 0$ such that

$$((e^{is_1 t} - 1)^{-1} - (is_1 t)^{-1})f(t)$$

is bounded for $s_1 t \neq 0$.

In order to prove that

$$\frac{1}{(2\pi)^n} \int \left(\iint e^{i(y(p-r+z)+(z-r)z)} v(x, y+p-r, s)\,dx \right) f(r)\overline{h(p)}\,dr\,dp)\,dz \qquad (49)$$

equals to I_1, we have to use the fact that the Hilbert transform of the function f of one variable satisfies the identity

$$P.V. \int \frac{f(t_1)dt_1}{t_1 - x_1} = \frac{1}{2}\left(\lim_{\epsilon \to 0^+} \int \frac{f(t_1)dt_1}{t_1 - x_1 + i\epsilon} + \lim_{\epsilon \to 0^+} \int \frac{f(t_1)dt_1}{t_1 - x_1 - i\epsilon} \right).$$

We only give the details of the proof of the following formula

$$\begin{aligned}
&\int \lim_{\epsilon \to 0^+} \iint \frac{\int e^{izz}\alpha(x+p)\overline{\alpha(x+r)}\,dx\, f(r)\overline{h(p)}}{p - r - i\epsilon}\,dr\,dp)\,dz \\
&= \int_0^\infty \left(\int_{-\infty}^\infty \alpha(p)\overline{h(p)}e^{-ipz}\,dp \int_{-\infty}^\infty \overline{\alpha(r)}f(r)e^{irz}\,dr \right) dz
\end{aligned} \qquad (50)$$

for $\alpha \in L^2(\mathbf{R})$ and smooth functions $f(\cdot)$ and $h(\cdot)$ of one variable with compact supports.

Using (50) and the fact that $\sum |\lambda_j| < \infty$ it is easy to prove that (49) equals to I_1 if the function (48) is replaced by

$$v_1(x, t, s) = \sum \lambda_j e_j(t+x)\overline{e_j(x)}e^{-is_1 x}(ist)^{-1}. \qquad (51)$$

for the case $n = 1$.

Besides, the part corresponding to $v(x, t, s) - v_1(x, t, s)$ can be treated much easier since the function $(x, t) \mapsto v(x, t - x, s) - v_1(x, t - x, s)$ belongs to $L^2(\mathbf{R}^{2n})$, and the generalization from $n = 1$ to the general n case is also easy. Therefore (49) equals to I_1.

Now, let us prove the formula (50). It is obvious that

$$\frac{1}{2\pi i} \int \frac{\xi(r)\overline{\eta(p)}\,dr\,dp}{p - r - i\epsilon} = \int_0^\infty \tilde{\xi}(y)\overline{\tilde{\eta}(y)}e^{-\epsilon y}\,dy, \qquad (52)$$

for $\xi, \eta \in L^2(\mathbf{R}^1)$ and $\epsilon > 0$ where $\tilde{\xi}$ is the Fourier transform of ξ. For fixed x, denote $\xi_x = \overline{\alpha(x + \cdot)}f(\cdot)$ and $\eta_x = \overline{\alpha(x + \cdot)}h(\cdot)$. Then

$$\tilde{\xi}_x(y) = \frac{1}{\sqrt{2\pi}} \int \tilde{g}(y + u)\overline{\tilde{\alpha}(u)}e^{ixu}du$$

$$\tilde{\eta}_x(y) = \frac{1}{\sqrt{2\pi}} \int \tilde{h}(y + u)\overline{\tilde{\alpha}(u)}e^{ixu}du.$$

By (52), the left hand side of (50) equals

$$i\int \Big(\lim_{\epsilon\to 0^+}\int_0^\infty e^{-\epsilon y}\int e^{ixz}\int \tilde{f}(y + u)\overline{\tilde{\alpha}(u)}e^{ixu}du \int \overline{\tilde{h}(y + v)}\tilde{\alpha}(v)e^{ixv}dvdxdy\Big)dz$$

$$= -i\int \Big(\lim_{\epsilon\to 0^+}\int_0^\infty e^{-\epsilon y}\int f(y + u - z)\overline{\tilde{\alpha}(u - z)}\overline{\tilde{h}(y + u)}\tilde{\alpha}(u)dudy\Big)dz. \tag{53}$$

It is easy to see that the function

$$(u, y, z) \mapsto \tilde{f}(y + u - z)\overline{\tilde{\alpha}(u - z)}\overline{\tilde{h}(y + u)}\tilde{\alpha}(y)$$

belongs to $L^1(\mathbf{R}^3)$, since f and h are smooth functions with compact supports. Therefore we may exchange the order of integrations and the operation $\lim_{\epsilon\to 0^+}$. Thus (53) equals to

$$i\int_0^\infty \Big(\int \tilde{f}(y + z)\overline{\tilde{\alpha}(z)}dz \int \overline{\tilde{h}(y + u)}\tilde{\alpha}(u)du\Big)dy,$$

which proves (50).

By the same method, we may prove that

$$\frac{1}{(2\pi)^n} \int \Big(\iint \Big(\int e^{i(z(t-r+x)+yx+s((t+x)^m - x^m)/m!)}v(x, y + p - r, s)dx\Big)f(r)\overline{h(p)}drdp\Big)dz$$

equals to I_2, which proves Lemma 6.

For $s \in \mathbf{S}_{m,n}$, let $V(s) = g_0(s, 0, t, 1)$. Then it is easy to see that

$$(V(y)U(t)(K(s) - W(s))V(s)^{-1} - (K(s) - W(s))V(s)^{-1}V(y)U(t))f = 0 \tag{54}$$

for $f \in L_C^\infty(\mathbf{R}^n)$. From (54) it is easy to prove that there is a constant $c(s)$ such that

$$(K(s) - W(s))V(s)^{-1}f = -c(s)f$$

for all $f \in L_C^\infty(\mathbf{R}^n)$. Therefore $K(s)$ extends a unique bounded linear operator from $L^2(\mathbf{R}^n)$ to $L^2(\mathbf{R}^n)$ which will be denoted by $K(s)$ again. Thus we have the following.

THEOREM 2. *Let W be a continuous mapping from $\mathbf{S}_{m,n}$ to the group of bounded linear operators on $L^2(\mathbf{R}^n)$ endowed with weak topology satisfying the conditions that $W(s_1 + s_2) - W(s_1)W(s_2) \in \mathcal{L}^1(L^2(\mathbf{R}^n))$ for $s_1, s_2 \in \mathbf{S}_{m,n}$, and*

$$g_0(\xi, t, 1)W(s) - W(s)g_0((0, \xi) + \mu(s, t), t, \nu(s, t)) \in \mathcal{L}^1(L^2(\mathbf{R}^n))$$

for $s \in \mathbf{S}_{m,n}, \xi \in \mathbf{S}_n^{m-1}$ and $t \in \mathbf{R}^n$.

Then there is a function $k(x, t; s)$ satisfying the conditions that the function $(x, t) \mapsto k(x, t; s)(e^{is_1(t-x)} - 1)$ belongs to $L^2(\mathbf{R}^{2n})$ for any s_1 and that the operator $K(s)$ defined by

$$(K(s)f)(t) = \int f(x)k(x, t, s)dx$$

for $f \in L_C^\infty(\mathbf{R}^n)$ satisfies

$$W(s) = K(s) + c(s)g_0(s, 0, 1), \tag{55}$$

where $c(s)$ is a complex function. Furthermore, the function k is determined by the trace formula

$$\mathrm{tr}\,(g_0(0, t, 1)[g_0(y, 0, 1), W(s)]) = (e^{iyt} - 1)\int e^{iyx}k(x, t + x; s)dx,$$

for $s \in \mathbf{S}_{m,n}$ and $y, t \in \mathbf{R}^n$.

7. In this section we consider the perturbation of a partial differential operator with constant coefficients. For the simplicity of notation, let $\frac{d}{dx} = (\frac{\partial}{\partial x_1}, \cdots, \frac{\partial}{\partial x_n})$. Any homogeneous polynomial $P(\cdot)$ of degree m of the form

$$P(t_1, \cdots, t_m) = \sum P^{i_1, \cdots, i_m}\frac{1}{m!}t_{i_1}\cdots t_{i_m}$$

with constant coefficients $P^{i_1\cdots i_m}$ is corresponding to a tensor $P = (P^{i_1\cdots i_m}) \in S_{m,n}$. Therefore the partial differential operator

$$P(i\frac{d}{dx}) = P\tilde{v}_m = \sum P^{i_1\cdots i_m}\tilde{v}_m^{i_1\cdots i_m},$$

where $\tilde{v}_m^{i_1\cdots i_m}$ is the differential operator defined in §2.

THEOREM 3. *Let $P(i\frac{d}{dx})$ be a homogeneous partial differential operator with constant real coefficients of order m. Let Q be a bounded self-adjoint operator on $L^2(\mathbb{R}^n)$ satisfying the following two conditions that*

$$QD(P(i\frac{d}{dx})) \subset D(P(i\frac{d}{dx})), \tag{56}$$

and

$$e^{iR(i\frac{d}{dx})}e^{i(\cdot)t}Qe^{-i(\cdot)t}e^{-iR(i\frac{d}{dx})} - Q \in \mathcal{L}^1(L^2(\mathbb{R}^n)) \tag{57}$$

for every real polynomial R with constant coefficients of degree $\leq m$ and $t \in \mathbb{R}^n$. Then there is a function $k(x,t;\rho)$ satisfying the condition that the function $(x,t) \mapsto k(x,t,\rho)(e^{is_1(t-x)} - 1)$ belongs to $L^2(\mathbb{R}^{2n})$ for any $s_1 \in \mathbb{R}^n$ and $\rho \in \mathbb{R}^1$ such that the operator $K(\rho)$ on $L^2(\mathbb{R}^n)$ defined by

$$(K(\rho)f)(t) = \int f(x)k(x,t;\rho)dx$$

for $f \in L_c^\infty(\mathbb{R}^n)$ satisfies

$$e^{i\rho(P(i\frac{d}{dx})+Q)} = \mathcal{F}^{-1}K(\rho)\mathcal{F} + c(\rho)e^{i\rho P(i\frac{d}{dx})}. \tag{58}$$

for $\rho \in \mathbb{R}^1$ where $c(\rho)$ is a complex function. Furthermore the function k is determined by the trace formula

$$\mathrm{tr}\ (e^{ixt}[e^{-y\frac{d}{dx}}, e^{i\rho(P(i\frac{d}{dx})+Q)}]) = (e^{iyt} - 1)\int e^{iyx}k(x,t+x;\rho)dx \tag{59}$$

for $y,t \in \mathbb{R}^n$.

Besides, if Q is not a multiple of identity, then the function $c(\cdot)$ is determined by the function $k(\cdot,\cdot;\cdot)$ in the following sense that $c(\cdot)$ is the only function satisfying the condition that $K(\rho) + c(\rho)e^{i\rho P(\cdot)}, \rho \in \mathbb{R}^1$ is a one parameter group.

PROOF. Let P be the tensor in $S_{m,n}$ corresponding to the polynomial P. For every $s \in S_{m,n}$ there is a unique decomposition $s = s_0 + \rho P$ satisfying $\rho \in \mathbf{R}^1$ and $s_0 P = 0$. Define

$$W(s) = \mathcal{F} e^{i\rho(P(i\frac{d}{dx})+Q)} \mathcal{F}^{-1} g_0(s_0, 0, 1).$$

We have to prove that

$$[W(s), g_0(\varsigma, 0, 1)] \in \mathcal{L}^1, \quad \text{for } s \in S_{m,n} \quad \text{and} \quad \varsigma \in S^{m,n}, \tag{60}$$

and

$$g_0(0, t, 1)W(s) - W(s)g_0((s, 0) + \mu(s, t), t, \nu(s, t)) \in \mathcal{L}^1, \tag{61}$$

for $s \in S_{m,n}$ and $t \in \mathbf{R}^n$.

For the simplicity of the notation, let $P = P(i\frac{d}{dx})$ and let $R = R(i\frac{d}{dx})$ where $R(\cdot)$ is a real polynomial of degree $\leq m$. It is easy to see that

$$e^{i\rho(P+Q)} = e^{i\rho P} + \int_0^\rho e^{iy(P+Q)} Q e^{i(\rho-y)P} dy.$$

Therefore, by (57), we have

$$e^{iR} e^{i\rho(P+Q)} e^{-iR} - e^{i\rho(P+Q)} = \int_0^\rho e^{iy(P+Q)}(e^{iR} Q e^{-iR} - Q)e^{i(\rho-y)P} dy \in \mathcal{L}^1,$$

which proves (60). Similarly, (57) implies (61).

From (60) and (61), it is easy to see that W satisfies the condition of Theorem 2, therefore there is a function $k(x, t; \rho)$ satisfying equations (58) and (59).

By the same method of proving Theorem 3 of [6], we may prove that the function $c(\rho)$ is determined by $K(\rho)$ if Q is not a multiple of identity.

From (58), we have

$$\mathcal{F} e^{i\rho(P(i\frac{d}{dx})+Q)} \mathcal{F}^{-1} = K(\rho) + c(\rho)e^{i\rho P(\cdot)}$$

If there is another function $c_1(\rho)$ such that $K(\rho) + c_1(\rho)e^{i\rho P(\cdot)}, \rho \in \mathbf{R}^1$ is also a one parameter group, then

$$b(\rho_1)e^{i\rho_1 P(\cdot)} K(\rho_2) + b(\rho_2)K(\rho_1)e^{i\rho_2 P(\cdot)} = b(\rho_1, \rho_2)e^{i(\rho_1+\rho_2)P(\cdot)}, \tag{62}$$

where $b(\rho) = c(\rho) - c_1(\rho)$ and $b(\rho_1,\rho_2) = c(\rho_1 + \rho_2) - c_1(\rho_1 + \rho_2) - c(\rho_1)c(\rho_2) + c_1(\rho_1)c_1(\rho_2)$.

Exchanging ρ_1 and ρ_2 in (62), we get

$$[e^{i\rho_1 P(\cdot)}, K_1(\rho_2)] = [e^{i\rho_2 P(\cdot)}, K_1(\rho_1)] \tag{63}$$

where $K_1(\rho) = K(\rho)b(\rho)^{-1}$, if $b(\rho) \neq 0$. We may assume that there is a kernel $k_1(\cdot,\cdot;\cdot)$ such that

$$(K_1(\rho)f)(x) = \int k_1(x,y,\rho)dy,$$

where the integral is in the distribution sense. Then (63) implies that

$$k_1(x,y,\rho_2)(e^{i\rho_1 P(x)} - e^{i\rho_1 P(y)}) = k_1(x,y,\rho_1)(e^{i\rho_2 P(x)} - e^{i\rho_2 P(y)}).$$

Thus there is $k_1(\cdot,\cdot)$ such that $k_1(x,y,\rho) = k_1(x,y)(e^{i\rho P(x)} - e^{i\rho P(y)})$. Therefore

$$b(\rho_1)b(\rho_2)K(\rho_1 + \rho_2) = b(\rho_1 + \rho)b(\rho_1,\rho_2)e^{i(\rho_1 + \rho_2)P(\cdot)}, \tag{64}$$

provided that $b(\rho_j) \neq 0, j = 1,2$ and $b(\rho_1 + \rho_2) \neq 0$. From (64), it is easy to prove that Q is a multiple of identity which proves Theorem 3.

It is easy to see that there is a non-trivial class of operators Q (e.g. some singular integral operators) which are not compact and satisfy the conditions (56) and (57).

REFERENCES

[1] Carey, R. W. and Pincus, J. D., Almost commuting pairs of unitary operators and flat currents, Integral Equations and Operator Theory 4(1981), 45-122.

[2] A. Cones, Non commutative differential geometry. Publ. Math. I. H. E. S. No. 62, (1985), 41-144.

[3] D. Xia, Trace formula for almost Lie group of operators and cyclic one-cocycles, Integral Equations and Operator Theory 9(1986) 570-587.

[4] D. Xia, On the almost unperturbed Schrödinger pair of operators, Integral Equations and Operator Theory 12(1989), 242-279.

[5]   D. Xia, Principal distribution for almost unperturbed Schrödinger pair of operators, Proc. Amer. Math. Soc. **112**(1991), 745-754.

[6]   D. Xia, Some trace formulas for almost unperturbed Schrödinger pair of operators, Proc. Amer. Math. Soc. **112**(1991), 755-764.

* The author gratefully acknowledges the support of the National Science Foundation, grant no. DMS-9101268.

Department of Mathematics
Vanderbilt University
Nashville, TN 37240
U.S.A

Titles previously published in the series

OPERATOR THEORY: ADVANCES AND APPLICATIONS
BIRKHÄUSER VERLAG